J. C. MARTIN ON PULSED POWER

ADVANCES IN PULSED POWER TECHNOLOGY

Series Editors:
A. H. Guenther, *Sandia National Laboratories, Albuquerque, New Mexico*
M. Kristiansen, *Texas Tech University, Lubbock, Texas*

Volume 1	OPENING SWITCHES Edited by A. Guenther, M. Kristiansen, and T. Martin
Volume 2	GAS DISCHARGE CLOSING SWITCHES Edited by Gerhard Schaefer, M. Kristiansen, and A. Guenther
Volume 3	J. C. MARTIN ON PULSED POWER Edited by T. H. Martin, A. H. Guenther, and M. Kristiansen

A Continuation Order Plan is available for this series. A continuation order will bring delivery of each new volume immediately upon publication. Volumes are billed only upon actual shipment. For further information please contact the publisher.

J. C. MARTIN ON PULSED POWER

Edited by

T. H. Martin
A. H. Guenther
Sandia National Laboratories
Albuquerque, New Mexico

and

M. Kristiansen
Texas Tech University
Lubbock, Texas

PLENUM PRESS • NEW YORK AND LONDON

Library of Congress Cataloging-in-Publication Data

Martin, J. C. (John Christopher), 1926-
    J.C. Martin on pulsed power / edited by T.H. Martin, A.H.
Guenther, and M. Kristiansen.
        p.   cm. -- (Advances in pulsed power technology ; v. 3)
    Includes bibliographical references and index.
    ISBN 0-306-45302-9
    1. Pulsed power systems.   I. Martin T.H. (Thomas H.)
II. Guenther, Arthur Henry, 1931-   . III. Kristiansen, M. (Magne),
1932-   . IV. Title. V. Series.
TK2986.M37 1996
621.31'2--dc20   96-11646
                                                                  CIP

ISBN 0-306-45302-9

© 1996 Plenum Press, New York
A Division of Plenum Publishing Corporation
233 Spring Street, New York, N.Y. 10013

10 9 8 7 6 5 4 3 2 1

All rights reserved

No part of this book may be reproduced, stored in a retrieval system, or transmitted in any form or by any means, electronic, mechanical, photocopying, microfilming, recording, or otherwise, without written permission from the Publisher

Printed in the United States of America

J.C. "Charlie" Martin
Atomic Weapons Research Establishment
Aldermaston, England

Presents the first Marx Award Lecture at the Pulsed Power Conference
Albuquerque, New Mexico, June 2, 1981.

FOREWORD TO THE SERIES

Pulsed power technology, in the simplest of terms, usually concerns the storage of electrical energy over relatively long times and then its rapid release over a comparatively short period. However, if we leave the definition at that, we miss a multitude of aspects that are important in the ultimate application of pulsed power. It is, in fact, the application of pulsed power technology to which this series of texts will be focused. Pulsed power in today's broader sense means "special power" as opposed to the traditional situation of high voltage impulse issues related to the utility industry.

Since the pulsed power field is primarily application driven, it has principally an engineering flavor. Today's applications span those from materials processing, such as metal forming by pulsed magnetic fields, to other varied applications, such as psychedelic strobe lights or radar modulators. Very high peak power applications occur in research for inertial confinement fusion, Ballistic Missile Defense (formerly the Strategic Defense Initiative), and other historical defense uses. In fact, it is from this latter direction that pulsed power has realized explosive growth over the past half century. Early thrusts were in electrically powered systems that simulated the environment or effects of nuclear weapon detonations. More recently it is being utilized as prime power sources for directed energy weapons, such as lasers, microwaves, particle beams, and even mass drivers (kinetic energy weapons). Consequently, much of the activity and growth in this field is spawned by government-sponsored research. In fact, the present activity is a result of research and results by J.C. Martin and his group at the Atomic Weapons Research Establishment at Aldermaston. This activity is of considerable benefit to the public and private sector through diverse areas such as bio-medical; inspection by flash x-rays; commercial ozone production; radio chemistry; food processing; surface modification; environmental clean up, and hazardous waste treatment, etc. Pulsed power is truly a multifaceted technology of diverse and surprising applications.

Pulsed power today is perhaps at once a flourishing field, yet one treated in perception as if it were of an arcane nature. Its exponential growth and development since World War II has taken place, in the large, without the benefit of text books, archival journals, or, for the greater portion of this period, recorded meetings or conferences devoted to pulsed power technology in an interdisciplinary manner. Only since the initiation of the IEEE Pulsed Power Conferences in 1976 have we been able to adequately delineate the requirements and focus pulsed power research from an engineering and physics standpoint. The physical, chemical and material sciences are now developing an appreciation and understanding of the processes involved in this field of pulsed power. Several interesting technology areas are involved, such as high voltage, high current, high power, repetitive discharges, particle accelerators, magnetic insulation, the effects of materials in adverse environments, instrumentation and diagnostics, as well as the always apparent problems related to switching, insulators, and breakdown. These events direct attention to the need for a strong and vigorous educational program. This need is not just a result of an increased awareness of pulsed power as an opportunity, but of the significance and potential of pulsed power to solve many of today's and the future's high technology problems.

The aforementioned explosive growth of this technology and the lack of recorded technical information has as well inhibited the development of pulsed power standards. We are now beginning to rectify this oversight. Our recent increase in understanding of fundamental pulsed power data has benefited greatly from the development of high quality diagnostic instrumentation techniques which have been perfected and applied to the field within the last 10-15 years. These techniques include lasers as applied in diagnosing plasmas or their use as light sources to visualize electric field distribution of even as voltage and current sensors using electro-optic effects. From the materials world the alphabet soup of surface analysis techniques, SEM, TEM, XPS, AES, LEED, SIMS, SNMS, ESCA, LIMA, etc., to name a few, have enhanced our understanding of not only surfaces but the interplay of materials and insulators either under the influence of electric fields or as a result of discharges. Importantly, we now have at hand rather elegant fast time resolution (approaching 10^{-15} second) diagnostics to employ in unraveling the processes limiting the performance of complex systems.

This series, therefore, is intended for physicists and engineers who have occasions to use pulsed power. They are not texts devoted to mathematical derivations but a thorough presentation of the present state-of-the-art. They will also point the reader to other sources when further in-depth study is necessary. We will address pulsed systems as distributed parameter systems in which electrical characteristics and propagation are considered in a

FOREWORD TO THE SERIES

continuous sense. Comparisons will be made, where appropriate, to lumped parametric analyses, such as in pulse forming networks. However, we will avoid a long list of inductance and capacitance formulae. Furthermore, we will assume that the readers of these texts are familiar with the basic materials which address those aspects of the pulsed power field. One should be able to apply a general knowledge of dynamic processes in the transient world and have little difficulty in understanding the material covered in this series and its application and importance in a systems sense. We also assume that the reader has familiarity with normal circuit elements and principles as they apply to the pulsed power world. It will than be seen that our purpose is to draw attention to the important aspects of system design, performance, and operation.

The editors intend that this series consist ultimately of about a dozen volumes. The first two volumes concentrate on those aspects of the field which usually are of the greatest concern as regards performance and reliability, to wit, switching technology. Thus, volumes I and II cover opening and closing switch technology and we refer you to the foreword in each volume for an in-dept analysis of its coverage. Other topics to be covered in later volumes will include Energy Storage, capacitor, inductor, mechanical, chemical (batteries, explosives, etc.), Voltage Multipliers; i.e., Marx generators, LC lines, Blumlein generators, etc.; Pulse Forming; Electrical Diagnostics; Components; and a volume on Supporting Analytical Tools such as field plotting codes, magnetic insulation codes, etc.; and certainly one or more volumes on Pulsed Power Applications.

With this array of intended publication on Advances in Pulsed Power Technology, we hope to lay the foundation for the field for many years to come. Certainly as this series of "Advances" are released, new directions and aspects of the field will be revealed. It is expected by the series editors to dutifully address new "special power" issues such as space-based power, inductive storage systems employing repetitive opening switches, materials in adverse environments, etc., as they reach appropriate stages of development or interest. Actually, we anticipate that the need for pulsed power information will outrun our ability to publish the appropriate volumes. This phenomenon is indicative of a vital and practical technology.

Certainly as this series of "Advances" are released new directions and aspects of the field will be revealed. It is expected by the series editors to dutifully address new "special power" issues such as space-baced power, inductive storage systems employing repetitive opening switches, materials in adverse environments, etc., as they reach appropriate stages of development or interest. Actually, we anticipate that the need for pulsed power information will outrun our ability to publish the appropriate volumes. This phenomenon is indicative of a vital and practical technology.

It has been our extreme pleasure to have been part of the pulsed power scene to date, a rather close knit community. Our undertaking this effort to document a chronology of pulsed power advances has been our pleasure. We express our sincere appreciation to the individual volume editors both presently identified and for those to come. If the readers have suggestions for subjects to be included in this series or would like to contribute book chapters or serve as volumes editors, we would be most receptive to talking with them.

A.H. Guenther
M. Kristiansen

Series Editors

Lubbock, Texas
1996

PREFACE

As indicated in the Foreword to this series on <u>Advances in Pulsed Power Technologies</u>, the pioneering roots of modern pulsed power as related by J.C. "Charlie" Martin and his co-workers of the Atomic Weapons Research Establishment, Aldermaston, Reading UK is an important if not essential record of the experiential history of the major developer of pulsed power advances during the post-World War II period. It finds great utility as an instructive accounting of the trials, tribulations and, finally, an almost chronological walk through their thoughts as they diligently and happily travel the yellow brick road to success.

It is recounted in the inimitable style of "Charlie" Martin as only he can relate, with some insightful perspectives by Mike Goodman, a constant companion, and collaborator who shares his unique view of "Charlie" and the Aldermaston Group.

This collection of selected articles is unique, for in large part, the documentation of their struggle and final triumph have not been formerly published in any archival manner. One reason, we suspect, was the defense-related application and significance of their work, compounded by the constant need for progress which did not allow for the time consuming preparation of formal submission to the literature. This also explains the "urgent" and sometimes terse manner of their writings. Yet the material remains remarkably current because we are dealing, in large measure, with pulsed systems less sensitive to those factors involved in slower pulsed scenarios.

The primary source of our selections in this book are Charlie's notes or, later, his lectures on high voltage topics, where he brings his knowledge to focus. We have left most of his writings as we found them (they are what makes the reading so enjoyable and illustrative). However, we have occasionally redrawn his original graphs.

The papers in this book are approximately 50% of the unpublished reports from the pulsed power group at Aldermaston. Those papers selected for inclusion, focus, in a hopefully coherent manner, on those of a more fundamental nature useful in the design and operation of large pulsed power systems rather than application.

As this volume was being assembled, careful attention was given to consistency between written text, figures, and tables to ensure coherence. The editors deemed it desirable in a few instances to recalculate, regraph, or retabulate data for clarity. Where some significant alterations from the original were made, it has been so noted by a footnote. Those instances were limited to maintain as faithful as possible a reproduction of J.C. Martin's writings and style. In addition, one paper by George Herbert, heavily referenced in a key paper by JCM, is included for ready reference.

J.C. "Charlie" Martin's place in history as a pioneer in high voltage endeavors is well established. He received the initial Erwin Marx Award in 1981 (Albuquerque, NM) of the IEEE International Pulsed Power Conference for *"Two decades of consistent yet extraordinary pioneering innovation in pulsed power technology."* Please read both his biography and autobiography, which, more than any other comments, will prepare you, the reader, for what is to follow in J.C. "Charlie" Martin's matchless style.

To enhance the cohesiveness of the writings we have organized this volume as follows:

Introduction	Chapters 1, 2, and 3
Technology Overview	Chapters 4 and 5
Gas Breakdown	Chapter 6
Liquid Breakdown	Chapter 7
Solid Breakdown	Chapter 8
Vacuum Flashover	Chapter 9
Switching	Chapter 10
Beams	Chapter 11
High Voltage Design Considerations	Chapter 12

The editors wish to acknowlege and thank all of the Atomic Weapon Research Establishment (AWRE became AWE) contributors.

Again, we are indebted to Marie Byrd, our friendly and consistent recorder and typist for this series of volumes on Pulsed Power Technology. She is not only able, but also committed to this endeavor. Thanks much!

Read and enjoy!

T.H. Martin
A.H. Guenther
M. Kristiansen

PREFACE

CHAPTER AND SECTION SUMMARIES

The papers naturally separated into main chapters and sections. Even though many sections were written at different times, grouping by date was not needed or done.

The Chapters 1-3 give an introduction and the development flavor of the rapidly moving Pulsed Power field. The fun of laboratory experiments, the best management help, and the wet blanket impact of regulations are provided by the AWRE (Atomic Weapons Research Establishment) group whose efforts finally went critical and brought Pulsed Power Technology into existence. A refined history in Chapter 1 is provided by M. (Mike) J. Goodman's article which was written for the AWRE corporate magazine. In the next chapters, J.C. Martin reveals all in his vivid style in an autobiography and then he tells the real story behind the rapid progress at Aldermaston.

Chapters 4 and 5 consist of two groups of articles that acquaint the reader with general pulsed power principles. First, and foremost, is the **Nanosecond Pulsed Techniques (NPT)** that is first presented as an overview and, later, can be used as a handbook by the reader. The fifth chapter is a collection of 5 lectures, given by Charlie at Hull University to introduce students and faculty to pulsed power.

Chapters 6-9 present information on breakdown in gases, liquids, solids, and on vacuum flashover. These chapters present AWRE's understanding of the basic breakdown phenomena along with experiments and explanations in each of the four major subheadings.

The Gas Breakdown Chapter starts with **The Pressure Dependency of the Pulse Breakdown of Gases** where the advantages of using short pulses is first shown. Next, is the **D.C. Breakdown Voltages of Non-Uniform Gaps in Air** that was written for the occasional designers of air gaps that did not breakdown at the expected electric fields. The third section **High Speed Pulse Breakdown of Pressurised Uniform Gaps** provides some theoretical background for fast pulse breakdown. The seemingly odd action of surface discharges is discussed in **Pulsed Surface Tracking in Air and Various Other Gases** where extremely long tracks were obtained with modest voltages using techniques that should be avoided in most pulsed power equipment. The last Section, **High Speed Breakdown of Small Air Gaps in Both Uniform Field and Surface Tracking Geometries**, provides data and discussion of these phenomena.

The **Liquid Breakdown Chapter** continues the breakdown theme with liquids. **Comparison of Breakdown Voltages for Various Liquids Under One Set of Conditions** shows the advantage of using less than one microsecond electrical pulses. **A Possible High Voltage Water**

Streamer Velocity Relation provides an interpretation and breakdown data for large multi-megavolt water gaps. An appendix to this section, **Velocity of Propagation of High Voltage Streamers in Several Liquids** by H.G. Herbert, provides the supportive data on oil, carbon tetrachloride, glycerin, and water. **Large Area Water Breakdown** provides data and direction for further experiments that are reported in the next section. **Interim Notes on Water and Breakdown** provides data and records a valiant attempt to conduct the next logical sequence of experiments. The results were not those expected and further research is needed. With privilege we acknowledge **T.H. Storr** as a co-author of this section. With regret we recognize his premature death. We will all miss the other sections that he would have contributed. The concluding section in this chapter is **Point Plane Breakdown of Oil at Voltages Above a Couple of Megavolts**. This provides data and analysis for large oil gaps.

The **Solid Breakdown** chapter contains three sections that provide insight into solid dielectric breakdown phenomena. Solids have the largest electrical field hold-off capability with practically no time dependence. This is shown in **Volume Effect of the Pulse Breakdown Voltage of Plastics**, and **Pulse Life of Mylar**. The volume effect is shown to exist and to be an intrinsic characteristic of solid electrical breakdown. **Pulse Breakdown of Large Volumes of Mylar in Thin Sheets** provides data and adds credibility to the defect hypothesis of breakdown.

The final breakdown chapter is on **Vacuum Flashover**. The section **Fast Pulse Vacuum Flashover** summarizes the AWRE work. The weak link in the power flow chain is generally the vacuum flashover limit. The vacuum power flow response to submicrosecond pulses is summarized and a formula predicting flashover is given.

The **Switching Chapter 10** contains data and information on pulsed power switching techniques. As contrasted to the Breakdown chapters, the intent of this chapter is to define the "breakdown process." Energy loss, inductance, rise times, and jitter in voltage and time are part of this process. The first section, **Solid, Liquid and Gaseous Switches**, summarizes and reviews the state of simple pulsed power switches. The next section, **Duration of the Resistive Phase and Inductance of Spark Channels**, outlines how to calculate energy losses and inductance of channels to provide switch risetime estimates. Recent results by one of the Editors (T.H. Martin) indicate that this method underestimates the actual losses by at least a factor of three (liquids) or more (gases) and the recent literature should be reviewed. The next section, **Multichannel Gaps (MG)**, along with **NPT** were the editors' favorites. **MG** is a classic that overflows with data, interpretation and results. Reading this section and browsing the data is recommended. **High Speed Breakdown of Pressurized SF_6 and Air in**

PREFACE

Nearly Uniform Gaps describes spark gaps for some of the AWE fast rise time pulsers. For futher analysis, one of the Editors (T.H. Martin) has used these results in a Pulsed Power Conference Proceeding. The section **Four Element Low Voltage Irradiated Spark Gap** outlines a 7 kV output fast breakdown gap that can be triggered by voltages from 0.5 to 1 kV. This low voltage trigger can initiate the larger systems.

The **Beams** Chapter 11 contains two sections on the first loads for pulsed power drivers. The first section is **Performance of the Tom Martin Cathode** which outlines a simple cathode design that provides a relative unaffected impedance using a cylindrical geometry cathode shank inside a closed end cylinder. This electron trapping geometry also forms the basis for magnetically insulated power transmission over much longer distances. In the next section, the electron beam is analyzed. The beam is formed in vacuum and then strikes a flat anode. By measuring the level and distribution of the radiation output as shown in **Electron Beam Diagnostics Using X-Rays** peak voltage levels can be obtained.

The final chapter is Chapter 12, **High Voltage Design Considerations**. This chapter contains information on constructing pulsed power devices. This chapter shows how to build simple and inexpensive high voltage systems using readily available materials. The first section is **Measurement of the Conductivity of Copper Sulphate Solution**. This section describes how to mix and measure copper sulphate water solutions that can be used for inexpensive high power resistors. The next section **Electrostatic Grading Structures** details how to design low field enhancement factor structures using simple materials. The final two sections **Pulse Charged Line for Laser Pumping** and **Notes for Report on the Generator "TOM"** details design, construction, and testing of two systems.

This work has been carried out with the support of the UK Ministry of Defence

T.H. Martin
A.H. Guenther
M. Kristiansen

CONTENTS

Chapter 1
HIGH SPEED PULSED POWER TECHNOLOGY AT ALDERMASTON
    M.J. GOODMAN 1

Chapter 2
J.C. "CHARLIE MARTIN, C.B.E. 15

Chapter 3
BRIEF AND PROBABLY NOT VERY ACCURATE HISTORY OF PULSED POWER AT
ATOMIC WEAPONS RESEARCH ESTABLISHMENT ALDERMASTON 21

Chapter 4
NANOSECOND PULSE TECHNIQUES 35

Chapter 5
HULL LECTURE NOTES
Section 5a
    Hull Lecture Notes No. 1 **Dielectric Breakdown and Tracking** . 75
Section 5b
    Hull Lecture Notes No. 2 **High Current Dielectric Breakdown
                                       Switching** 95
Section 5c
    Hull Lecture Notes No. 3 **Marx-Like Generators and Circuits** . 107
Section 5d
    Hull Lecture Notes No. 4 **Fast Circuits, Diodes and Cathodes
                                       for e-Beams** 117
Section 5e
    Hull Lecture Notes No. 5 **Odds and Sods** 125

xvii

Chapter 6
GAS BREAKDOWN
Section 6a
    Pressure Dependency of the Pulse Breakdown of Gases 135
Section 6b
    D.C. Breakdown Voltages of Non-Uniform Gaps in Air 139
Section 6c
    High Speed Pulse Breakdown of Pressurised Uniform Gaps . . 145
Section 6d
    Pulsed Surface Tracking in Air and Various Gases 155
Section 6e
    High Speed Breakdown of Small Air Gaps in Both
    Uniform Field and Surface Tracking Geometries 169

Chapter 7
LIQUID BREAKDOWN
Section 7a
    Comparison of Breakdown Voltages for Various
    Liquids under One Set of Conditions 177
Section 7b
    A Possible High Voltage Water Streamer Velocity Relation . . 181
Section 7c
    Large Area Water Breakdown 191
Section 7d
    Interim Notes on Water Breakdown 201
Section 7e
    Point Plane Breakdown of Oil at Voltages Above a Couple
    of Megavolts 225

Chapter 8
SOLID BREAKDOWN
Section 8a
    Volume Effect of the Pulsed Breakdown Voltage of Plastics . . 227
Section 8b
    Pulse Life of Mylar 235
Section 8c
    Pulse Breakdown of Large Volumes of Mylar in Thin Sheets . . 245

Chapter 9
FAST PULSE VACUUM FLASHOVER 255

CONTENTS

Chapter 10
SWITCHING
Section 10a
    Solid, Liquid and Gaseous Switches 261
Section 10b
    **Duration of the Resistive Phase and Inductance
    of Spark Channels** 287
Section 10c
    Multichannel Gaps 295
Section 10d
    **High Speed Breakdown of Pressurised Sulphur Hexafluoride
    and Air in Nearly Uniform Gaps** 335
Section 10e
    Four Element Low Voltage Irradiated Spark Gap 351

Chapter 11
BEAMS
Section 11a
    Performance of the Tom Martin Cathode 367
Section 11b
    Electron Beam Diagnostics Using X-Rays 375

Chapter 12
HIGH VOLTAGE DESIGN CONSIDERATIONS
Section 12a
    Measurement of the Conductivity of Copper Sulphate Solution . . 413
Section 12b
    Electrostatic Grading Structures 417
Section 12c
    Some Comments on Short Pulse 10 Terawatt Diodes 425
Section 12d
    Pulse Charged Line for Laser Pumping 439
Section 12e
    Notes for Report on the Generator 'TOM' 489

Index 537

CHAPTER 1

HIGH SPEED PULSED POWER TECHNOLOGY AT ALDERMASTON*

M.J. Goodman

INTRODUCTION

Interest in devices capable of generating many millions of volts and passing hundreds of thousands of amps for tens of nanoseconds has grown considerably in the last few years. At the present stage machines have been built in the United States which can produce X-ray doses in excess of 7000 roentgens (R) at 1 metre in 100 nanoseconds with voltages as high as 10 million volts and electron beam currents around 250,000 amps (1).

At Aldermaston the problem was first investigated eight years ago (1961) by a division of S.S.W.A. led by J.C. Martin, when methods of improving the radiographic facilities existing in the bomb chambers were outlined and tackled. At that time the largest machine was a 30 MeV electron linear accelerator capable of production 0.3R at 1 metre in 450 ns. The beam energy is rather high for good radiographic contrast and the dose too low except for the most sensitive X-ray film. The machine was used to radiograph explosively driven systems and the exposure time was rather too long and resulted in significant motion blur. Some improvement occurred by reducing the pulse to 150 ns. but this brought the dose down to 0.1R at 1 metre, which reached the limit on film speed available with the objects then being studied.

---

* First Printed as "High Speed Pulsed Power Technology at Aldermaston" in AWRE News, Vol. 16, No.5 pp. 18-23, March 1969.

There seemed to be two reasonable ways of designing improved systems. The first involved a super linear accelerator concept where the beam current is increased from the usual 100 mA during the pulse to 100 amps and such a device, called Phermex (2), has been built by Los Alamos Laboratories. The principal objection to this method is cost: it involves the use of very large resonant cavities and high power R.F. valves. The second approach was to produce a generator capable of driving, say 50,000 amps at 6 MV for 30 to 50 ns through an X-ray tube. The second approach had the merit that it was capable of being made with relatively cheap materials, but as a technology was nowhere near as well developed as the first. We decided to try the second course of action and concentrate on making cheap fast pulse generators and hope the X-ray tube would't have time to develop serious trouble.

A series of machines were constructed over the next few years to study the design problems of generators, X-ray tubes, cathodes, electron beam transport and focusing, targets, monitoring and many other aspects of the work. The largest of this sequence was a machine called S.M.O.G. of which two are still in use in the bomb chambers in H. Area. This can give between 5 and 10 roentgens/pulse at 1 metre with a 25 ns X-ray pulse length and operates with 3.5 to 4 MV on the X-ray tube and a load impedance around 100 ohms.

These early machines were quite sophisticated in design and operation and enabled us to do a great deal of work: for instance, they produced the concept of multi-layer X-ray tubes capable of holding electric fields of over 2 MV/foot which are now in use in virtually all present-day generators here and in the States.

Most of the high-speed X-ray generators produced so far can be split into three parts:

(A) D.C. capacitor store.

(B) Fast charged high speed pulse forming section.

(C) X-ray tube.

The reason for fast charging item (B) is that liquids and gases will sustain considerably greater electric fields for periods of a few microseconds than they will D.C. For both oil and water long duration pulses approaching seconds can cause breakdown at fields as low as 10 kV/cm, whilst pulses of microsecond duration can enable both oil and water to be used at fields over 300 kV/cm, in certain circumstances. As the energy stored goes as the square of the field strength, this means that the pulse forming section can be several orders of magnitude smaller than if charged D.C.; indeed, due to ionic conduction, water could not be charged D.C. at all. As mentioned above, a high speed (tens of nanoseconds) delivery is required to avoid premature breakdown of the X-ray tube envelope and to prevent gross plasma formation on the cathode and anode which would also cause serious collapse problems.

In addition to delivering the stored energy quickly to the load, at some stage the voltage has to be considerably increased. Whilst it is possible to D.C. charge compressed gases up to several million volts directly (Ion Physics Corporation has used this technique successfully) we lacked the engineering techniques and expertise to follow this idea. The generators built so far can be split into two basic classes; those with large voltage multiplication in (A), and those in (B). Some small voltage gain normally occurs at each stage at which energy is handed on, but it is necessary to have one stage at which the main voltage gain occurs. Those systems which use a Marx generator do it essentially in (A) but generators like S.M.O.G. and similar stacked line devices do it in (B). Before describing some of the generators in more detail, it might be worth while talking a little more about some of their common characteristics.

D.C. CAPACITOR STORE

In most contemporary generators the energy store is also the voltage multiplication section in the form of a Marx Generator. This, in essence, is just an arrangement of condensers which can be charged in parallel and then, by causing a series of switches (usually gas) to breakdown, connecting them in series to give nV volts where there are n stages of condensers at V volts each. In the case of a Marx, it pays to use the highest voltage condensers easily available and here we have been lucky to have had access to a large supply of 0.5 μF 100 kV low inductance capacitors destined for the canceled I.C.S.E. experiment which would have been the successor to Zeta. Figure 1-1 shows a typical, small, 45 KJ, I.C.S.E. condenser Marx (Mini 'A') prior to installation in an oil tank. Most of the Marx generators built here are installed under oil to prevent voltage breakdown. the output appears between the plate at the bottom and the condenser body connected to the round plate at the top of the perspex switch column. To reduce the number of switches yet again, the condensers are joined together and charged plus and minus. Hence a nominal 2 MV Marx would contain 20 condensers and 10 switches rated at 200 kV each. In the simple theory given in the textbooks, the Marx is a self-triggering system such that when the first gap breaks down the second gap sees double the volts normally existing across it and hence breaks down and this process rapidly runs up the stack, breaking all the gaps. In fact most of the older generation of Marxes had to operate within a few percent of the self break level because, due to the arrangement of stray capacities, the second gap only saw a few percent of the overvolt pulse, certainly not double. A new layout has been

Fig. 1-1 A typical Marx.

devised which arranges for the stray capacities to be such that the generator can be fired at 25% or less of the self break level. Again, in the conventional Marx a great deal of the cost lies in the spark gaps and the charging resistors, which we have reduced by using standard 6 in. Perspex tubing and ordinary 2 in. balls for the gaps, pressurised to no more than one atmosphere with SF_6, and standard garden hose P.V.C. tubes filled with copper sulphate solution for resistors. These techniques save not only money but also time, being very simple to make. If the output exceeds 1 MV, the generator is put under transformer oil to prevent flashover problems.

An alternative way to obtain high voltage gain is to use a pulse transformer fed by some relatively low voltage condensers (3). This approach requires the use of low inductance condensers if the transformer is not to get very large, and similarly, iron cores are virtually impossible because of breakdown problems. Figure 1-2 shows a 3 MV air cored transformer driven by a 1.5 μF. 100 KV condenser system sitting in its oil tank (empty). It is being used here to test the breakdown strength of transformer oil between the two circular plates located in the bottom right of the

Fig 1-2   A 3 MV air cored pulse transformer

picture. the tube on the right of the test plates is a copper sulphate dump resistor and the two tubes under the beard are part of a graded safety switch to prevent the volts lasting too long on the transformer.

   Hence all pulse transformers built by this section are air-cored. In order to prevent turn-to-turn flashover from the sharp edges of the thin copper foil windings, the transformer is impregnated with dilute copper sulphate solution of a carefully calculated conductivity. The conduction currents through the solution spread the electric field out as the pulse rises and relieves the otherwise impossible fields existing on the sharp edges. The highest voltage to which these transformers have been taken is 4 MV; this was from a unit 12 ins. diameter and 18 ins. high with a pulse lasting a few microseconds. At this voltage the transformer has to be under oil and is usually put in a cube of oil 1 metre on a side; at 4 MV the cube is completely covered in sparks, but this is more a strain on the nerves than on the system. In essence the transformer system is very simple, but it is very sophisticated in practice and will only handle about 20 kilojoules of energy before it starts to tear its leads out, and similar problems.

FAST CHARGED HIGH SPEED PULSE FORMING SECTION

This part of the machine can be in many various physical forms but electrically they are nearly all the same. Considering the main function of this section, that is, it should be capable of delivering its stored energy in a very short time; this requires that the circuit have a band width of tens of megacycles. This usually implies that it is in the form of a transmission line, either coaxial or as a strip line, which is usually two strips of conductor spaced either side of a piece of dielectric where the separation is less than the width.

The simplest circuit is one where a transmission line is pulse charged and then a switch connects its inner conductor to a load, usually the X-ray tube. A characteristic of this circuit is that under normal load conditions half the volts on the line appear on the output load. If the circuit is designed for high-current, low-voltage operation there may be some advantage in this, but for X-ray work where the X-ray production goes as the cube of the voltage, it is inefficient to lose so many volts after struggling so hard to get them.

The most common circuit which is used to overcome this problem is the Blumlein circuit, named after Mr. A.D. Blumlein, the brilliant young British Scientist who was killed in a bomber crash during the war (4). This basically consists of two transmission lines in series, one of which is switched at the end remote from the load. The load is put across the two outside parts of the lines which are nominally earthy during the charging phase, thus avoiding volts across the tube except during the main pulse. With this circuit the volts developed across a matched load is the same as the volts applied to the lines during charging; in practice, for X-ray production, the load is about three times the output impedance of the circuit and this results in 1.5 times as many volts across the load as were applied to the line. Most of the modern circuits use a single Blumlein line charged to a high voltage but it is possible to couple many such circuits together to produce high voltage gain as well as high speed pulsed and such a machine is, in fact, S.M.O.G., referred to earlier.

The dielectric used in this section of the machine can take many forms: for instance, polythene, perspex, mylar, water and transformer oil have all been used at some time or another. The high speed switch needs to have low inductance and resistive phases to ensure that the pulse rise time is shorter than the pulse, not always an easy job. In addition to the materials mentioned above which have also been used as switch components, freon gas and highly compressed SF_6 have also been employed.

THE X-RAY TUBE

One of the most important steps forward was to show that, providing the pulse was short enough, the vacuum in the X-ray tube could be rather poor and not affect its performance significantly (5). Until this time, every effort had been made to achieve very high vacuua in flash X-ray tubes and this had automatically meant glass envelopes, prolonged baking and a consequently drastic limitation on the materials that could be employed in the tube's construction.

Having established that the pressure could sometimes be as high as 10^{-2} mm Hg, we built a simple tube using a standard fork and spoon as the cathode and anode. Figure 1-3 shows one of our earliest attempts at an X-ray tube, at 300 KV and passing 5000 amps for 25 ns it gave several mR at 1 metre and would have done even better if we could have afforded a gold spoon!. This tube worked quite well up to the voltage of 300 kV. Apart from the considerable expertise and time required to produce the older, very high vacuua tubes, their size was strictly limited by the supply of suitable glass and if any small stray currents of a few hundred amps of 2 MV electrons went into the walls - (something that happens quite frequently) - then you either had a very poor vacuum or a broken envelope. However, with the relatively poor vacuum now allowable, plastics could be used to make big, stable and quite cheap tube envelopes.

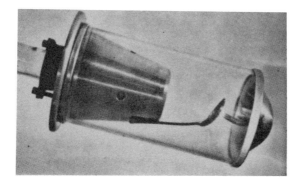

Fig. 1-3 An early X-ray tube that proved a point.

We were also able to show that a multistage design with inclined inside surfaces could support gradients well over 2 MV per foot, even when the surface of the insulator had suffered considerable dendriting from stray electron beams hitting them.

An extensive study was made of the effects of various angles and electric field configurations on different materials to discover the optimum for X-ray tube performance and tests were performed on perspex, polythene, glass, silicone rubber, ordinary rubber, and even ice was tried (5).

With regard to the cathodes used, anything partially works, and a surprising array of objects work well. We have used balls, small balls on balls, tubes, pointed rods, roughened spheres, razor blades, and human hair. One of the facts that has emerged about all these is that they are really covered in plasma blobs and a rather more sophisticated cathode we have developed acknowledges this fact and consists of plastic inserts in a curved metallic surface. Due to the initial field gradients across the plastic inserts, or possibly the pulse occurring prior to the main pulse, breakdown occurs over their surfaces and electrons are then extracted from these plasma covered surfaces. The virtue of this cathode is the electron trajectories are much better defined and enable better focusing to be obtained under some conditions. The production of the X-rays is an almost trivial part of the business, the electrons being allowed to hit a more or less flat anode plate of a high Z material. For voltages over 1 MV the X-rays pass through the anode and irradiate the volume beyond it: if less than 1 MV, the target is viewed through a window mounted in the side of the anode cap to avoid self-absorption by the target.

BRIEF REVIEW OF SOME OF THE SYSTEMS DEVELOPED
IN THE LAST FIVE YEARS (1964-1969)

Table I summarises the salient characteristics of the AWE machines selected for reference and in addition there is a brief comment about them in the following text. Some of the figures are estimates for performance not necessarily realised; where this is so, it is noted as such. For instance, the maximum open circuit voltage is not a level to which the generator is taken because it could cause breakdown for no good reason. Similarly some machines have been built for purposes other than X-ray production and in this case the figures refer to the doses that would be achieved if a reasonable cathode anode set-up had been used. The rise times and dose figures quoted are about 20% accurate. As far as numbers of machines are concerned, some three SMOGs, seven WEWOBLs and two PLATOs have been built and only single versions of the rest exist at the moment.

HIGH SPEED PULSED POWER AT ALDERMASTON

TABLE I
AWE Pulse Power Machines

	SMOG	WEWOBL	PLATO	MINI 'A'	MOGUL	MINI 'B'
Slow (μs) Pulse Charging Circuit	Ringing Capacitor	Capacitor Feeding Pulse Transformer	Capacitor Feeding Pulse Transformer	Marx	Marx	Marx
No. and Type of High Speed Circuits	20 polythene folded strip Blumlein	1 water strip Blumlein	1 oil strip Blumlein	1 coaxial oil Blumlein	1 coaxial oil Blumlein	1 coaxial oil Blumlein
Commissioning Date	February 1963	July 1966	July 1966	September 1967	September 1967	March 1968
Volume of Dielectric	385 gals	12 gals	250 gals	2,600 gals*	17,500 gals	3,400 gals
Total Energy in Store	15 kJ	2.25 kJ	9 kJ	42.5 kJ	95 kJ	45 kJ
Max. Line Voltage	0.22 MV	0.35 MV	2.5 MV	2.5 MV	4.2 MV	2.5 MV
Peak Current carried by High Speed Switch	3,600 kA	100 kA	50 kA	340 kA	200 kA	270 kA
Max. $\frac{dI}{dT}$ at Switch	$30.\ 10^{13}$ amps/sec	$2.\ 10^{13}$ amps/sec	$1.\ 10^{13}$ amps/sec	$2.5\ 10^{13}$ amps/sec	$1.4\ 10^{13}$ amps/sec	$2.2\ 10^{13}$ amps/sec
e-Folding Rise Time	12 ns	5 ns	4 ns	14 ns	14 ns	12 ns
Output Impedance	30 ohms	7.5 ohms	80 ohms	12 ohms	30 ohms	16 ohms
Max. O/P Voltage Open Circuit*	5.4 MV	0.55 MV	3.2 MV	3.4 MV	7.6 MV (tube 6.1)	4.1 MV
O/P Pulse Length	30 ns	45 ns	13 ns	20 ns	48 ns	50 ns
Max. Energy Delivery Rate (into Matched Load)	$25.\ 10^{10}$ watts	$1.\ 10^{10}$ watts	$3.\ 10^{10}$ watts	$25.\ 10^{10}$ watts	$50.\ 10^{10}$ watts	$28.\ 10^{10}$ watts
Typical Tube Impedance	120 ohms	12 ohms	200 ohms	35 ohms	100 ohms	50 ohms
Max. Dose at 1 M.	12 R	.06 R	1 R	25 R	300 R	Estimated 50 R

* possibly not achievable with X-ray tube

SMOG

In many ways the most sophisticated generator yet built at A.W.R.E., it used 20 or 40 triggered solid dielectric switches which switched 3.4 million amps in about 12 ns, giving a rate of current rise at the switches of some 3.10^{14} amps/second. The high speed section was 20 stacked strip Blumlein circuits contained under water in a large fibreglass tank. The dielectric in these lines was polythene and they were pulse charged to a maximum of 220 kV by ringing from two sets of ICSE condensers which were plus and minus charged 100 kV in some cases and minus 200 kV in others. The output was at the end remote from the switches and the fact that the generator was completely shorted out by numerous switch and other cross connections made no difference to its output. This was because the finite velocity of light is such that the pulse is over before the output end can discover the input end exists. Whilst the system is complex and time consuming, e.g. changing a set of switches takes 15 minutes and cleaning a tube an hour, SMOG did enable a lot of work to be done on the tube and enabled us to show that a large amount of current could be focussed into a small spot for radiographic uses. A pair of these generators are installed alongside the H.1 bomb chamber and have taken many radiographs both singly and together.

WEWOBL

This is a fairly small system and uses a pulse transformer to charge the high speed section via a long cable. This is to enable the storage condensers to be located outside the bomb chamber in somewhat cleaner conditions, whilst the high speed section within the chamber has only to withstand the pulse volts. The transit energy store uses water as the dielectric and demonstrates the very compact and low impedance systems that result from the use of this high dielectric constant material where the velocity of light is nearly a tenth of that in free space. Because the energy is well below a million volts, a knife edge cathode is used which produces a line X-ray source on an anode only a couple of millimetres away from it. The radiographs are taken along the axis of this line source, so giving a spot of small dimensions.

PLATO

This is a parallel strip line generator using oil as the dielectric. It is fed by a pulse transformer and uses a high speed gap made of compressed SF_6. This required some development but finished up as a very reliable component capable of quick and simple adjustment. Because of the nature of the parallel strip lines used in the high speed section, access is particularly easy in this generator and in addition to a lot of tube and cathode development it has been used to put high speed pulses on materials and objects and test them. This is also one of the cheapest systems made.

MINI 'A'

This was designed to be a cheap quickly built generator with a modest output but a short pulse. It was designed, constructed and commissioned with the help of another Division in a period of two months. It has proved quite reliable in operation and fired over one thousand times to date. The Marx generator uses ICSE condensers and pulse charges the oil filled coaxial Blumlein in the

HIGH SPEED PULSED POWER AT ALDERMASTON

manner pioneered by Physics International. The high speed gap is a self-closing oil gap located between the inner and intermediate cylinders. The cylinders are separated from each other by short polythene legs which are cylinders with a flared out section near the positive electrode. Other than a little trouble with dirt, this machine has been very reliable.

MOGUL

Mogul is a larger oil filled coaxial Blumlein system shown as Figure 1-4. This figure shows a cross sectional view of Mogul. The Marx is located, under oil, on the left, the high speed Blumlein section is in the centre mounted on wheels and the X-ray tube is located in the end of the outer cylinder on the right. The high speed switch is at the left hand end of the Blumlein section mounted between the inner and intermediate cylinders.

The line is charged with a Marx with an output voltage, open circuit, of 3.8 MV. Considerable development work was put into the Marx before the rest of the system was assembled. The triggering range of the Marx is down to 50% of self-break and can go even lower with extra triggering points added. The system is rather more engineered than other generators built by SSWA and has been incredibly reliable, the oil gap, for instance, having a much lower scatter on breakdown points than would be predicted, being about 2%. The tube has performed quite well, after revealing one or two new points about electron transport and has given nearly 300 roentgens at 1 metre in a pulse and could go higher. Some radiographs are shown in Figure 1-5. This radiograph was taken with Mogul producing about 20 R at 1m. Crytallex film was used at 3 metres from the anode. This is a static picture but the results would be the same with a rapidly moving system. The oil filled cylinder is the bottom end of an I.C.S.E. capacitor and shows the air filled expansion bellows used for temperature compensation.

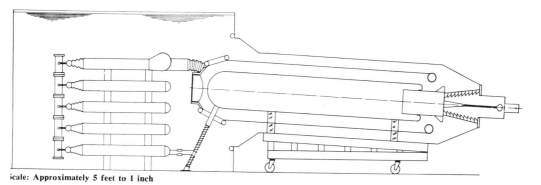

Scale: Approximately 5 feet to 1 inch

Fig. 1-4 Scaled cross section of Mogul.

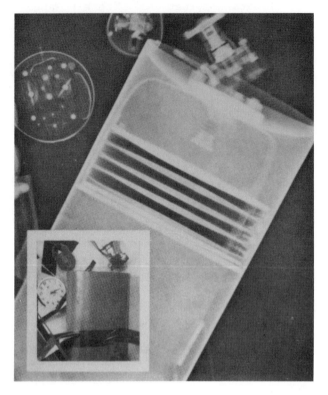

Fig. 1-5 Radiographs from Mogul.

This machine has had some six hundred firings and most of the work has been devoted to reducing the spot size and investigating the problems of transporting a beam of electrons a metre or so and then bring them down to a focus. The radiographic requirements on this machine can nearly all be met at this moment and work on this machine has now been suspended to enable the Marx to be used to charge a new variety of test lines involving water.

MINI 'B'

This is a later version of Mini 'A' with a longer pulse length to use the stored energy more efficiently. Some work has been done with this machine on electron beams and cathode design and it is now being connected to a larger Marx (Mini 'C') to drive the Blumlein a bit harder.

SUMMARY

As can be seen from the above bewildering display of devices, this field is hardly what one would call settled and it seems to have the promise of many exciting developments in the future as even higher rates of energy transfer become possible. Finally I would like to acknowledge the help of Mr. J.C. Martin, who contributed greatly to this article, and to the other members of his section who have been an inspiration to me.

REFERENCES

1. New Scientist, 1 Aug. 1968, p. 240; Financial Times, 29 July, 1968; Sandia Lab News, 20 15.1.

2. LA. 3241/TID-4500. Phermex. Douglas Venable (Ed).

3. Pat. App. No. 50127/64; Notes on Construction of a High Voltage Pulse Transformer. J.C. Martin, P.D. Champney, D.A. Hammer, Cornell University, CU-NRL/2.

4. Pat Appl. 589127, Oct. 1941; Electronics and Power, June, 1967, p. 218.

5. "Pulse Breakdown of Insulator Surfaces in a Poor Vacuum", I.D. Smith, International Symposium on Insulation of High Voltages in Vacuum, M.I.T., Oct. 19-21, 1964.

Chapter 2

J.C. "Charlie" Martin, C.B.E.

At the 3rd IEEE Pulsed Power Conference convened in Albuquerue, New Mexico on June 1-3, 1991, J.C. Martin was presented the initial Erwin Marx Award in recognition of "two decades of consistent, yet extraordinary pioneering innovations in pulsed power technology". From the proceedings of that conference, we reproduce both a biography and an autobiography, which more than any other comments will prepare you, the reader, for what is to follow in J.C. "Charlie" Martin's matchless style. A photograph of the presentation is included in Chapter 2.

Conference Biography*

During the early 1960's, John Christopher Martin and his group at the Atomic Weapons Research Establishment (AWRE), Aldermaston, England proposed and demonstrated a unique combination of ideas and hardware which allowed building terawatt pulsed power accelerators. Concepts and basic data concerning Marx generators, fast pulse liquid dielectric breakdown, switching, and vacuum flashover generated by him are still the standards in this field.

J.C. Martin confers regularly with pulsed power groups throughout Europe and the United States. The large accelerators now operational show the impact of this collaboration.

---
* First printed 3rd IEEE Pulsed Power Conference Proceedings, 1981, then updated 1991, 1995.

In 1977, he was awarded the United States of America Defence Nuclear Agency Exceptional Public Service Gold Medal in recognition of his many contributions. In 1990, he was made a Commander of the Order of the British Empire by HM Queen Elizabeth the Second. In 1946, Charlie graduated from Kings College, London with Honours in Physics. He joined the UK atomic weapons project (later to become AWRE) when it was established. He was for many years a Deputy Chief Scientific Officer with special merit at the AWRE where he continued to generate novel and innovative concepts. He then worked part time at AWE (as it is now known) for several years. Charlie is presently retired in London where he has purchased a flat and, according to rumor, a computer.

AUTOBIOGRAPHICAL DETAILS*

J.C. Martin

John Christopher Martin, born 21 September 1926 of middle-class parents who still loved him despite his early decision to call himself Charlie. This was partly because he could not spell Christopher and partly because he loathed A A Milne's poem "Christopher Robin". Attempts were made to educate him at a minor public school (i.e. private school) and Kings College London. After getting a war time degree and doing a years post graduate research into bridge and life in Soho, he joined the UK atomic weapon project a week before it started, at the mature age of 20.

During the first 18 years in what was to become the Atomic Weapons Research Establishment, he was regularly moved from field to different field of research every 3 or 4 years. This may have been because his services were widely required, but a more likely explanation is that people just wanted to get shot of him. As a result of this he knows nothing about a lot of things rather than everything about nothing. Some 17 years ago he strayed from the field of flash radiography into what was to become modern short pulse power technology and was trapped there by minor success. At this time he had two great strokes of good luck, a management that allowed him to produce nothing very concrete for 3 or 4 years and a small but very good group of scientists who did all the real work. He was also fortunate in not having anyone around who knew anything about high voltage technology and hence what could not be done.

---

* First printed 3rd IEEE Pulsed Power Conference Proceedings, 1981 as written by Charlie.

His major present fear is that since he is now deemed an expert, he too could act as a brake on progress. Early co-operation with scientists in the USA proved to be both stimulating and personally very enjoyable to him and his group. The last fourteen years has been largely spent back in the field of flash radiography where some modest success has attended the group's efforts.

Being of limited ability he has had to adopt approaches of such stunning simplicity that they border on the moronic. A mania for cheap constructional techniques has led to vast expenditures on black tape and simplex cement which have however sometimes resulted in rather economical machines.

He has been blessed by a supportive management who have never failed to find the money to support the group's work. He believes that the group have done their work nearly as economically as possible and hopes the poor British taxpayer has had value for money.

He stands in awe of the present developments of pulse power technology in the States and has been known to express fellow feelings with Baron Frankenstein. His major hope for its future is that it will occasionally work almost as well as intended. His major wish for those working in it is that they will have as much fun as he has had and make as many friends, although he doubts if this is actually possible.

J. C. "CHARLIE" MARTIN, C.B.E.

A.H. Guenther presents the first Erwin Marx Award to J.C. Martin while the Conference chairman, T.H. Martin (one of the editors) look on. Our other usually visible editor of this book. M. Kristiansen apparently is taking the photo. The award recognized J.C. Martin's initiation of and technical support of the pulsed power community.

Chapter 3

BRIEF AND PROBABLY NOT VERY ACCURATE HISTORY OF PULSED POWER AT ATOMIC WEAPONS RESEARCH ESTABLISHMENT ALDERMASTON

J.C. "Charlie" Martin
11 December 1991

As I get older and older exactly what happened and especially when it happened seems less important than why it happened. The following is an outline of the events leading up to the work done by the pulsed power group at Aldermaston and a brief resume of what I think we may have achieved.

As far as I am concerned modern short pulse power technology at AWRE started with an experiment done in building H36. This was a meter wide 3 meter long mylar insulated DC charged Blumlein generator. This was built in about a week including supporting research concerned with tracking margins, etc. The line was in a shallow perspex box under freon and was DC charged to 50 kV. It was switched by a hammer-operated blunt tin tack (US usage thumb tack) which was estimated to switch 2 MA with a rise time of perhaps 5 ns. The objective was to take the 50 kV pulse in the 1/20 of an ohm output line and stack this in a transit time isolated pulse adder and generate a 1 or 2 MV output pulse into a hundred ohms or so.

I can well remember the result of the first test. The line was charged with absolutely no trouble and only a little crackling to the full 50 kV. Having charged it to the full voltage so easily there seemed to be little point in not firing the machine. There was the usual noise of the hammer hitting the tin tack and, in addition, a sharp crack. About 50 small columns of smoke rose through the freon all round the edge of the lines. Far from being depressed at the apparent failure of our first test, I was delighted that we had accidentally stumbled on a multi channel solid dielectric switch.

Indeed this was a constant theme of the early work of the group. If the experiment worked, that was fine. If it did not, which was usually the case, you simply changed the objective of the experiment and considered exploiting the unexpected occurrence.

I am mildly embarrassed by the fact that while I can remember exactly what happened, I cannot for the life of me recollect when we did it. It seems likely that it was during 1958, but this is not certain. Anyway if anyone wants a date this is as good as any other. The third of a century anniversary will thus have been sometime this year (1991).

I would now like to back track to explain how we arrived at this initial experiment.

I had the very good fortune to be shifted from job to job during the first ten years of my work in the UK Atomic Weapon project. In one of these jobs, I had been involved in neutron detecting systems which counted at rates of up to 100 million counts per second and had efficiencies of detection of 100% or thereabouts (in addition to neutrons there were some gamma rays as well as X-rays from the inelastic scattering of the neutrons). I had also dabbled in a related technique where a gamma ray source is used in conjunction with moving metal components. Changes in the apparent output of the source can be correlated with the location of the metal. However, I was not very enamoured by the technique as its spacial resolution was poor and the results liable to misinterpretation.

At AERE Harwell a group had built what was then a powerful 25 MV linear accelerator. The accelerated electrons so produced hit a high-Z target, generating X-rays which in turn were used to produce bursts of fission neutrons via the giant photoneutron resonance at about 12 MeV in uranium. In a subsequent development, the uranium was made out of a slightly subcritical assembly and the neutron pulses multiplied by a factor of about 20 or so. These sub-microsecond large neutron bursts were used to measure neutron cross sections by time of flight methods, etc.

Anyway, the very able group were building a new larger linear accelerator and had offered us the old one for free. I got to hear about this and also heard that my site would probably decline the gift. I then went to see Dr Pike (in those days a junior staff member would not address him as Herbert even though he was the very nicest of people and also very interested in archaeology). I said we should take up the offer and radiograph models of explosive systems, etc. and that it could be very useful. A few days later I was called to a meeting with the Director (then Sir William Cook, one of the best directors we ever had). After a half-an-hour discussion, we decided to acquire it. I cannot forebear to compare this with the endless meetings, committees, proposals, studies, quality assurance assessments, environmental impact statements, etc. that would now be de rigour.

I returned to the various jobs I had in hand and in due course the accelerator arrived and was installed along side one of 3 bomb chambers. The job of installing and running it was given to the Safety Section for historical reasons. Because of the large load of work on the Safety Section (which in my opinion they did superbly) only two junior staff could be spared for what was indeed a very challenging job. There the accelerator languished (understandably), largely ignored by safety and as yet unappreciated by the owners of the bomb chambers, the Hydrodynamics Division.

In yet another job change, I was later moved into said Hydrodynamics Division (H Area) partly as a result of some suggestions I had made. This involved six months of relative tedium with Ted Walker and Sid Barker in a theoretical study of whether it was possible to light a small amount of DT with explosives. In retrospect, I learned a lot which was later of significant use, but at the time our small group wisely decided to see if there were any approaches which might have even an outside chance of success. Hence the work was purely theoretical and I am an experimentalist. At the end of half-a-year we wrote a paper outlining lots of approaches but concluded that the project was so difficult that no practical work be done until some new approach surfaced. During this period I learned I was slated to become the head of a new superintendency. I was horrified and I count as one of my major achievements to have abolished this latent superintendency. Also looking back at nearly 40 years unsuccessful ICF efforts I think our conclusions at the time were sound.

The result of this abolishment was that I was now located in the H area without a job to do. So was the non functioning linear accelerator for which I felt a measure of responsibility. Nemesis struck and I got stuck.

There were a number of reasons why the accelerator had languished, but the major one was the klystron that powered it. This was a UK copy of the Stanford 20 MW klystron. It had the mildly unfortunate characteristic that the cathode needed 100 hours running in and its average life was 110 hours. There was a spare but this was a mirror image of the original and using it was impracticable under the cramped conditions the accelerator now found itself.

The first job of the small group I had was to replace the 20 MW klystron by two 6 MW commercial klystrons. Also steps were taken to focus the 0.2 amp current as well as possible.

It was during this time that Ian Smith joined site. In an initial interview we had waved our hands at him and suggested he take on the focusing problem. When he actually arrived on site, he was given some other job. Quietly but firmly he declined this and said he would leave if he did join us. Anyway, after about a year's work we had an 0.8 mm spot and about 0.2 R at 1 meter of X-rays from 25 MV electrons in a short pulse. Alan Stables had been a tower of strength during this phase and subsequently. Unfortunately, having a small X-ray spot was only of use if located close to the explosive charge. I had gaily proposed blowing up the end of the machine on every shot. Indeed two fast acting valves had been installed to this end. At this point Mike James came to the rescue and developed the Snorkel. This was like a re-entry body in reverse, sticking into the large bomb chamber.

The focusing method which Ian developed had a high quality DC magnet followed by a coil which produced a pulsed 50 Kilogauss field. In order that the field could get to the electron beam there was a glass tube section very close to the anode.

I do not recall this tube ever cracking although the anode was frequently within 6 inches of several pounds of explosive, a fine tribute to Mike James.

Now having a workable flash X-ray machine of reasonable output and excellent spot size, a period of remarkable progress followed. In part this involved making fairly small lumps of explosive behave as if they were much bigger. In addition, odd geometries were evolved to hide the weaknesses of radiography and exploit its strengths.

It is still inadvisable to report this work but one day I hope it will be possible. Much of it was elegant and all of it was highly informative. While many people contributed to the development of flash radiography on the hydrodynamic side, Peter Moore and Mike James made outstanding contributions. As a result of the rapid development of novel (at least to us) and useful techniques, we were asked what new machine developments were possible.

To any pulse power readers who have understandably nodded off it _may_ just be worth while waking up now as I am approaching the genesis of AWRE pulse power work.

We considered building a more powerful linear accelerator. Tentatively we thought we might get 1 amp of electrons to 100 MV. This would produce more dose (possibly at the expense of the spot size). However, the development would be beyond the capability of the small group that had accumulated and the machine would be expensive and complicated.

We also considered the possibility of an air cored betatron which might have a kilo amp of circulating current (with however a limited charge unless the machine was large). We briefly considered a plasma filled betatron, shuddered and moved on. Finally, the decision was made to go to 20,000 amps and 5 MV. The reason for the large current was that the current goes roughly as the inverse cube of the voltage for a constant dose. A subsidiary advantage of the high current approach was that the resulting X-ray spectrum was better for certain work than that from a 100 MV accelerator. I remember that I added that the generator should be the size of a Principle Scientific Officer's desk (I was then a PSO and was sitting at such a desk). For those unused to the opaque ways of the British Civil Service, each grade had various entitlements as regards to office furniture, hat stands, etc.

Thus was AWRE pulse power work born.

Looking back at the beginnings we had six substantial advantages.

Several of the small group had experience of short (sub microsecond) pulse technology at a few hundred volts. There was also an excellent bible ("Millimicrosecond Pulse Techniques" by Lewis & Wells, Publisher Pergamon Press) of which most of us had read.

We fortunately did not have a high voltage expert who would have known that the dimensions we were thinking of were impossible. Such an expert would have immediately ordered huge ceramic bushings etc.

We were all experimentalists with a well-grown philosophy of make it yourself in a few weeks, designing it as you went along.

We communicated almost all the time and, on occasions, even listened to what the other person or persons were saying.

As we were in a hurry, we raided component recovery stores and, on occasions, liberated locally stored material and equipment. We had a workshop containing a pillar drill and a clapped out lathe. In the same building there was the local, official engineering workshop with whom we maintained good relations and who were invaluable in knocking-up the more elaborate bits and pieces we always wanted yesterday.

Finally our senior superintendent, Dr Arthur Bryant, while interested in what we were doing, left us to get on with it. As we initially spent little money, apart from salaries, we were not a large financial item. Consequently, we did not have to spend half of our time justifying our work in front of committees reporting bi-monthly, inventing milestones, etc., etc. We were most fortunate, we just got on and did it.

Dr Bryant was a remarkable person, he knew what was going on in the division in detail by occasionally wandering around and chatting to the people. I recall one occasion about 10 years after we started. Dr Bryant appeared when Mike Goodman was about to fire MOGUL A and look at the propagation of a 30 kA relativistic electron beam. This involved open shutter camera work, so before the firing the laboratory was blacked out. Unbeknown to Mike a valve had not been completely closed and the oil level in the Marx generator tank was lower than it should have been. Anyway Mike fired the machine, there was a shattering bang as the Marx earthed itself in a long track over the surface of the oil. This was followed by the sound of tinkling glass for what seemed to be half a minute or so. The cameras were closed and the laboratory lights turned on. Dr Bryant kindly thanked us and wandered out, I am sure, aware that something untoward had happened. We were collectively astonished, since we knew there was absolutely no glass in the machine. Two minutes of a fault finding disclosed the explanation. Two fluorescent lamps had been mounted on the end of the machine and the shock had spalled these into some equipment racks.

The role of good senior management: to provide just adequate resources, to discretely monitor progress, to back an able group that appears to be going in a potentially fruitful direction, and finally, to encourage them by enlightened interest is under appreciated. The late Dr Bryant was an outstanding boss and a first class scientist to boot, as well as being an extraordinarily nice human being.

In due course, Dr Bryant retired and Mr Schofield took his place. Once again we had all the backing we could ever expect and it is a great pleasure to record this.

One of the consequences of this trust and support is that we all felt obligated and motivated not to wander off into self-indulgent by-paths. Over the years there must have been several hundred occasions when we made observations which could well have merited further study. It is very difficult to quantify but I guess about 2% of our time was spent in academic or old style university research. Indeed we could reasonably be faulted for not having done more. However, we all felt our work was opening possibilities for solving difficult problems for AWRE and this was what we were being paid to do.

While I am dealing with overall direction issues, it is probably worth-while recording how the group functioned. While everyone could do everyone elses job, certain people had or acquired an extra degree of expertise or interest in some field. George

Herbert tended to get vacuum work, Tommy Storr electronic recording, etc., etc. They certainly did not specialise in this field it just was that their advice was always sought or if there was a big job they would make the major decisions and do much of the work. This was never really organised, it just seemed to happen.

With regard to more formal management, at irregular intervals (maybe two on average a year) there was a full group meeting. Once this was held in the car park because it was a (rare) beautiful day. These happened when there was some substantial policy decision to be made and/or some official site responsibility to be allocated.

Corralling all the group together in one room was a major exercise and I sometimes felt like a dog with a particularly recalcitrant flock of sheep.

Where an official administration of management requirement had to be met, the group would (with much moaning) eventually agree we would go along with it. The delicate phase as to who was going to be lumbered next ensued. Fairly rapidly the group would decide who was prime victim. At this point the prime victim usually started to back track and withdraw his support. However, gentle persuasion combined with group blackmail usually prevailed. Bribery was also used and I can remember Mike Goodman being offered his own marked car parking space to become building manager. I think he even got it for a few months.

More interestingly was the resolution of technological disputes as to the way forward to meet some challenging objective. Mostly the general approach would be agreed, but sometimes there was substantial disagreement. When this happened an agreement was reached that the two competing ideas should be worked on in parallel for say two months. At the end of this time we would have a second meeting to review progress and make a final decision. The two competing groups would work like mad to tackle the outstanding difficulties and resolve them, thus justifying the selection of their approach. As the group communicated continuously, the progress of two parties was know on an almost daily basis. Strangely enough the second meeting never took place, the losers drifted off to other work, still proclaiming that their approach would have worked "if only it had been given a fair crack of the whip . . .".

The end of these meetings was always the same, people just started drifting away. Formal decisions were never made, but in practice it was all resolved and, nearly always, in an ultimately amicable way. However, the passage to this end was far from the quiet reasoned discourse of savants. I regret to record that on occasions the closest analogy was "The Storming of the Bastille".

For this deplorable state of affairs I am sure I was solely to blame.

Returning to technical matters, our initial efforts tended to some form of pulse stacking. This was motivated by the belief that as the pulse duration became 10's of ns, many of the known breakdown processes could not occur. Hence it was hoped, that the breakdown and tracking fields would rise substantially. This would have two consequences. Firstly, the dimensions of the feed lines, etc would shrink and the energy feeds would become much lower inductance. Secondly, spark channels would become shorter in switches, again decreasing their inductance (at that time we did not know of the resistance phase but this also decreases).

Looking backwards in time (always a dangerous thing to do), it now appears that we were too worried about breakdown, surface tracking, etc. Hence there was a distinct tendency to attempt to get too much gain in the high speed (~ 10 ns time scale) section rather than in the slower (~ μsec time scale) section that, early on, was used to pulse charge the fast section. While our first few efforts (such as the initial test described above and various cable and line stacking devices) were DC charged, we rapidly moved on to pulse charged systems. In these, transformers or Marx generators were used to charge the fast sections in about one μsecond.

It should be remembered that 30 kV was a fairly high voltage for general laboratory usage at the time we started, but because we had access to free 100 kV capacitors (a left over from an aborted plasma physics machine at Culham), we initially used 100 kV stages in our Marxes or transformer drivers and then pretty quickly went over to ± 100 kV charging, that is 200 kV on the Marx gaps. Even so, many of our initial generators of any size went for about equal voltage gains in the slow and fast sections. While this was aesthetically satisfying in this case it was not the optimum way to go.

Over the first two or three years the group built a bewildering array of machines (mostly in the 1 or 2 MV range and typically storing a few kilojoules). The high speed sections used polythene, mylar, oil, or water as the main insulating and storage dielectric. These were charged with air cored transformers or Marxes.

It would probably be impossible now to detail these attempts and very tedious if I did, however, a list of their names might be amusing. The people who <u>built</u> the generator got to name it.

BRIEF HISTORY OF PULSED POWER AT ALDERMASTON

<u>Partial List of (some) Honour</u>

Agamenon	Moomin	It
Dagwood	The Water Marx	Ace
Gusher	Elsatron	Splatt and Splattlet
Dreadnaught	Triton and Snail	Pebble
WOBL	Lark	Snoopy
Wee WOBL	Rebel	Tom and Jerry

Some of these (notably Dagwood) gave rise to a family of generators, others were without issue (see Mike Goodman's article).

In parallel with this prototype building activity was a lot of other work. Principally this was concerned with two main areas of technical interest, breakdown and switching (which is nothing more than intentional breakdown). We had early on developed a simple air cored transformer which could be used on a laboratory bench and gave an output of up to 1 MV on a pair of large crocodile (US usage alligator) clips. These were invaluable in investigating the above fields of interest.

Versions of the air cored transformer were developed which gave 3 MV under oil in about a 1 meter cube perspex box. I remember Ian Smith doing some tests on a 3 MV pressurised SF_6 gap when it tracked outside over the gap body. The energy locally deposited was small but that stored in the pressurised gas was not and this demolished the cube, leaving Ian paddling around in a few inches of transformer oil.

Another major area of effort was monitoring the output. Here the late Tommy Storr laboured mightily and successfully. We probably devoted 20% of our effort to this activity and it is arguably the case that we should have invested more.

To digress a bit, the monitoring issue was a part of the reason we did not publish. We were happy that our measurements of say voltage and current were not badly adrift and adequate for what we needed. However, to publish data we would have needed to spend twice the time in order to be reasonably sure of our data. In fact a few years ago we were able to establish that our voltage scales were probably accurate to a couple of per cent. This was because we had developed a solid dielectric switch (a sheet of polythene with 50 needle produced stabs in it) which was very reproducible and worked up to 300 kV in a single switch and twice this in a double (triggered) version. Having recently stumbled on one of the old stabbers in a heap of debris we were able to establish that at least our initial voltage scales were satisfyingly accurate.

In addition to the major areas of interest listed above, there were a number of other (<u>not</u> in order of importance) power packs, quick and dirty methods of construction, and safety. The latter we took very seriously. Pick up suppression was subsumed in the work on monitoring and safety. The latter because we worked in an area which handled explosives and, in particular, various forms of detonators.

In parallel with the pulse power work, we also worked on cheap infrequently used high-current capacitor banks. This work culminated in the 2 MJ bank built by Angus MacAulay, Rex Bealing, Dave House and others, which operated at a maximum voltage of 30 kV on the capacitors. It used several unusual techniques which included a multi-channel-strip, solid-dielectric switch which broke in several hundred channels in some 10 ns. The short circuit current of the bank just down stream of the switches was 60 megamps. Each capacitor (made by BICC for us) stored 30 kJ.

As we moved on to prototype 2 MV, 40 ns generators we started work on the diode (or tube UK) and studied vacuum flashover. Field Emission Corporation was making very nice flash X-ray tubes involving very high vacuum technology and field emission cathodes. These of necessity involved glass envelopes and high technology processing. It seemed to us very important, if not imperative, that we get away from glass. Hence the work on vacuum flash over and multi-stage plastic tubes.

I recall quite early on we made a ~ 300 kV tube where the cathode was a fork, the anode a spoon and used the X-rays to take a radiograph of a knife. If (as rumour had it) AWRE had solid gold cutlery, we would have got twice the dose.

Once we could take the fast pulse into a vacuum we started work on cathodes and the production of tens of kilo amp relativistic electron beams and, in particular, their transport and focusing.

About this time we began to have close contact with the USA. As I recall the first people to visit us were Dave Sloan and Walter Crewson (then at EG&G). However, contact rapidly broadened to include Dr Alan Kolb then at NRL, Sandia Corporation at Albuquerque, Ion Physics, Physics International, and DNA. We already had a sort of communication link to Don Martin, Bernie Bernstein and Jim Lyle at Livermore Laboratory concerned with flash radiography. Hence, when the first two left Livermore to set up PI, this link remained.

While our main concern was still flash radiography, it had not been lost on us that the techniques we were helping to develop were potentially capable of raising the rate of energy deliverable by 2 or more orders of magnitude and the second time differential by 4 or 5 orders of magnitude. Our thoughts had therefore turned to a number of possible applications. A far out long term possibility was ICF. More immediate possibilities were really fast Z pinchs and the dense plasma focus. However, the interchange (particularly via DNA) alerted us to much more immediate applications in the general field of simulation, vulnerability and above ground testing. As AWE was going through a fairly long period where we had no major tests, the latter was particularly relevant to the work of the site. I (and occasionally other members of the group) began to commute across the Atlantic.

It was both a duty and a pleasure to give what help we could to others in the States. It was also very stimulating and satisfying. I do not think we gave what help we could with any other motive at the time, we just cast our bread upon the water. In this case with typical American generosity and drive, it came back buttered, covered with jam, and with a thick layer of cream on top.

It is invidious to select one memory out of so many, but perhaps the reader will indulge me. I recall having my first flight directly into San Francisco. This flight was both long and tiring and also involved an 8 hour clock change. I staggered off the plane, wishing I was dead, to be met by Don Martin. He hauled me over to San Leandro (rather as you would a sack of potatoes) and there Bernie Bernstein showed me the first coaxial oil Blumlein charged by a Marx. Even as I slumped against the machine sliding gracelessly to the floor and oblivion, I realised that this was the way to go for flash radiography.

I have occasionally wondered why we had not arrived at that solution too. Part of the reason has been explained earlier, that is, our unspoken belief that there should be a fairly big voltage multiplication in the fast section. George Herbert has rightly suggested a second reason was my statement that it should be the size of a PSO's desk. It is very easy to calculate at the high fields pulse break down permitted that the necessary energy (~ 10 + kJ) could be stored in much less than a cubic meter of water or mylar. Hence my off the cuff remark. It is also true that some of our initial attempts were very compact, and this probably over inclined us to seek tricky solutions (plus they were lots of fun). Anyway I am solely to blame if anyone is.

Our early interchanges were not limited to Government Agencies and Defence Contractors, but also most enjoyably involved a group at the Stanford Linear Accelerator group on streamer chambers,

Cornell and Maryland Universities on plasma physics experiments (thanks to Alan Kolb) and later other organisations in lasers, etc.

However from AWRE's point of view the defence aspects were the most significant and enormously worthwhile.

Before concluding this already overlong history, I had better bring the pulse power group's story rapidly up to date. The group worked seriously in the field for about 20 years. We made a series of home grown generators for simulation purposes. We also provided generators for other Government organisations in the UK. We puttered around in pumping gas lasers for a short while. We tossed off a number of flash X-ray machines from time to time but in general these were not properly engineered and fairly regularly disintegrated in use. About 12 years ago we decided we had to go back into flash radiography seriously, partly to provide a range of good flash X-ray machines working at different voltages. However, the main reason was that we could see our way to taking a radiograph of a weapon at an important time. This involves a great deal more than just having a suitable machine. There are many other tricks that have to be played and if good data is to be obtained, lots of other advances made. This has kept the relatively small pulse power group gainfully occupied and hopefully the British taxpayer has had value for his money.

Speaking of money I have done a _very_ crude estimate of the cost of pulse power at AWRE over the 22 or so years we worked seriously in the field. This comes out at around 9 million present day Pounds, which is about 0.4 M£ per annum. This is a lot of money but I can think of a number of times when we have saved more than we cost. It is also very nice to have helped pioneer a field in which maybe 5,000 people now work. I personally believe that pulsed power has modestly contributed to maintaining the balance of deterrence during the period that it was vital so to do.

What I am certain of was that it was all great fun and most enjoyable. We have gained a large number of good friends all over the world but particularly in the States. What also struck me very forcibly was the remarkable absence of the "not invented here" syndrome. This is not to say there were not warm and occasionally heated discussions about which way the field might evolve. However, when a new concept or approach was proposed within a year or two it had either been adopted where it was useful or regretfully discarded if it failed to live up to expectation. It is almost as if the field went from the Wright Brothers first flight to the 747 in 20 years with only a small fraction of the people in aviation contributing ideas and concepts. Partly it was that there were so many possibilities to exploit. However, I personally think that much of the reason for the N.I.H. syndrome being barely detectable was the fact that mostly the generator was only a stepping stone to the real objective. The machine builder was also its user in the early days and what turned the user on was the electron beam or the laser output or whatever it was designed (hopefully) to produce. It also did no harm that the field seemed almost over supplied with exceptionally nice and interesting people. I can only express my personal thanks for having been associated with the amazing developments of the last 3 decades and with all the friends I have made over that time.

In conclusion I have now arrived at the part that I have been dreading, that is acknowledging the people in the pulse power group who in this country kicked it off. Many people at Aldermaston have contributed to its work. In addition people (usually junior scientists) have come and gone after a couple of years. Also we have had excellent support from many non scientists and support groups. However to create a definitive list would be time consuming and boring to others. So I will compromise and list the core members and apologise to those I leave out.

These are:
Peter Carpenter, Phil Champney (deceased), Dave Forster, Mike Goodman, George Herbert, Ted Russell, Ian Smith, and Tommy Storr (deceased).

What we did, we did as a group.

Chapter 4

NANOSECOND PULSE TECHNIQUES*

J.C. Martin

INTRODUCTION

This review of pulsed high voltage techniques is of a rather personal nature and mainly reports on work done by the high voltage pulse group at AWRE, Aldermaston. I would like to acknowledge the efforts of this group, which over the past seven years has managed to elucidate some of the features of the design of high voltage pulsed systems. Reference will be made to the work of others and, where I fail to do so I would like to apologize in advance. Published references are not extensive and the lecturer must confess a weakness in not having searched diligently for these. There are a few general references at the back, Nos. (1) to (5) and in the note specific references are made to some sources of more recent material. This field has expanded rapidly in the last four years or so and very big, fast, high voltage systems are now in existence.

In general, high voltage pulsed systems may be DC charged or they can have their high speed section pulse charged. In the case of DC charged systems, where the voltages can rise to 12 million volts or more, the dielectric media employed for energy storage are gaseous or, at lower voltages, solid. Ion Physics Corporation, of

---
* Circuit and Electromagnetic
  AFWL System Design Notes
  Note 4
  April 1970

Burlington, Massachusetts, are the prime practitioners of DC charged, pressurised gas insulated pulse systems. Such systems require very great engineering expertise and much experience and it is not in general within the ambit of the experimental worker to build his own. DC solid insulated systems take the form of a cable, fast Marx or strip transmission line systems and can be fairly easily built in the laboratory. The simplicity of the DC charged systems is obvious but this simplicity can sometimes be bought at a very considerable price and, in general, for the more demanding applications a pulse charged high speed section makes sense.

In a system which pulse charges a high speed section, much greater energy densities can be stored and higher breakdown gradients transiently achieved. The energy storage medium can now be gaseous, liquid, or solid, and indeed one of the best media is water, which would be quite impossible to use DC. In addition to the energy store there is an output switch or switches, which deliver the energy rapidly to the load and these, too, can have a higher performance when pulse charged. The pulse charging supply can be relatively slow, typically charging in times of about 1 μs, which in this note is referred to as the "slow" time scale. The high speed store needs a fast, very low inductance switch or switches and must be designed using transmission line concepts so as to feed the energy into the load in a few tens of ns or less. This time scale is generally referred to as the "fast" one.

In addition to having breakdown criteria which can be applied to prevent breakdown in the high speed section and to achieve breakdown in the switch, it is necessary to have theoretical relationships which enable the rise of the pulse to be estimated. It is also necessary to monitor the pulses at various parts of the system and as these systems can have rates of change of current up to 10^{15} amps/sec, pick-up is a serious problem. However, even in these conditions monitoring can still be fairly simple. It is also necessary to have additional components such as load resistors for test purposes, or deflector gaps and loads which will prevent the system ringing on until the time that it tracks somewhere. In addition to this, it is necessary to have dump resistors so that the system can be safely discharged in the event of a malfunction.

After briefly outlining the various items mentioned above, a few systems will be described, to give some idea of currently achievable items. In general, small systems store less than 1 kJ and provide test pulses for breakdown studies, to drive spark or streamer chambers, or to operate other low energy absorbing items. Such systems can go up to a few million volts and occupy 100 square feet of laboratory, or so. Systems involving, say 100 kJ and above, will need a floor area ten times bigger and stand typically 10-15 feet high. These can give towards 10 million volts, but are

NANOSECOND PULSE TECHNIQUES

obviously rather outside the scope of a general laboratory. However, they do not require the enormous hangar-like buildings that the slower, older systems needed, largely because they employ pulse charged dielectric media which can stand rather greater stresses.

PULSE CHARGING OR SLOW SYSTEMS

Air Cored Transformers

These are deceptively simple systems which can give up to 3 or more million volts. However, a surprising amount of expertise is required to make them properly and, in general, a small Marx, even though it contains many components, is more reliable. At the low voltage end - 200 or 300 kV, say, - they work well and are very simple to construct. However, for completeness, the more advanced type will be described as well, (Refs. (6) and (7)).

The high voltage dictates that in general no magnetic materials can be used in the core and consequently the primary inductance is low. It is therefore necessary to keep the leakage inductance very low so that gains approaching that of a lossless transformer may be achieved. As a consequence of this, it is also necessary to have a low inductance bank to feed the primary turns and

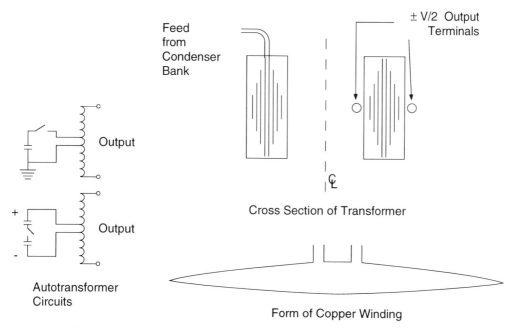

Fig. 4-1 High Voltage Pulse Transformer.

care has to be exercised to use reasonably low inductance capacitors and low inductance switching. At low voltages an isolated primary can be employed but at high voltages even small transients can break the insulation between the primary and secondary turns, so that an auto transformer design is employed. Figure 4-1 shows the basic circuit used, with a single condenser. Also shown is the preferred version which used plus and minus charging. In general, in all the applications in this note plus and minus DC charging is greatly to be recommended. The first reason for this is that the DC flashover conditions are eased, since flashover distances go as a fairly high power of the voltage. Secondly, in any power pack the addition of a second set of rectifiers doubles the output voltage cheaply. Thirdly, pick-up is reduced in a balanced system.

Typically, high voltage pulsed transformers will have up to 70 : 1 turns ratio and into small loads have gains a little more than this. They have winding thicknesses which are of the order of 20 per cent of the mean radius and Fig. 4-1 shows a cross-section through the winding. By tapering the winding, the transformer is reasonably well macroscopically graded, but in the absence of further palliatives there are very large fields developed at the edges of the winding, since the copper is only a few thou. thick. These high fields are avoided by impregnating the transformer with a dilute $CuSO_4$ solution, which forms thin resistive films at the edges of the copper. The resistivity of this solution is chosen so that the voltage pulse diffuses along the distributed resistor capacitor network so formed, a distance of a few mm. during the pulse. This technique of grading is effective but great care is needed so that the impregnation is complete. This is achieved by vacuum impregnating and by letting the solution into the transformer slowly so that the films of $CuSO_4$ solution can flow into all the windings. The output is of course ± V/2 and leads from the inside and outside of the transformer take the voltage to the test region.

For voltages in the region of 1 million to 3 million volts, the transformer sits in a cube of oil about 1 metre on a side, but for voltages up to about 1 million, a bench top model can be employed, with the working volume under oil or water. In the latter case, heavy capacity loading can be avoided by blocking most of the water volume out with perspex. Such a transformer should not be allowed to ring on at high voltages for long periods of time, as a track will eventually occur, and auxiliary water or oil gaps are used to dump the energy into a $CuSO_4$ resistor, after the pulse has gone over peak. The typical pulse rise time depends on the loading, but for a high voltage system is in the region of 1/2 μs. Such systems are particularly useful for doing scaled breakdown studies, tracking studies across interfaces, and experiments on switch design. In general, in pulsed voltage application, voltages scale, but some care and judgment has to be exercised. However, if

NANOSECOND PULSE TECHNIQUES

the model tests show that tracking will occur, then higher voltage full-scale tests are obviously necessary.

A very elegant transformer design which used the same volume as the DC capacitor has been invented by R. Fitch and V. Howell (Ref. (8)). This is known as the spiral generator and only needs a low inductance switch to generate high voltage pulses. Unfortunately, insulation flashover problems have limited it to voltages of under 1 MV or so.

In triggering gaps, fast and reliable pulses of 100 or 200 kV are very useful and these can be simply obtained by using much simpler pulse transformers. Typically, gains of 10 are all that are required and by using a one or two turn primary fed from a fractional μF capacity and spark gap, and by having a 10 to 20 turn secondary made out of the inner core of cable, this can be achieved. Additional insulation is placed between the secondary and the primary, but care is still taken to keep the leakage inductance low. Typically, such a transformer is 8" diameter and 8" high, and can work in air. The insulation is arranged to increase towards the high voltage end or ends of the secondary, which reduces the self capacity of the transformer as well, at a modest cost in self inductance. Such systems can have pulses rising in tens of ns when used into small loads.

Marx Generators

There has been a considerable development in the field of generators which charge capacities in parallel and then stack them in series by means of a number of switches. The prototype of such systems is the Marx generator, but the term has now been extended by usage to cover systems quite different from the original one, but still having the feature of stacking the charged condensers. Such generators are more complex than the auto transformer to make and involve many more components, but with care they can be made reliable and, using modern materials and techniques, can be constructed cheaply. Versions storing megajoules are being developed and systems giving up to 10 million volts and delivering their energy in a μs or two have also been made. (Ref. (9)). Many of the early Marx generators had an extremely limited triggering range and for large systems involving many gaps, care has to be taken that the mode of erection of the Marx is reasonably well defined and consistent. I personally would call a Marx healthy if it could erect satisfactorily at voltages down to 60 per cent of its self breakdown and under these circumstances there is a region, above 70 per cent, say, of the self breakdown voltage over which the erection time does not decrease much and is typically in the region of 1 μs or less. The reason for requiring a reasonable firing range is that small changes in gap performance, line voltages, etc. have little or no effect. In general, the Marx generators are built with fixed gaps and the operating range covered by changing the gas pressure and/or nature of the gas. At AWRE we have had satisfactory service using gaps charged up to ± 100 kV, these gaps being simply made out of 2" ball bearings inserted at intervals along a column of perspex tubing, 6" in diameter. To achieve a reasonable working range, the pressure can be changed from rather under one atmosphere to 2 1/2 atmospheres absolute of air and then the air replaced by dried SF_6.

Figs. 4-2 and 4-3 show a plot of the best achieved output inductance and erection time for large Marx generators. These are ones storing 100+ kJ. The parameters are quoted per million volts of output and it must be repeated are the best that can be achieved at the approximate date quoted. However, such graphs may be of use in showing what has been achieved and what can reasonably be expected to be do-able in the near future, for systems where the inductance of the capacitors is low.

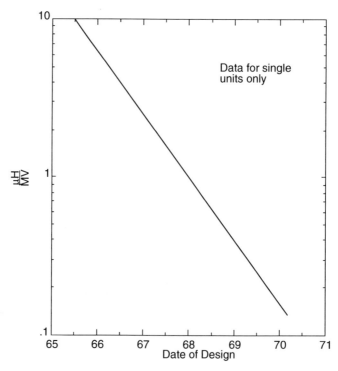

Fig. 4-2 Inductance per MV for best design for large Marx like systems E > 100 kJ.

NANOSECOND PULSE TECHNIQUES

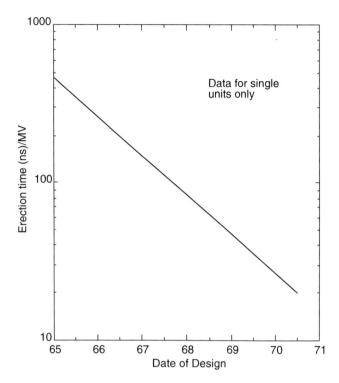

Fig. 4-3 Erection time per MV for best design for large Marx like systems E > 100 kJ.

In addition to developing single Marxes, they can be paralleled, thereby achieving a further reduction in output inductance, but care has now to be taken that erection of these sub-Marxes is consistent and reliable, and consistency is usually achieved by making the erection fast.

Versions now exist which will trigger down to zero volts, such systems having essentially a triggered pulse generator in each switch. Others employ triggered gaps which are fed back across a number of switches by resistors, so that the trigger electrode stays roughly at constant potential while the gap erects past it. The additional complexity of putting trigger electrodes in the Marx gaps enables Marx generators to become more compact and still have reasonable triggering ranges. Where the capacitors do not have to be very closely packed, then the original approach of Marx modified to give increased triggering range can be employed, and apart from a few gaps at the bottom of the Marx, the gaps can be untriggered.

There is a novel version of the Marx due to Mr. R. Fitch which is called the "L.C. Marx". In this, a switch and inductance is placed across every other condenser and when these switches are closed simultaneously, every other condenser rings and reverses its polarity. This leads to very low inductance systems, but has quite a bit of additional complexity in getting all the switches to go reliably and has some interesting fault modes. Yet another novel Marx has been developed by Mr. Fitch and his associates at Maxwell Laboratories Inc., San Diego, where a small, lightly loaded Marx erects very fast and cross-connections from this fire up a number of parallel Marxes, again with triggered gaps, in between each condenser.

Most of the Marxes, referred to here have particular virtues and these are, in general, of course bought at increased complexity. However, quite elementary Marx generators can be made healthy in quite simple ways. In the case of untriggered Marxes (that is, apart from the first few gaps), stray capacity built into the system can be made to link across every third gap. The magnitude of this capacity has to be considerably greater than the capacity across each gap and if two gaps have fired, nearly three times the stage voltage appears across the unfired third gap. These strays are usually built in by the capacity between the cases of the condensers, which are conveniently stacked in three columns. If it is not possible to do this, three sets of charging resistor chains can be employed, these chains coupling to every third capacitor. This discharges the unfavourable strays quickly for values of the resistors of a few kilohms. These can be made very conveniently and cheaply by using $CuSO_4$ resistors.

In the case of the triggered Marx, again the trigger electrodes are tied back by a resistor to the trigger of the switch 3 below.

These systems can of course be extended to couple across more than three switches but experience has suggested that, with care, coupling over three switches is adequate.

In the case of multimillion volt compact Marxes, they usually operate under oil, but Marx generators up to say 1 million volts can be built compactly in air, providing care is taken to prevent flashover. Such a Marx need not be more than 5 ft. high for a 1 MV output and can still have quite a low inductance.

The Marx generator can be a very reliable and consistent system, when properly made, and some versions can achieve a low enough inductance that they can feed a high speed pulse directly into the load. However, in general they are used to pulse charge a high speed section, some forms of which are mentioned in the next section of this note.

NANOSECOND PULSE TECHNIQUES

HIGH SPEED SECTIONS

As the high speed section has to be treated as a transmission line, the impedance of the simplest forms are given in Fig. 4-4. Since the local velocity of light is given by

$$\sqrt{LC} = \sqrt{\epsilon} \times 3.3 \times 10^{-11} = \frac{1}{v}$$

and the impedance by

$$\sqrt{\frac{L}{C}} = Z$$

if the latter is known, the inductance and capacity per cm length can be easily derived. The formula for the strip transmission line strictly only applies for thin lines but the impedance of fat lines can be obtained to about 10 per cent accuracy by calculating the

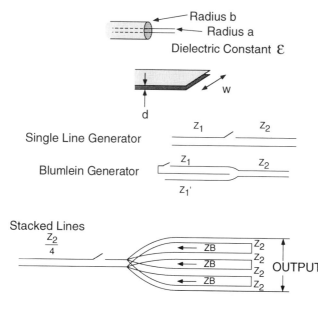
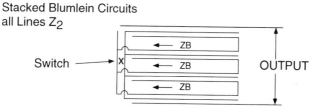

Fig. 4-4 Examples of transmission lines.

impedance given by Z_{trans} in parallel with $200/\sqrt{\epsilon}$ ohms. The extra impedance is roughly the energy flowing in the fringing fields. The impedance of a line of square cross-section in air is given as 200 ohms, which is correct. When the width of strip lines becomes very small compared with their separation, the impedance of parallel wires applies.

Two basic transmission line generators are also shown in Fig. 4-4. Firstly the case where a charged coaxial cable or transmission line (impedance Z_1) is switched into a load. This load can either be a resistance or another transmission line of impedance Z_2. The gain of the system is given by

$$G = Z_2 / (Z_1 + Z_2)$$

and is equal to 1/2 when the system is matched. This can be painful where great efforts have been expended in order to charge the original line to several million volts.

The second circuit shown in Fig. 4-4 was invented by A.D. Blumlein. The two transmission lines form the generator and these usually, but not necessarily, have the same impedance. For a duration equal to the two-way transit time the circuit acts like a generator of twice the charging voltage with an internal impedance of $Z_1 + Z_1'$. Thus the gain is

$$G = 2Z_2 / (Z_1' + Z_1 + Z_2)$$

which is equal to unity when the generator is matched ($Z_2 = Z_1 + Z_1'$).

For X-ray generation the optimum X-ray output is obtained when the load is about three times that of the generator, as the efficiency of X-ray production is a high power of the voltage. Under these conditions the gain of lossless Blumlein circuit is 1.5, so that 6 MV can be put on the X-ray tube from a system pulse charged to only 4 MV. If the high speed section is pulse charged from a Marx and the capacity of the Marx can be made twice as big as that of the Blumlein (considered now as a lumped circuit component), then in this case a further lossless ringing up gain of 4/3 occurs, so the Marx need only have an open circuit voltage of 3 MV. Various practical factors reduce the gains to be had at each stage, but approximately the calculations given above apply.

Both of the simple circuits mentioned above can be stacked to give voltage gain. Fig. 4-4 shows a four-stage generator using a simple transmission line to drive it. The generator only adds pulses for a duration equal to the two-way transit time of the stacked lines and hence the driving line is usually made the same electrical length. If the pulse rise time of the switch is comparable to this time, the pulse falls before it has finished rising and serious drops in gain may occur. Also, as is shown in Fig. 4-4, the volume between the stacked lines represents transmission lines going back to the switch. By making the impedance of the charged line very low compared with that of the stacked lines in parallel, a lossless open circuit gain of 4 is possible. However, real gains of 2 are more likely. In Fig. 4-4, the circuit of four stacked Blumleins is shown in the version where the simultaneously fired individual switches are replaced by a single switch. The lossless open circuit gain of the system is 8 and real gains of 6 may be achieved in practice. However, a very low inductance switch is needed to achieve an adequate fast pulse rise. A system of this form is described in Ref. (10).

In addition to the above circuits, tapered transmission lines, as described in Ref. (1) have been employed to obtain gain. However, such a system no longer produces square output pulses, even for ideal switching. Intrinsically the line produces a pulse which droops and such a shape is particularly susceptible to loss of gain with real life switches. A version of the tapered line which does not suffer from this disadvantage is a series of lines, each of which has an impedance twice that of the preceding one. In an ideal case a gain of 4/3 is obtained for each stage, with a small but acceptable energy reflection at each cable. However, real life small losses can reduce this gain substantially and in my experience this circuit is factually of not much interest.

In practical systems the coaxial version of the Blumlein circuit is particularly useful for generating high voltage pulses, an approach pioneered by Mr. D.F. Martin of Physics International, San Leandro. For smaller, laboratory bench top systems, stacked Blumlein cable systems can have open circuit gains approaching 10 when using careful construction and 8 or so stages. These use quite a lot of cable, but can have outputs of over 1/2 MV when employing reasonably sturdy coaxial cable. The impedance at the switch will be in the region of 2 or 3 ohms and a solid dielectric switch is usually the best way to get an adequate rise time, although advanced pressured gaps can be used. The system would of course be DC charged and can be simple and reliable. The size will be of the order of 1 metre cubed and hence is a bench top fast pulse system.

BREAKDOWN STRENGTH OF DIELECTRIC MEDIA

Before starting to consider the individual media of use in high speed sections, it is worth making a general point about the degradation suffered in the breakdown field when the area or volume of the system is increased greatly. These are quite general considerations and refer to all media and indeed refer to other areas, such as strength of large structures, etc. The basic point is that if a major parameter, such as the breakdown field for a given volume of solid, has an intrinsic scatter, then there must be a degradation of the mean field as the volume is increased. This can be shown by considering an ideal experiment where the mean breakdown field of a large number of samples is known as well as the intrinsic standard deviation. Ten samples are then stacked or laid out and the question is asked what the mean breakdown voltage of this and other sets of ten units will be. Obviously in any large group the unit with the lowest strength will break as a rising voltage is applied and the new mean field will correspond to the voltage which broke only about 6.5 per cent of the samples when they were tested one at a time. This means that the mean breakdown field must decrease as the volume of the sample increases and that the local rate of decrease can be calculated from the standard deviation of the unit sample breakdowns. This, of course, assumes that there are no measurement errors involved. If the standard deviation is σ, then for a change of 10 the shift in mean field is just about 2 σ. Thus for a σ of 12 per cent, which is typical of solid breakdown intrinsic scatter, an increase of volume of a factor of 10 will decrease the mean breakdown field to 80 per cent and the mean breakdown field will halve for an increase of volume of 1,000. In the case of solids, the breakdowns originate within the volumes for samples (other, possibly, than in thin films) and the effect is a volume one. However, for liquids and gases the breakdowns originate from the electrodes in general and hence the relevant parameter is the area of the electrodes. The reason the area is not quoted for, say, uniform field spark gaps is that, when conditioned, these have standard deviations of well under 1 per cent and an increase of 35 orders of magnitude would be needed to halve the measured breakdown field, so essentially the data is independent of electrode area in this case. However, for liquids and solids the standard deviations are frequently more than 10 per cent and the effect is quite important. Even for gases an area effect has been found in large high pressure DC systems, such as are made by High Voltage Corporation.

The distribution of breakdown voltages is not Gaussian but is a skew distribution which can be generated by replicating the Gaussian distribution a few times.

With regard to the way the breakdown data for media is presented, the point should be made that the designer only needs data to 10 or 20 per cent, since he will include a safety factor in general or be prepared to enlarge the size of the system slightly. In what follows several relationships are quoted and most of these are only approximately correct. However, in the absence of anything better they may be quite worth while, but they are not intended to have a high degree of accuracy. In any case, as there is an intrinsic scatter of solid and liquid breakdowns of the order of 10 per cent, accuracies much better than that are a bit meaningless.

In the next portion of this section, the breakdown of gases, liquids, and solids will be covered and in general one would like to be able to calculate the breakdown voltage of any of these for any area or volume of dielectric and for pulses of a given duration in any geometry. The pulse durations covered by the data are from a few μs to a few ns and in the case of the geometrical arrangement this is simplified to uniform, mildly diverging, or point geometries. In the case of gases and solids these are largely self-healing, although conditioning (both up and down) can occur. However, to the first approximation these materials do not have a life criterion, but for solids there is a definite life when operating at fields close to the intrinsic breakdown field and this, too, is covered briefly below.

Gases

As has been mentioned, Ion Physics have developed compressed gas insulated DC charged systems up to generator voltages of 12 MV. However, laboratory usage DC insulation in gases does not go much above 200 KV, but data in this range is useful for designing gas switches, etc., so some brief remarks on the DC breakdown of uniform and mildly non-uniform gases may be useful.

In uniform field gaps in air, the breakdown field can be expressed in the form

$$E = 24.6 \, p + 6.7 \, p^{1/2}/d^{1/2} \text{ kV/cm}$$

where p is in atmospheres and d in centimetres. For gaps in the region of 1 cm at 1 atmosphere, the field has the usually used rough value of 30 kV/cm, but this can rise substantially as the gap is decreased into the millimetre range. The expression also shows that the field does not go linearly with pressure when the gaps are small.

When spheres or rods are used as electrodes in gaps, unless these have radii large compared with the gap, the mean breakdown field can decrease considerably. The method of calculating the breakdown voltage, in the case of air, is to use the field enhancement factor (FEF) which is the ratio of the maximum field on the electrodes to the mean field. The FEF for spheres and parallel cylinders is given on page 6 of Alston's "High Voltage Technology". For instance, for spheres separated by their diameter, this factor is 1.8. It is now necessary to calculate the maximum breakdown field and this involves a length characteristic of the field fall off. Experiments show that this distance corresponds to the point at which the field is down to about 82 percent of the maximum on the electrodes. This gives an effective distance:

$d_{eff} = 0.115 \ r$ (spheres)

$d_{eff} = 0.23 \ r$ (cylinders)

For example, for a 1 cm diameter sphere, $d_{eff} = 0.057$

and $E = 24.6 + 28.4 = 53$ kV/cm

and using the FEF of 1.8, the mean field $\bar{E} = 29$ kV/cm.

As the gap is 1 cm in the example considered, the breakdown voltage is 29 kV at 1 atmosphere. However, at 3 atmospheres the breakdown voltage is 68 kV, significantly less than three times as great.

This treatment gives values within a few percent of the observed conditioned breakdown voltages, except when the gap is small compared with the radius, when the uniform field approximation should be used.

In addition to the non-linear dependency mentioned above, an additional non-linearity sets in when the air pressure rises above 10 atmospheres or so.

Another effect can change the breakdown voltage. If corona has set in, because the diameter of the electrodes is small compared with the gap, then the field on the electrode changes and the breakdown voltages can be significantly affected.

When hard gases such as freon and SF_6 are used, the dependency of breakdown voltage with pressure is considerably non-linear. From a practical point of view, freon is very much cheaper and is useful in flooding equipment to help to raise its breakdown. However, in gaps, dried SF_6 is much preferable but it must be changed reasonably frequently because the breakdowns cause its dielectric strength to degrade.

For pulse breakdown in gases in uniform or near uniform fields, there are two effects to consider. One is the statistical delays caused by waiting for an initiating electron to occur, to start the Townsend avalanche. However, for reasonable areas of rough electrodes, such as are used in pulse charged gas insulated lines, this effect only leads to increases of a few per cent. The second and more important effect is the fact that the avalanche process and the subsequent streamer takes time to occur and if the pulse is short enough, the breakdown field rises. Curves for various gases are given (Ref. (11)) which show, for instance, that for 1 atmosphere of air, the breakdown field has increased by a factor of 3 for a 10 ns pulse. The increase is only a few per cent for a 1 microsecond pulse and for high pressures the improvement decreases, so the effect is only useful for rather short pulses.

However, for hard gases the fall-off of the DC breakdown field with moderate pressures mentioned above does not set in until much higher pressures are reached. Thus pulse charged SF_6 is fairly linear up to 10 atmospheres, at which pressure fields above 600 kV/cm can be obtained with the proper electrode materials.

For very divergent fields such as apply to points or edges, a rather approximate relation applies for gaps greater than 10 cms or so. This rough expression is

$$(F_\pm)(dt)^{1/6} = (k_\pm) p^n$$

where F is the mean field (V/d) in kV/cm, d is the distance in centimetres, t the time in microseconds and p is in atmospheres.

Table 4-I gives the values of k and n for three gases.

The pressure dependency power only applies from 1 to 5 atmospheres or so. For air the time dependency disappears for times longer than of the order of 1 μsec for negative pulses and for several hundred μsec for the positive polarity. Thus for air at 1 atmosphere, a point will require about 1.5 MV to close across 100 cm in 100 ns.

For short pulses, mean fields comparable with 30 kV/cm can be achieved.

From the above integral relations differential velocity relations can be obtained but these are rather inaccurate and must be used with considerable discretion. However, lengths of incomplete streamers have been calculated which agree reasonably with experimental observations for high voltage, short pulses applied to wires.

Table 4-I
Values of K_+, K_-, and n for gases

	Air	Freon	SF_6
k+	22	36	44
k-	22	60	72
n	0.6	0.4	0.4

Liquids

For liquids, the breakdowns originate from the electrodes and usually from the positive one. Thus an area dependency would be expected and is indeed observed. The smoothness of the electrodes is not critical, provided gross roughness is avoided. In addition, impurities and additives have little effect on the pulse breakdown field strength. For instance, several per cent of water or carbon introduced into transformer oil has less than 20 per cent effect on the breakdown fields, in complete distinction to their effect on DC breakdown.

For uniform fields the breakdown field is given approximately by

$$Ft^{1/3} A^{1/10} = k$$

where F is in MV/cm, t in microseconds, and A is the electrode area in square centimetres. For transformer oil $k = 0.5$ and for water $k = 0.3$. However, experiments with diverging field geometries show that water has a considerable polarity effect, so that this k corresponds to positive breakdown streamers, while negative breakdown fields have the value $k_- = 0.6$. The expression is not very accurate and it is not clear that other liquids obey exactly the same relation, but if they do, methyl and ethyl alcohol have about the same value of k as transformer oil, while glycerine and castor oil have values of k which are about 1.4 times as big as transformer oil.

In general, for pulse charged liquid lines, transformer oil and water have been the principal media used. Water is particularly useful because its dielectric constant is 80 and remains constant up to about 1 gigacycle. Water has to be deionised, not because impurities affect its pulse breakdown strength but just to prevent it providing an ohmic load on the generator. Resin deionisers give water with resistivities above 1 megohm centimetre when it is recycled through them and this is usually good enough, as the self-discharge time is then 8 microseconds. Thus for a microsecond charge time some 10 per cent of the pulse charge energy is lost by ionic conduction.

For mildly diverging fields, the breakdown field is applied to maximum field on the electrode (making allowance for any polarity effect) and using the area of the electrode which is stressed to within 90 per cent of the maximum field. A small correction can be applied to allow for the fact that the streamers are moving into a diverging field and this typically increases the breaking voltage by 20 per cent or less.

For point or edge plane breakdown conditions, the mean streamer velocity has been measured for a number of liquids for voltages from about 100 kV up to 1 MV and the mean velocity is given over this range by

$$\bar{U} = d/t = kV^n$$

where $\bar{U}$ is in centimetres per microsecond and V is in MV. Table 4-II gives the values of k and n for transformer oil and two other liquids.

For water the relationships are rather different, being best fitted over the range mentioned by

$$\bar{U} t^{1/2} = 8.8 V^{0.6} \text{ positive}$$

and $\bar{U} t^{1/3} = 16 V^{1.1}$ negative.

Thus for oil, a negative point or edge will break down at 1 MV in 100 ns over a distance of about 3 centimetres. The fundamental mode of streamer propagation seems to be the negative one and for voltages in the region of 1 MV the slow positive streamers in water and transformer oil speed up and move with the velocity of the negative streamers.

For voltages from rather over 1 MV to 5 MV, both polarities in oil obey a relationship of the form

$$\bar{U}_{\pm} d^{1/4} = 80 V^{1.6}$$

A similar relation seems to apply to water, but with a lower value of k. Data is lacking in this case.

Table 4-II
Values of k_+, k_-, n_+, n_- for Liquids

	k_+	n_+	k_-	n_-
oil	90	1.75	31	1.28
carbon tetrachloride	168	1.63	166	1.71
glycerine	41	0.55	51	1.25

Solids

For solids the streamer transit times are very short and down to a few nanoseconds the breakdown field is independent of the pulse duration. The breakdown field is given by the expression

$$E\,(vol)^{1/10} = k$$

where the field is in MV/cm and the volume is in cc. Table 4-III gives the values of k (which is the breakdown field for 1 cc) for various plastics.

For thin sheets, the standard deviation of the breakdown field decreases and the breakdown strength becomes almost independent of volume. For 1/4 thou. sheets this occurs at 5.5 MV/cm, for 2 thou. sheets at 4.0 MV/cm and for 10 thou. sheets at about 3.0 MV/cm.

For diverging fields, the maximum field on the electrode is calculated and the breakdown field used which corresponds to the volume which is stressed to about 90 per cent of the maximum field.

For solids, there is a reduction in breakdown field for repeated pulses. This life is given by the expression

$$\text{Life} = (E_{BD}/E_{op})^8$$

where E_{BD} is the breakdown field and E_{op} is the field at which the life is wanted. The power in the life relation seems to be related to the standard deviation and for thin films of mylar the power is 16 or higher.

Table 4-III
Values of k for Plastics

	k
polythene	2.5
tedlar	2.5
polypropylethylene	2.9
perspex	3.3
mylar (thick)	3.6

NANOSECOND PULSE TECHNIQUES

As an example of the calculations applied to polythene, a volume of 10^4 cc has a mean pulse breakdown field of 1 MV/cm for a single pulse. However, this decreases to about 0.5 MV/cm if a life of 1,000 pulses is required. Thus a 1/16" sheet of polythene can be used in uniform field conditions at about 80 kV for a reasonable life.

For DC charged solids, several effects combine to alter the breakdown strength and usually, but not invariably, lower it. For instance, in some plastics conduction currents can heat the plastic and cause run away thermal degradation. Chemical corrosion from surface tracking can cause degradation of the breakdown fields, as can mechanical flow under electrostatic forces. All of these effects vanish in pulse charged systems, of course. In polythene, DC charging can give fields some 20 to 30 per cent higher than the pulse values. This is caused by enhanced conduction in the regions containing the defect that originates the breakdown. However, if the voltage is rapidly reversed, this "annealing" charge separation now adds to the field on the defect and polythene can be made to break with a pulse reversal of only 30 per cent when it has been DC charged. The time scale of this charge and annealing is of the order of milliseconds. Mylar and perspex show little or none of this effect and hence are to be preferred for DC charged strip transmission lines.

In addition to the above DC effect, both polythene and polypropylethylene show a polarity effect when pulse charged in diverging geometries. The negative point, or small sphere, is over twice as strong as the positive one.

Interfaces

This whole area is very complex and cannot be simply treated. In general, interfaces and legs can be designed to support fields as good as the main lines in which they are placed. In the case of diaphragms, the metal surfaces are recessed, so the field is reduced at the triple interfaces. In general, it is possible to avoid interface tracking in transformer oil, but in water, with its very big dielectric constant, the problem is much more difficult. However, acceptable solutions have been developed for pulse voltages up to 4 MV or so in a couple of systems. In this area more than any other, scaled experiments are of great help in developing promising solutions which can then be proved out at full voltage.

Edge Grading of Lines

When building strip transmission lines, if relatively thin copper sheets are employed, very large field enhancement occurs at the edges of the lines. This can be reduced to acceptable levels by using pseudo-Rogowski contours. These can be quickly made by using a file on thick plywood to solve Laplace's equation and then covering the thick rounded line with copper foil a few thou. thick. Another technique in pulse work is to use dilute copper sulphate films to resistively grade the edges during the pulse rise. A third technique of use for solid lines DC charged is to surround the edge of the metal foil with blotting paper, which helps to suppress surface flashover. Flashover voltages can be doubled, or more, by using this technique with care.

Vacuum Breakdown

This subject is rather beyond the scope of this note, but pulse fields can be very much greater than the DC ones, providing care is taken with the finish of the metal electrode surfaces. If there is no prepulse, oiling with transformer oil can raise these fields at which significant current flows to levels of the order of 1/2 MV/cm. In addition, self-magnetic fields can be used to suppress currents in vacuum insulated coaxial lines. As regards flashover across vacuum-insulated interfaces, Ref. (12) describes some work which has enabled X-ray tubes working at gradients of up to 2 MV per foot length to be made.

Switching

Before dealing with some aspects of switching, a general topic will be briefly covered. This relates to estimating the rise time of the pulse produced by a given switch. As was mentioned earlier, values good to 20 per cent are more than adequate and in order to be concise, various aspects which can occasionally be relevant will be omitted.

The rise time of the pulse, which is the same as the fall of voltage across the gap, of course, is largely controlled by two terms: the inductive term and the resistive phase term. The inductance of the gap is larger than that of the spark channel itself but in a well designed gap it is usually the major term. This inductance changes with time as the conducting channel expands and on occasions this can be important. However, in this note only a constant inductance will be considered. The second component of the rise time is the resistive phase. This is caused by the energy absorbed by the plasma channel as it heats and expands. In the case of channels working at reasonable high initial voltage gradients and driven by circuits of tolerably low impedance (i.e., less than 100 ohms or so), the plasma channel lowers its impedance mainly by expanding to become larger in cross section. The energy used up in doing this is obtained by the channel exhibiting a resistive impedance which falls with time. This expansion can be measured optically and correlated with the energy deposited in the plasma itself. A fairly wide range of experimental results have been collected and summarised in a semi-empirical relation which is easy to use.

The two terms in the rise time expressions both refer to e-folding times. For the case of a constant inductance L joined to a generator of impedance Z the voltage falls exponentially with a time constant $\tau_L = L/Z$.

The accurate calculation of the inductance of the channel is difficult and would require accurate knowledge of the channel radius. Fortunately it is not necessary to do this, because an adequate approximation is to use that of a wire of radius "a" fed by a disc of radius "b" in which case the inductance is

$$L = 2 \ell \ln b/a \text{ nanohenries.}$$

In this expression b >> a in practice and the value of L is only weakly dependent on a. The rate of expansion of the channel is of the order of 3×10^5 cms per sec. for air at one atmosphere and maybe half an order of magnitude greater for liquid and solid gaps, depending on the field along the channel when it forms. In the above expression ℓ is the channel length and is in centimetres. For rise times of a few ns, the log term is of the order of 7.

Incidentally, when care is taken in selecting approximations to the various sections of an array of conductors, simple inductance relations can be used to obtain the total inductance to an accuracy of 20 per cent or so. The way of getting the inductance per centimetre of a chunk of an equivalent line was given at the beginning of the High Speed Sections discussion.

The relation for the resistive phase is given in two forms, for convenience. For gases, it is

$$\tau_R = \frac{88}{Z^{1/3} E^{4/3}} (\rho/\rho_0)^{1/2} \text{ ns}$$

where E is the field along the channel at its closure <u>in units of 10 kV/cm</u>, ρ/ρ_0 is the ratio of the density of the gas to air at NTP, Z as before is the impedance of the generator driving the channel.

For solids with unit density

$$\tau_R = \frac{5}{Z^{1/3} E^{4/3}} \text{ ns},$$

where E is now in <u>MV/cm</u>.

To obtain the effective rise time of the pulse τ_{tot} the two times are added, i.e.

$$\tau_{tot} = \tau_L + \tau_R.$$

This time is the e-folding time, where the waveform is exponential and where the pulse rise is not, this parameter approximates to the time obtained when the maximum slope of the voltage waveform is extended to cut the 0 and 100 per cent values of the voltage, i.e.

$$\tau_{tot} \simeq V/(dV/dt)_{max}.$$

Where the waveform is exponential, the e-folding time is multiplied by 2.2 in order to obtain the 10 to 90 per cent time. However, where a number of equal time constants are operating, one after the other, this factor falls rapidly and eventually tends to a little less than unity. Then there is no unique relation between the e-folding time and the 10-90 per cent rise time and I would suggest that for many applications, the one used here is of more use than the standard parameter. In particular, it gives the value of di/dt_{max} reasonably well and is the main parameter of use in calculating the energy deposited in the spark channel.

After the main voltage fall has occurred, small voltages persist across the plasma channel as it cools, moves, bends, etc., as well as the more usual drops at the electrodes, but most of the energy is deposited in a time of the order of the resistive phase. In most gaps both terms are important, but for circuits with impedances of tens of ohms and higher working at modest gradients, the resistive phase may be several times as large as the inductive one.

Various approximate treatments exist to cover branching, non-uniform conditions down the channel, etc., but these are beyond the scope of this note.

Table 4-IV gives an example of a calculation for a plasma channel of length 5 cms with a starting voltage across it of 100 kV, in air at one atmosphere pressure. The calculations are shown for three different values of Z.

Table 4-V shows a representative approximate calculation for a solid switch with 200 kV across of 3 mm solid dielectric, such as polythene.

Table 4-IV
Rise Time (ns) Calculations for Air*

Z	τ_L	τ_R	τ_{tot}
100	.7	7.5	8
10	7	16	23
1	70	35	105

Table 4-V
Rise Time (ns) Calculations for Solid Dielectric*

Z	τ_L	τ_R	τ_{tot}
100	0.04	1.8	1.8
10	0.4	4.0	4.4
1	4	8.6	12.6

All the times are in ns.

As can be seen, calculations based on the inductive term alone can be badly wrong. The combined effect rise time is also used below in some brief comments on multichannel switching.

General Comments on Switching

This field is very extensive, so here, even more than in the rest of these notes, the comments are of a personal nature and in particular the next few paragraphs are given purely as my views, views, however, that I am prepared to defend to the bitter end.

In order to reduce the vast field to more manageable proportions, I consider that, apart from mechanical closure gaps (which can be very useful - see below), the main kinds are:

---

* Recomputed by Editors from original manuscript.

trigatrons
cascade gaps
field distortion gaps
UV triggered gaps
laser triggered gaps.

Disposing of the last two versions first, the UV triggered gap has a very limited operating range and while the laser triggered gap can have an acceptable operating range, it cannot do anything that cannot be done in a tenth of the time and cost by one of the preceding gaps (see Editor's Note).

The trigatron gap has been extensively used and, properly designed, can be reliable. It tends to require a smaller trigger voltage (with, however, more drive) and was much favoured when 10 kV was a high voltage pulse. Nowadays this aspect is almost irrelevant. When used in a reliable mode, however, it needs about as large a trigger pulse as the other gaps and as the discharge goes to a limited area, erosion can be a serious problem.

When large capacitor banks had to be reliably triggered for plasma physics work, the cascade gap was extensively employed and for many applications it is still of considerable use. It can be triggered with jitters of tens of nanoseconds, but in most versions it needs a small irradiation gap in the trigger electrode and is also fairly bulky and expensive.

For most applications I prefer the field distortion gap which, because of the small radius on the trigger electrode, does not need any extra irradiation, as field emission from the rough trigger provides the necessary initiating electrons. In addition, the length of the gap, which is closely related to the inductance, is a minimum in this gap and it can be cheaply and quickly made. It can also lead to the spark channels being well distributed along or around the main electrodes, so erosion is a smaller problem where this is a factor of importance. Figure 4-5 shows a cross section of a field distortion gas gap (which also goes by the name mid-plane gap). The gap can use either ball bearings as spherical electrodes or rods for a cylindrical electrode gap. The trigger electrode need not be in the middle of the gap and, indeed, in one mode of operation it should be offset. The field distortion or mid

---

* Editor's Note: (T.H. Martin)
For specifics on applications since this note was written, the reader should consult recent literature. Laser triggered gaps although expensive, are probably the better way for high voltage (5 MV), higher impedance gaps. Electrical trigger lines have considerable circuit loss for high voltage systems in water.

NANOSECOND PULSE TECHNIQUES 59

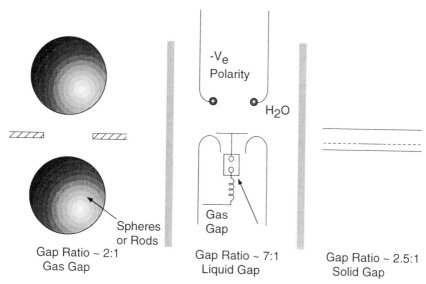

Fig. 4-5 Field distortion spark gap.

plane gap can be easily modified for use with liquids or solids and indeed for solids this form of construction is so simple as to be positively moronic.

Consideration will now be given to a few of the factors involved in gas, liquid, and solid gaps, concentrating mainly on the field distortion gap as a personally preferred type.

Gas Gaps

In general, gaps become easier to trigger and operate faster the higher the voltage at which they work. The difficulty is usually to stop them working in unwanted ways such as tracking. In general, brass is an excellent material to make the electrodes out of and this material is equally good with air or SF_6. It can be easily worked and its erosion rate is as low as any other material in almost all circumstances*. With regard to the gases to be employed, air and SF_6 can cover a wide range and in general are all that we use. The gas flow can be significant in the case of air at

---

* Editors Note (Kristiansen):
  For high average power, repetitive operation many electrode materials have considerably lower erosion than brass (e.g. W+Cu).

no cost and while SF_6 is quite expensive, a modest flow is necessary to obtain reliable operation. The gas employed should be dried, especially if the gap has a large coulomb usage. In the case of gaps using spheres as electrodes, these can be made out of steel or phosphor bronze ball bearings, with some very minor corrosion effects. The spark gap body can be made out of perspex, either in standard tube lengths or, for a high pressure gap, turned out of solid. Nylon has also been used with success, as well as other more exotic materials, but perspex has the advantage that it can be quickly joined with Simplex cement. This material is a powder monomer with a catalyst and is extremely useful in building high voltage systems cheaply and quickly.

The field distortion gap is usually employed in a cascade mode. In this, the trigger electrode is central and its potential (which is earth in a balanced system) is changed by the trigger pulse or by shorting it to one of the main electrodes via a small inductance. One half of the gap fires and the gap volts are then imposed on the second half. Such a gap will trigger fairly quickly down to something like 60 per cent of its self-breakdown voltage (i.e. it has a triggering range to 0.6). However, the gap can be easily used with a displaced electrode and then it can be made to trigger down to 0.4 or less. Such a gap, when used at 80 per cent of self-break, can trigger in a few ns and can have a jitter of a few-tenths of a ns. These figures apply to gaps working at tens of kV and above. The trigger pulse needed is of the order of the self-break voltage of the gap, but as the capacity it is driving is small, this is easy to achieve. The gap with a displaced trigger electrode is arranged to break both portions of the gap simultaneously and the ratio of the gaps is of the order of 1:2, but depends a little on the gas and pressure.

In spherical electrode gaps the trigger electrode takes the form of a disc with a hole cut in it and this hole can be of the order of the gap separation, so that small errors in the DC potential of the trigger do not have any effect on the self-breakdown voltage. In general, the thickness of the trigger disc can be in the 1 mm range.

Such gaps work down to 15 to 20 kV well, but below this various factors contribute to making their operation more difficult. To bridge the gap between a few kV and these potentials, a cheap, simple gap exists (Refs. (13), (14)), which can be triggered quickly by a 300 volt pulse in a 70 ohm cable. This pulse is fed via a ferrite cored transformer and the resulting 1 1/2 kV trigger pulse applied to the gap. These gaps do not in general handle much energy and are best in the low voltage range, but versions working at 200 kV and handling 10 kilojoules have been built. Their main virtue is their cheapness and ease of construction. They can also have a jitter of firing on the trigger pulse of 1 or 2 per cent and

NANOSECOND PULSE TECHNIQUES 61

have been used to make pick-up resistant trip, delay, and output circuits. In these, the internal pulses are all at the 10 kV level and the output pulses are of low impedance and have rise times of 50 ns or so. In conditions where electromagnetic pick-up is bad, they can be very useful.

As an example of a good low inductance gap, a rod or rail gap has been built with an outside diameter of a little over 3 inches. Strip line feeds to it were well insulated with doubled-over mylar and under freon it did not track at 210 kV. Its inductance, used as a single channel gap, was under 50 nanohenries and a triggered multichannel version was calculated to have an inductance of about 5 nanohenries. It took about three days to build and the materials cost a few tens of pounds.

Liquid Gaps

Figure 4-5 shows the general form that has been used for pulse charged liquid gaps. Water gaps have been operated up to 3 MV and have shown modest multichannel operation at these levels. They have also been operated with oil by Mr. I.D. Smith of Physics International, San Leandro, at levels up to 5 MV, where again they have performed well. Mr. Smith has contributed greatly to many aspects of pulsed high voltage technology and it is a pleasure to record this.

In the version shown, the trigger pulse is simply derived from the energy stored within the gap, but better multichannel performance would be obtained by using an external pulse to operate the gap. In the case of water, the gap ratio is more like 7:1 but significant expertise is still required to operate liquid gaps at the multimegavolt level.

For simple, over-volted, untriggered gaps, both oil and water perform well and fast rising pulses can be simply produced. With care, jitters in the voltage breakdown can be 3 or 4 per cent and closure gradients of 400 kV/cm in oil and 300 kV/cm in water obtained for microsecond pulse charge times.

Liquid gaps represent an attractive approach for switching at the multimegavolt level and the only other type gap probably worth looking at is pressurised SF_6 or similar gas mixtures.

Solid Gaps

Mechanically operated solid gaps have been used for a long time and, for many DC applications, a slightly blunt tin tack and a hammer is by far the best approach. Indeed, this switch probably has the fastest rise time of any when used in a low impedance circuit. The reason is that both the thin copper or aluminum sheet top electrode and the insulating film flow and intrinsic breakdown eventually occurs in a very small volume at fields in excess of 8 MV/cm. The deformed electrodes also form a good feed to the very short plasma channel; all in all, it is quite a sophisticated gap. Several versions of mechanically broken gaps have been developed at Culham, as start and clamp gaps, and these are operated with modest jitters by means of exploding wires or foils to cause the mechanical deformation when required.

Two versions of intrinsic breakdown solid gaps (Refs. 15, 16) were developed at AWRE some eight years ago. The first type uses an array of 50 needles to stab holes in, say, a 60 thou. polythene sheet. By varying the depth of stab, the operating range could be changed from 40 to 150 kV with the stabs positive. As there is a strong polarity effect on polythene with the stabs negative, the range was from 120 to 250 kV or more. This switch could be stacked for higher voltages and a pair of them made a simple triggerable switch. The standard deviation of the breakdown voltage could be as low as 2 per cent and the switches had very similar characteristics when used from DC down to 10 ns charging times. In one quite old system, 40 such switches were regularly fired simultaneously at 200 kV, producing a 4 MA current rising in 8 ns or so, i.e. a di/dt of 5×10^{14} amps per second. Such switches are still in use in various systems and have been operated up to 1/2 MV pulse charged.

A second and even lower inductance solid dielectric switch was developed along the lines shown in Figure 4-5. Two mylar sheets of different thicknesses enclose a trigger foil whose edges are sealed with a thin film of transformer oil or silicon grease. Copper or aluminum main electrodes are added to complete the switch. The inductance of a single channel switch of this kind is very low and the resistive phase is small as well. For DC charged high current banks a multichannel version of this switch is easily made and by injecting a pulse rising to, say 40 kV in a few ns along the line formed by the trigger foil, hundreds of current carrying channels can be made to occur. Currents of tens of megamps can be switched by a couple of such switches and their inductance is extremely small. Pulse charged versions of this gap have been operated up to 1/2 MV.

Solid dielectric gaps have to be replaced each time, of course, but, in high performance banks with a relatively low rate of useage, they can be a very cheap substitute for tens or hundreds of more orthodox gaps.

Multichannel Operation of Gaps

Single channel gaps, transit time isolated, have been operated in parallel for several years, but these do not come within the rather strict definition of multichannel operation that I use. For me, multichannel operation of a gap occurs when all the electrodes are continuous sheets of conductor. As such, while transit time isolation may play a small part, the main effect is that before the voltage across the first channel can fall very much, a number of other channels have closed. As such, it is the inductive and resistive phases which are important in allowing a very brief interval in which multichannel operation occurs. This time Δt is given by

$$\Delta T = 0.1 \, \tau_{tot} + 0.8 \, \tau_{trans}.$$

τ_{tot} is as defined in the previous section and τ_{trans} is the distance between channels divided by the local velocity of light. Both of these terms are functions of n, the number of channels carrying currents comparable to the first one to form. ΔT is then placed equal to 2δ where δ is the standard deviation of the time of closure of the gap on a rising trigger pulse. This is related to the standard deviation of voltage breakdown of the gap by the relation

$$\delta(t) = \sigma(V) \, V \, (dV/dt)^{-1}$$

where dV/dt is evaluated at the point on the rising trigger waveform at which the gap fires.

Figure 4-6 gives the jitter of a negative point or edge plane gap as a function of the effective time of rise of the trigger pulse. For ordinary gaps (both gaseous and liquid) a good jitter is 2 per cent or so, but for edge plane gaps charged quickly the jitter can be down to a few tenths of a per cent. Typically ΔT is a fraction of a nanosecond - ordinary gaps require trigger pulses rising in 10 ns or so. However, for edge plane gaps, the trigger pulse can rise in 100 or more ns and still give multichannel operation. In one experiment 140 channels were obtained from a continuous edge plane gap. The expression has also been checked for liquid gap operation and more approximately for solid gaps, and gives answers in agreement to some 20 per cent for ΔT. The way it is used is to define a rate of rise of trigger pulse and as such, this sort of accuracy is ample. Fast trigger pulse generators are required to obtain multichannel operation and these must have impedances of 100 ohms or less. Of course the pulse length produced by the trigger generator need not be very great and the generator can be physically quite small. Mr. T. James of Culham has done much work on high speed systems and gaps in general and Ref. (17) gives an example of a high current low inductance multichannel gap he has developed.

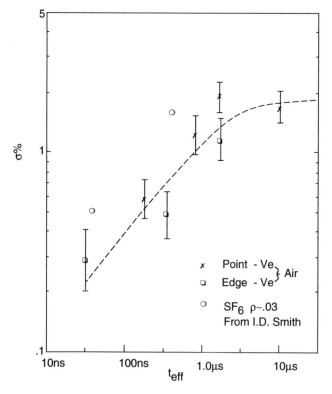

Fig. 4-6 Jitter of pulse charged point and edge/plane gaps.

MONITORING AND OTHER GENERAL ITEMS

Monitoring

The subject of monitoring is fairly intimately interwoven with the other problems of pick-up and the oscilloscopes available to display the attenuated pulses. In general it seems little sense to me to attenuate a signal by a factor of a million and then display it on an oscilloscope with the sensitivity of a few volts per cm. Attenuation problems are extreme and the effort to reject unwanted signals considerable. Consequently, wherever possible, we have tended to attenuate the signal by a factor of a few hundred and to transmit it by cable at the level of 10 or 20 kV. When the signal is passed through the shielding around the oscilloscope, it may be reduced by attenuators using standard 1 W resistors without loss of band width. However, even here we have followed a policy of using insensitive oscilloscopes whenever possible, a matter briefly dealt with below. In this section the main techniques for doing the initial attenuation will be briefly discussed.

NANOSECOND PULSE TECHNIQUES

In the case of voltage monitors, the stray capacity is the component which tends to limit the response of the attenuator and the inductance of the monitor is usually unimportant. For current shunts the reverse is the case. It is usually worth distinguishing between the two time scales, since high voltage monitors for the pulse charging system can have impedances of 10 kilohms or thereabouts, while the faster system can usually accept attenuator impedances of a kilohm or two. Reducing the attenuator impedance, of course, helps to reduce the integrating effect of any residual stray capacities. In high voltage dividers, the effects of stray capacity can be largely mitigated by placing them in an environment where they have essentially zero stray capacity, as is shown in Fig. 4-7. The ideal place is in the uniform field between, say, two parallel transmission lines. However, the gradients are so large here that flashover might well occur. If, however, the monitor is moved into the fringing fields and located along a line which approximates but is not the same as the field line, this stray capacity can be canceled. By using $CuSO_4$ resistors in their simplest form, a tube of $CuSO_4$ will divide the potential along its length uniformly because of its resistance and also it will divide inductively in a uniform fashion. If now it is located so that it crosses the equipotentials uniformly as well, then to a first order its stray capacity is zero. This trick only applies for durations longer than a few times the transit of the signal across the gap and it does not apply to very high frequencies, when a complete theory would be very complicated. However, for lines 1 or 2 feet apart, these times are a nanosecond or two.

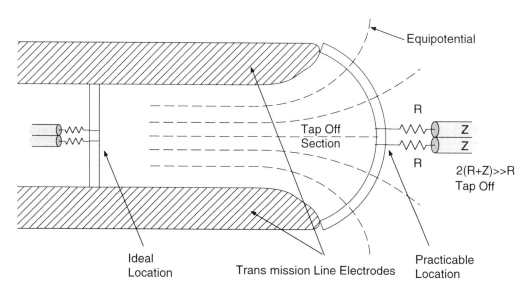

Fig. 4-7 Location of voltage divider.

In the situation shown in Fig. 4-7, a plus and minus signal is taken out of the centre tap of the monitor. Using a balanced monitoring system helps to reject pick up but the approach works quite well where the tap off section is at one end of the resistor column or chain. Uniformly graded resistor columns are not essential and in diverging fields the cross-section of a $CuSO_4$ resistor can be changed, or indeed its concentration, but these are elaborations which are just not worth while.

Where the divider is made up of standard resistors, these can be zigzagged back and forth in a low inductance fashion and then the stack bent to conform with the required layout.

In general, 1 W carbon resistors will stand pulse volts up to about 15 kV per resistor for short pulses and will absorb one or two joules with only a modest long term drift. At 5 J per pulse per resistor this change becomes larger but can be cancelled to a first order by tapping off across a representative resistor. At the level of about 10 J per resistor, they blow up with a satisfying bang!

An alternative approach to monitoring the high voltage pulse is to use capacitor dividers embodied in the walls of the transmission line. Several groups use this approach and, in particular, Mr. J. Shipman of the U.S. Naval Research Laboratory, Washington, D.C., has developed internally damped high frequency capacitor dividers to a high level of perfection.

For monitoring currents flowing in the system, low inductance shunts can be employed, or integrating Rogowski loops. We have tended to use current shunts made from thin Nichrome, steel, or brass, and, once again, since tens or hundreds of kiloamps are flowing, output signals can easily be at the level of a few kV. The shunt is usually made fairly wide, rather short, and insulated with a few thou. of mylar.

Pick Up

While it is easy and indeed convenient to have most of the pulses monitoring the functions of the generator produced at the level of several kV, certain signals such as those from strain gauges cannot be made very big. In these conditions serious problems with pick up are likely to be encountered. These can be minimised by careful attention to each stage at which the signals are handled and it is possible to monitor few milliamp signals in the presence of low voltage banks generating 10 million amps.

In general, pick up occurs because of capacitive coupling or inductive coupling. Inductive coupling is much the most difficult one to cope with and my remarks will be restricted to this mode. A lot of work has gone on in the last few years to rationalise the rejection of unwanted signals. The field is a very large one and is eminently susceptible to rational analysis. The first message is that there is no "black magic" in pick up: the laws of electromagnetism apply here just as anywhere else. Where many interacting metal loops have been allowed to occur, the situation may be too complex to analyse, but if care is taken from the beginning to simplify such loops, then the circuit can be analysed and answers to a factor of half an order of magnitude obtained. Such accuracies are entirely adequate because the aim is to reduce the pick up to very small levels. In general, there are three areas of importance in pick up - the system that is generating the unwanted electromagnetic signal, the monitor and its connections; and the oscilloscope, which is usually inside a screened room or box.

Every effort should be made to make sure that the bank or pulse generator is as low an inductance as possible. This is achieved by making sure that all return currents flow close to the outgoing current. Balanced strip line feeds radiate very little and, of course, help to speed up the pulse itself. Balanced charging and balanced output signals from attenuators can also reject common mode interference considerably.

With regard to the connectors between the attenuators and the monitoring oscilloscopes, two approaches are very useful. One is known as "tree wiring" where every endeavour is made to reduce the area of the pick up loop. For instance, should it be necessary to join a monitor to an oscilloscope and the 'scope is joined to other 'scopes and units, all the wires flow along the trunks and branches, as in a tree, and are closely taped together. This means that the area in which pick up can occur has been made as small as possible.

The second general approach is to put a complete break or, if impossible, a high impedance in each loop. Under these circumstances only small currents flow in the outer braiding of the coaxial cables in which the monitoring signals are transmitted. this is not difficult to do in practice. For instance, pulses can be sent between oscilloscopes by small isolating ferrite cored transformers. The bank or high voltage system can frequently be isolated from earth by heavy duty $CuSO_4$ resistors of 10 or 20 ohms impedance. Since the bank will have a capacity to the rest of the laboratory, you can calculate the sort of impedance this represents and make this equal to the earth isolating resistor. From a safety point of view it is probably better to have three or four easily seen large $CuSO_4$ resistors than to vaporise a copper conductor inside its outer insulation cover because of the large circulating currents introduced by multiple earthing. Battery operated oscilloscopes can also be employed, but a much simpler solution is to wind 50 yards or so of mains cable on a cardboard drum to make an isolating inductor out of it. Small capacities of the order of 0.1 μF are needed between (live and neutral) and earth at both ends of the inductor, to avoid producing transient voltage pulses in the mains wiring and the mains transformer of the oscilloscope. The oscilloscopes can move away from earth by a few kV during the operation of the generator and while the shocks experienced by touching them would be small, the 'scopes should be isolated during the operation of the system. When proper care is taken, large fast banks can be fired without blowing plugs out of wall sockets and ringing telephone bells in adjacent buildings.

With regard to building a shielded monitoring room, or individual shields around the oscilloscopes, these techniques are fairly well reported and the only point probably worth making is that the boxes can be cheaply made out of wire mesh, providing good low impedance joints are made frequently along the edges of the mesh. This is because even a few tens of milliohms will cause surface currents to flow non-uniformly and allow signals to leak in through these high resistance joints. Phosphor bronze draught excluder strips make good seals for the necessary access door. Any further attenuation that the signal needs is done as it comes in through the wall of the oscilloscope box. The 'scope itself should sit off the deck and it is desirable to employ 'tree wiring' within the box.

The two concepts mentioned above of 'tree wiring' and placing impedances in each loop are not mutually incompatible and, in general, both should be employed. Where any one scheme is consistently carried out, a drop in interference of some 40 db is usually easily obtained.

Safety

It is not possible in this note to deal adequately with safety, but it is a matter which should be taken very seriously. In general, condensers storing more than a few tens of joules should be in a metal enclosure with limited access. In order to ensure safety before entry, it is desirable to have three complementary methods of making sure that, in normal operation, the bank is discharged. One useful method is a voltage meter that monitors the potential on the condensers. These have to work each time or else it is not known how much the bank has been charged and after the bank or system has been fired, they should be watched to make sure that they have fallen to zero or near-zero.

The second line of defence is a weight which is lowered onto the top of the charged system and which is coupled to two $CuSO_4$ resistors in parallel. These are designed so that they will not leak and are placed in a position where they are clearly visible.

The third, and last-ditch, shorting technique is a dropping weight which places a short circuit across the condenser system. This should be observed each time that it is operated and should, of course, never cause any spark on closure. In the event that it does, both of the preceding systems are malfunctioning and the matter should be investigated urgently. In the event of a bank blowing up, or some obviously unusual occurrence taking place, by far the safest thing is to sit down for ten minutes or a quarter of an hour and to think hard. Then re-entry with care and with a shorting-stick well to the fore can be attempted.

$CuSO_4$ Resistors

$CuSO_4$ resistors are of great use in nearly all high voltage applications. They are very much cheaper and simpler to make than soldering together 500 1 W resistors. They can be flexible and disposed in elegant shapes. They can absorb vast amounts of energy, one cubic foot or so of solution taking, quietly and safely, the energy out of a 1 megajoule bank. As has been mentioned above, they are useful as voltage dividers and also act excellently as high energy terminators for transmission lines. They have a resistive response into the several hundred megacycles region and apart from the electrode-liquid interfaces obey a much better understood bit of physics than carbon resistors. The disadvantages are that it is quite easy to measure incorrectly their resistances. This is because insulating films form on the electrodes rather like electrolytic condensers and hence resistance measuring systems working at fractions of a volt see large capacitive components. However, these films can be easily stripped off with a DC current, or the voltage raised to the point at which the effect goes away. In use, of course, the fact that there is, say, a 5V drop at the electrode-liquid interface is irrelevant when a one million volt pulse is applied to the resistor.

In making $CuSO_4$ resistors, it is sensible to use deionised water and $CuSO_4$ only, and copper electrodes. The body can be either flexible PVC tubing or, if rigid resistors are required, perspex. The resistance of ionic conductors is a function of temperature and in long resistors temperature gradients can exist, but this is an easy matter to measure and should cause no problems, although it sometimes does. When large energy densities are dumped into $CuSO_4$ resistors the temperature rises and the resistance changes, but in addition a pressure wave develops because the liquid does not have time to expand. Consequently for dump resistors which may experience rapid rises of 10° or 20°C, flexible walled vessels should be employed.

In general, for low energy applications, resistors, or resistor chains, are quite adequate up to 50 or 100 kV but above this they become messy and for high energy absorbing applications and/or high voltages, $CuSO_4$ resistors are greatly to be preferred.

Low Sensitivity Oscilloscopes

One of the ancillary pieces of equipment developed at Aldermaston has been the low sensitivity oscilloscope, which has a deflection sensitivity of some 4 kV/cm on the film. It has a rise time in the region of 1 ns and apart from the rectifiers in the power pack, the only active element is a spark gap triggered off by a portion of the incoming signal. The oscilloscope is sufficiently immune to pick up to be used unshielded. The sweeps are exponential but this is an advantage since one is usually mainly interested in what happens to the front pulse, with a diminished interest in later reflections, etc. In one application where it was required to monitor the high speed pulse in a system pulse charged in about 1 μs, one of these 'scopes was floated up to 1 million volts and operated quite happily when the high speed pulse came along. The signal leads between the generator and the oscilloscope and the mains cable from the 'scope to earth were wound in the form of inductances and the scope located approximately in its proper potential place in the fringing field of the generator.

Power Packs

With modern plastics whose surface resistivities can be very high, modern power packs can be made light and compact. For instance a $\pm$ 125 kV power pack need not occupy a volume much greater than 3 cu. ft. and can be easily carried. Air insulation is obviously used and to avoid having to make a large number of metal domes, the simple expedient is used of wrapping perspex sheets, or mylar, around and over the sharp edges of the power pack. A brief burst of corona charges up these insulating sheets in such a way as to grade the potentials correctly and, a small AC ripple apart, no further corona-ing will occur. For lower voltages, commercially made RF power packs work well and are reliable.

In general, either carbon resistor chains or $CuSO_4$ resistors should be included in the output leads of the power packs so as to limit the rate of discharge of these when the short occurs in the main generator, or if the generator is fired without switching the power pack off. This simple precaution can save quite a lot of money in replacement rectifiers, at the same time as isolating the power pack so that it does not radiate pick up.

NANOSECOND PULSE TECHNIQUES

EXAMPLES OF PULSE SYSTEMS

As an example of a large system, Ref. (9) can be consulted. This machine is Hermes II and was built by Mr. T. Martin of Sandia Corporation, some three years ago. It works with up to 10 MV on the X-ray tube and the tube current is in excess of 100 kA. Physics International of San Leandro, California, Ion Physics of Burlington, Massachusetts, and Maxwell Laboratories of San Diego, all build a range of pulse power systems in the multimegavolt range and I am sure they will be pleased to supply literature on these. In the range 100 kV to 2 MV, Field Emission Corporation of McMinnville, Oregon, produce a wide range of systems. As a review of work at AWRE up to a few years ago, Ref. (10) lists some of the systems built and gives their characteristics.

In general, Marx generators of output voltages up to 10 MV and more are available and these can deliver energy at the rate of up to 1 GW (10^{12} watts). Fast systems have been built which provide

$di/dt = 5 \times 10^{14}$ amps/sec

$dV/dt = 5 \times 10^{14}$ volts/sec

and $dE/dt = 10^{13}$ watts.

However, such systems are large and expensive. In addition, their rate of operation is fairly limited, a matter of no great importance for their applications.

Turning to somewhat more modest systems, Table 4-VI lists a series of slow and fast systems and gives <u>very</u> rough cost estimates when built by the persons wanting to use them.

In addition to these systems, fast and slow pulse generators with outputs up to 1 MV and storing a kilojoule or less can be built for a few hundred pounds in a few man weeks.

Systems giving a hundred kV or so can be knocked up in a week and at a very small cost, when the construction techniques we employ are used.

Applications of the pulse generators include pulse X-ray generators, the generation of relativistic electron beams, various possible types of ion accelerators, spark chambers and, in particular, streamer chambers, dielectric breakdown studies and electromagnetic pulse generators of various kinds.

Table 4-VI
Pulsed Power System Examples

Speed	Type	V (MV)	i (kA)	t_{pulse} (ns)	t_{rise} (ns)	Approximate Cost (£s)
Slow	Transformer	1	2	500	-	500
		2 1/2	5	1,000	-	2,000
	Marx	1	10	1,000	-	3,000
		5	100	2,000	-	30,000
Fast	WeeWOBL	0.4	20	45	5	1,000
	Plato	2	20	13	4	3,000
	MOGUL	5	60	48	14	10,000

CONCLUDING COMMENTS

The field of high speed voltage generators has seen a rapid development in the last seven years or so. This trend is likely to continue and systems an order of magnitude bigger than those mentioned above are within the range of real possibility. Approximate treatments have been developed that enable these systems to be designed with some confidence and relatively recent developments have enabled multichannel gaps to operate at such a level that the rate of rise of the pulse is essentially limited by the properties of the dielectrics available and not by the switches.

There is a rich field of interesting work waiting in the development of dielectrics with improved characteristics, particularly in the field of liquids. Understanding of the physics of pulse breakdown of media is also a fruitful area waiting to be exploited. The generators can be used to produce large relativistic beams of electrons and the study of the behaviour of these is both complicated and fascinating.

In concluding I would like to repeat my thanks to my colleagues at AWRE and the many other very able workers in the field in the United States and the UK.

These notes of necessity have had to be very compressed and I am only too well aware of the drastic simplifications that this has forced upon me. I am afraid I have had to leave out many aspects of all the subjects touched on in these notes, some of which may be of considerable importance in any particular application.

My sincerest thanks again due to Mrs. V. Horne who has had to slave over a red hot typewriter, while disentangling my horrible syntax and impossible writing and spelling.

REFERENCES

Useful General References

1. Lewis and Wells: "Millimicrosecond Pulse Techniques". Pergamon Press, 1959.
2. M.J. Mulcahy (Ed.). "High Voltage Technology Seminar". Presented by Ion Physics Corporation, Sept. 19-20, 1969.
3. Frank M. Clark: "Insulating Materials for Designs and Engineer Practice". John Wiley and Sons, Inc., 1962.

Standard High Voltage Books

4. L.L. Alston (Ed.). "High Voltage Technology". Oxford University Press, 1968.
5. Frank Frungel: "High Speed Pulse Technology". Vols. 1 and 2. Academic Press, 1965.

Text References

6. J.C. Martin and I.D. Smith: "Improvements in or relating to High Voltage Pulse Generating Transformers". Pat. No. 1,114,713. May 1968.
7. J.C. Martin; P.D. Champney; D.A. Hammer: "Notes on the Construction of a High Voltage Pulse Transformer". Cornell University. CU-NRL/2.
8. R.A. Fitch and V.T.S. Howell: "Novel Principle of Transient High Voltage Generation". PRDC. IEE. Vol. III, No. 4. April 1964.
9. T. Martin; K. Prestwich; D. Johnson; "Summary of the Hermes Flash X Ray Program". Sandia Laboratories Report SC-RR-69-421. Radiation Production Note 3.
10. M.J. Goodman: "High Speed Pulsed Power Technology at Aldermaston". Reprint from AWRE News, March 1969.
11. P. Felsenthal and J.M. Proud: "Nanosecond Pulse Breakdown in Gases'. Tech. Report No. RADC-TR-65-142, Rome Air Development Center, Griffins Air Force Base, New York. June 1965.
12. I.D. Smith: "Pulse Breakdown of Insulator Surfaces in a Poor Vacuum". International Symposium on High Voltages in Vacuum. MIT. Oct. 19-21, 1966.

13. J.C. Martin: "Corona Triggered Spark Gaps". Pat. No. 1,080,211.
14. R.J. Rout: "Triggered Spark Gaps for Image Tube Pulsing". Journal of Physics E. 1969. Series 2, Vol. 2, p. 739.
15. J.C. Martin: "Switch for Fast Electrical Discharge having a Plurality of Electrodes in a Non-porous Dielectric Material inserted between Electrodes". Pat. No. 988,777.
16. J.C. Martin and I.D. Smith: "Electrical Switch having a Trigger Electrode whose Sharp Edges are sealed to suppress the Formation of Corona". Pat. No. 1,080,131.
17. T.E. James: "A High Current 60 KV Multiple Arc Spark Gap Switch of 1.7 nH Inductance". CLM-P.212. Culham Laboratories, 1969.

Chapter 5

HULL LECTURE NOTES

Section 5a

HULL LECTURE NO. 1*   DIELECTRIC BREAKDOWN AND TRACKNG

J C Martin

This set of notes was prepared in conjunction with a one-week course given by Charlie Martin at Hull University in June of 1973. The notes serve as an excellent supplement to his, "Nanosecond Pulse Techniques" and an introduction to other AWRE publications.

* SSWA/JCM/HUN/5
  June 1975

INTRODUCTION

These lectures will cover superficially the rapidly advancing field of high voltage pulse technology and a few of its applications. Many groups and individuals have contributed to the present state of the art and while I have given an occasional acknowledgment, I am afraid I have been far from diligent in the matter. This is partly because of the speed with which the following notes have had to be produced (three days), but mainly because nearly all the notes of which I had spare copies to bring to Hull have been our own. Thus these were the easiest to refer to and to make available during the lecture course. Consequently I want to apologise to all the other workers in the field for failing to acknowledge their efforts and for not referring to their papers in the open literature. If anyone at Hull wants to refer in more detail to any subject, we can provide copies on loan, or references to most, but not all, of the published work in any specific area.

With regard to the great bulk of the notes (an apt phrase) of which I have provided copies for Hull (in addition to these lecture notes), the majority come via Carl Baum of Kirtland Air Force Base, Albuquerque, New Mexico, who has done a magnificent job of reprinting them. Such copies are much clearer than the originals and are suitable for reproduction. This accounts for the fact that many have two reference numbers. With regard to the remainder, some are recent and of good quality (referring of course to their legibility, not their intelligibility) but a few date from the years BX (Before Xerox) and look as if they had been duplicated by the moist jelly process. Indeed this may actually have been the case and I am afraid there was neither time nor justification for having them retyped, but I apologise in advance to those who try to read them.

Finally I would like to acknowledge that most of the work reported was done by other members, past and present, of the SSWA pulse power group, to whom all the credit is due: any faults that exist are mine. It might be assumed from running an eye down the list of largely internal references given at the back of the later lecture notes that I am the only one at present in the group who can write: a perusal of any of the notes will rapidly disprove this. I can't write, either. However, despite the universal illiteracy of the group, it has managed to make modest contributions to the field from time to time and it is a great personal pleasure to me to acknowledge the fact that we worked as a group and that any success we have achieved was as a group; I just happened to be the one who could afford to buy the ballpoint pen refills.

Summarising, these lectures are a cursory survey of the field and are very largely based on the combined work of the SSWA pulse power group.

I would like to add one personal comment to those who have suffered through the lectures themselves. I could have dealt with half the number of topics in the seven and a half hours and left some degree of comprehension with the audience. However, I have elected to cover, however rapidly and inadequately, all of those areas of the field which are, or may possibly become, of use to the work of the group at Hull University. As such, I aimed to give a general survey, however condensed and confused it may have become. I hope that the stack of single copies of notes will enable those who wish to find out more clearly what I was trying to say on any particular topic and, having perused these, they can contact us to obtain further information and clarification, in the unlikely event that this is forthcoming. To the remainder, who have sat through long tracts of confused and hurried exposition, I can only offer my deepest sympathy and request that if, by accident, they understood any section, perhaps they would be kind enough to enlighten the lecturer as to what it was he was trying to say.

VOLUME/SURFACE AREA DEPENDENCY OF INTRINSIC BREAKDOWN

The treatment given in Reference (1), while being largely concerned with solid dielectrics, is generally applicable. Thus, while solids have a volume effect, most, if not all, of the breakdowns in liquids and gases originate at an electrode surface, in which case an area dependency applies, although, as is explained in the Nanosecond Pulse Techniques note ("NPT"), for conditioned electrodes in gas the area effect is so small as to be practically non-existent. For the case of a chain of many links, where the breaking strength of each individual link has an intrinsic finite scatter, there would be a linear dependency of strength.

As the note, although written eight years ago, is still essentially correct, I will merely update the data in it a little. The distribution which is graphically derived in the note is a Weibull distribution, to which reference should be made by those of a probablistic turn of mind. The applicability of this distribution function is very wide: indeed, I would hazard an opinion that in real life it is met up with much more frequently than the better publicised Gaussian one. Certainly it is a very good approximation to the distributions found in dielectric breakdown, spark gap firing voltages, mechanical strength, etc. In particular the distribution curve is considerably faster falling on the upper side, which is in intuitive accord with many real life situations - such as the distribution of salaries and other vitally important issues. As is explained in the note, if the intrinsic scatter has a constant standard deviation, then the slope of the breakdown strength against volume or area is constant, too. As is hinted in the note and was subsequently found for mylar in thin sheets, the standard deviation can, of course, change, in the case of thin mylar decreasing to a few percent. When this happens, the slope of the curve decreases very considerably and the volume effect of intrinsic breakdown essentially disappears, the breakdown strength becoming practically constant above some volume. However, even in this situation there is still a finite chance of coming across a sheet with a mechanical defect, or hole, in it, and a new non-intrinsic volume effect takes over. If some electrical or mechanical testing procedure is employed to weed out those areas with such defects in them, this volume effect can be largely removed. The remarkable thing about solid dielectric intrinsic breakdown is not that there is a volume effect, but that over an enormous volume range the standard deviation seems to remain essentially constant (with the exception of very thin films in large volumes).

A second example of the applicability of the note is to the self breakdown voltage of a spark gap. In this case, when the gap is carrying significant but not very large currents, a good gap will have a σ of $\pm$ 3%, or thereabouts, and the distribution is closely a Weibull one. With this distribution, even in thousands of firings there is no real chance of a breakdown occurring at one half of the mean breakdown voltage. However, in real life such low voltage breakdowns can happen, sometimes with percent-like probabilities. These "off distribution" breakdowns are called "drop outs" and visual inspection will show that they are nearly all or all associated with sparks originating well away from the axis of the spark gap, the arc following a long looping path. It is believed that close to the axis metal vapour does not have a chance to condense, or, if it does, the whiskers that grow get knocked by the hydrodynamic shocks from later sparks. However, at some distance from the near axis sites of the vast majority of the breakdowns, long whiskers can grow and these eventually lead to the drop out breakdowns. Thus the breakdown distribution has two components. The issue is one of importance, as in a Marx with twenty such spark gaps as many as 10% of the self breaks can be drop outs occurring at very low voltages. The frequency of drop out breakdowns is a complicated function of electrode material, gas, pressure, current through the gap, and even perhaps gravity, and it has to be controlled in systems using many gaps.

SOLID DIELECTRIC PULSE BREAKDOWN

Firstly there is little or no sign of a time dependency. Jumping ahead, the reason for this can be seen from Reference (13), which mainly deals with the mean velocity of transit of streamers in liquids but also quotes some values for polythene as well. For transformer oil at 100 kV the time to close over 0.1 cm, is about 70 ns from a point to a plane. In polythene the time is 5 ns. While the proper transit time has to be calculated by integrating as the streamer moves away from the metal electrodes, the ratio stays about the same. In addition to the much higher velocity in polythene, the breakdown field for polythene of a few cc volume is about 2 1/2 MV/cm, whereas in oil charged in about 1 μsec it is more like 1/2 MV/cm. Thus while there are transit time effects in good plastics these are only a percent or so of those in liquids and essentially the breakdown streamer propagation phase can be considered zero, except for large thicknesses of solid dielectric being broken in a few ns. Such streamer velocities were studied in 2 cm thick perspex slabs, where thin holes were introduced part way through and a voltage of 3 MV rising in 10 ns or so applied across the slab. While time resolved measurements were not attempted (it was in the very early days), it was shown that after a very regular growth of small positive bushes the streamer growth went over discontinuously to a much faster mode which travelled at a velocity of up to 1/30th of the local velocity of light. This phenomenon has also been observed in water, where for large gaps the relatively slowly growing, heavily branched bushes from positive edges changed their nature and their velocity of propagation to become much faster and appeared like the more straggly negative bushes. This effect was known as "bush on bush" and suggests that for two very dissimilar materials the negative streamer mode is ultimately the fundamental one and for large voltages or high fields the velocity of this can reach appreciable fractions of that of light.

Thus for solids in most practical situations breakdown is field dependent only and not significantly time dependent. Because there is typically an intrinsic scatter in breakdown fields ($\sigma \sim \pm$ 12%), there is a volume or area effect. The data supporting the contention that there is a volume effect is not extensive but it is reasonably sound and I know of no data that contradicts it. Thus for uniform, near uniform, or very diverging geometries, the volume stressed to 90% of the peak field is estimated and this gives the value of the peak breakdown voltage that the geometry will stand.

In addition to Reference (1) data for some plastics is given in References (2) and (3). One phenomenon that should be mentioned is that while pulse charging gives probably the best measure of intrinsic breakdown strength available, it is possible to achieve a field higher than this using DC with a few materials, typically, and most importantly, polythene. Here, using DC, breakdown fields 20% above those quoted in Reference (1) can be obtained in well arranged tests.

This has been studied and can be attributed to charge annealing around the microscopic, probably molecular, scale defect, due to body conduction in the plastic. Thus charge separation occurs along the weakness, reducing the field locally. If such a sample is then put in an LC circuit and rung so that the voltage is reversed, the sample will break at reversed voltages of only 20% of the original applied fields. This is because the annealing charge separation now adds to the reverse applied fields and leads to pulse breakdown at a very low level. The time scale over which this charge annealing takes place is of the order 1 to 10 milliseconds. Other plastics do not show this effect and mylar in particular can be rung nearly 100% without causing it to break after DC charging. DC in this context is 1 min: the effect might show up over longer periods.

Reference (4) gives some data on mylar and was written well before Reference (5), which advances the idea that for large enough volumes of thin sheets of mylar the standard deviation of the breakdown field decreases and the volume effect essentially disappears. The data on which this theory was advanced was notable more by its absence than anything else. However it has been vindicated by fairly recent results from Physics International, where stacks of many thin sheets of mylar of volumes of several hundred cc have shown standard deviations of ± 2% and have breakdown fields essentially on the original curves in Reference (4), thus displaying a much reduced volume effect.

While the breakdown of plastics shows no time dependance it does show a live dependancy which, like condensers, is proportional to $(F_{breakdown} / F_{operating})^8$. Thus to get a mean life of a thousand shots, a volume of plastic has to be used at half it single shot breakdown field. The pulse life of mylar was investigated as an example in Reference (6). In this it is shown that given the intrinsic standard deviation and the life law, the distribution of breaks with number of pulses at constant field could be calculated. It was also speculated in Reference (5) that there was a direct relation between the standard deviation of the single shot breakdown fields and the life law power. It was shown that for a volume of thin stacks of mylar only just into the volume independent region, the life law was more like a 16th power. This speculation received strong additional support when the life law found for the above tests at Physics International was a 50th power, or thereabouts. Thus from the single pulse breakdown data of a set of samples of a plastic of the same volume, the mean breakdown field and the slope of the volume dependency in the region of the sample volume can be determined. A guess can be made of the life law (an 8th power if $\sigma \sim \pm 12\%$) and then the distribution of breaks with number of shots at a constant pulse amplitude can be calculated: not too bad a return for some modest effort.

LIQUID PULSE BREAKDOWN

The following is a very brief description of what we believe are some of the stages involved in liquid breakdown; these ideas were developed about seven years ago. The model starts with whiskers of lengths in the range 10^{-5} to 10^{-4} cm on the electrode surface. Field emission from these either boils the liquid at the tip of the whisker or warms it so that its surface tension decreases to a low level. A bubble then develops at the tip, where it is hydrodynamically unstable in the large local electric field. The electrostatic pressure at the point is estimated to be in the range 30 to 300 atmospheres. This causes the bubble to elongate and become 10^{-3} to 10^{-2} cm in length. The velocity of this hydrodynamic phase is quite low but because the distances moved are small, the time can still be short. When the sharp bubble has lengthened enough, the field at its tip becomes big enough to start the streamer phase proper, which then accelerates out away from the shielding effect of the adjacent electrode.

The range of parameters given above applies for the span of breakdown fields usually encountered with reasonable pulse charging rates in the range several microseconds to tens of nanoseconds. One of the most important phases in this tentative model is the hydrodynamic phase. The existence of this was proved indirectly to our satisfaction by the following experiment. If the hydrodynamic phase is theoretically investigated, a viscosity of order 1 million cgs units will stabilise the bubble against distortion in the applied electric field. We therefore investigated the breakdown of a number of liquids which could be cooled so that their viscosity passed through the range required without crystallisation occurring. Among the liquids investigated were transformer oil and glycerine. For viscosities lower than the critical value, the breakdown had a time dependency and gave field values of the order of 1/2 MV/cm. However, when the temperature was lowered so that the viscosity was greater than about 1 million, the breakdown field rose to 2-3 MV/cm and the time dependency vanished. Thus these experiments showed that the difference between liquids and solids, from a breakdown point of view, is a viscosity of order of 1 million. At the time of doing these experiments there was a brief period of great optimism, as glycerine has a dielectric constant of about 45 and it looked as if (practical difficulties apart) we had a solid of high dielectric constant and high time independent field strength. It was, however, only a couple of hours before it occurred to both Ian Smith and myself to ask the obvious question, and it was quickly shown that the dielectric constant had fallen to about 2 by the time the strength had climbed to those of good solid dielectrics.

Quite recently, some Russian work has confirmed the existence of bubbles at the base of the streamers. In addition the above picture explains why pressurising water to several tens of atmospheres can cause the time dependency to disappear for relatively low fields, and the higher the pressure, the bigger the field at which the time dependency vanishes. This is because the externally applied pressure stabilises the microbubbles against deformation.

The picture also led us to coating experiments which for small areas gave dramatic improvements, when used with transformer oil. However, as the coating area increased, the effect became smaller and smaller, producing improvements of only 10% for areas of 10^5 cm^2 or so. This, too, might have been anticipated.

The NPT note summarises the time and area dependency of liquid breakdown and Reference (7) to (12) cover measurements of the pulse breakdown of liquids for various pulse lengths and electrode geometries and areas.

In the NPT note and Reference (9), the method of calculating breakdown fields in mildly diverging fields, as well as uniform fields, is given. For mildly diverging fields there is usually a polarity term which can be up to about 50% in transformer oil and a factor of 2 for water, with the negative electrode being the stronger. Further work on transformer oil breakdown by Ian Smith, when he went to Physics International has extended the area tested to over 2×10^5 cm^2 and investigated very thoroughly oil purity effects, coating, and surface finish. None of these have very much effect and the large area data lies pretty close to the extrapolated curve based on the earlier work at AWRE. The time dependency was also further investigated at Physics International and times out to 50 microseconds were used, and while a more complicated time dependency was found, the basic time dependency can still be expressed as $t^{1/3}$ to an adequate accuracy for design purposes.

For highly diverging fields, i.e. point/edge plane geometry, the data in Reference (13) is of use. NPT covers the case where higher voltage pulses are applied to point plane gaps in the range 1 to 5 MV with a relation based on data obtained by I.D. Smith at Physics International. This shows that, as we indicated for perspex and water, as the voltage is raised the mean velocity becomes polarity independent, when the applied voltage is large, presumably settling down in the fastest mode.

GASEOUS DIELECTRIC BREAKDOWN

The DC breakdown of near uniform pressurised air is covered in Lecture No 2, as is the breakdown of SF_6 to some extent, and will not be covered here.

The pulse breakdown of uniform gases over a very considerable range is effectively covered by Felsenthal and Proud in Reference (14). There may be delay effects due to awaiting the emission of an initiating electron from the electrodes, but for pressurised gases with rough dirty surfaces there is usually a plentiful supply provided by field emission. In the experiments of Felsenthal and

HULL LECTURE NOTES

Proud the electrodes were irradiated*. They essentially calculate the time the Townsend avalanche takes to build up from an initiating electron to the level at which the avalanche overrides the applied field because of its space charge. The avalanche then accelerates rapidly and significant current begins to flow, the time to do this phase being small compared with the first. The calculations of Reference (14) are in good agreement with the extensive measurements reported. However, the breakdown time that we would use is the time until the resistive phase begins (see Lecture 2 notes), i.e. the time at which the thin plasma channel first links the electrodes, and the current begins to approach the current that the driving circuit can apply. The measurements reported in Reference (14) take the time until a rather ill-defined current flows and a further time elapses on the records before the start of the resistive phase. However in many cases this time interval is not very long in fast pulsed uniform fields and the hydrodynamic and plasma streamer propagation phases which I believe occur add only a little time to that of the initial avalanche phase. Thus the measurements and calculations of Reference (14) give a lower limit to the time of breakdown, but usually one that is not grossly so. Over most of the range of parameters studied the breakdown field is a fairly low fractional power function of the time. Hence if this time is lengthened a little, the increase produced with calculated breakdown field is rather small. However, under certain circumstances the hydrodynamic and plasma column streamer phase may become the dominant term in the breakdown time. This was first observed by Laird Bradley of Sandia Corporation, Albuquerque. He was investigating the pulse breakdown of uniform pressurised nitrogen gases. These gaps were weakly irradiated by u.v. to provide a supply of initiating photoelectrons from the electrodes. The experiments were in the range of gap spacing 1/2 cm to 2 cm and went up to about 10 atmospheres. As such they were in a different range of parameters to those studied by Felsenthal and Proud and he used pulsed voltages of up to 400 kV. Laird Bradley found that the breakdown field-time relation was as given in Reference (14) for some of the range studied, but for smaller gaps and lower pressures there were significant departures from those calculated essentially only considering the initial avalanche phase. Reference (15) is a note applying the point plane data (discussed below) to these conditions and showing reasonable agreement.

---

* Editor's Note: (T.H. Martin).
Felsenthal and Proud data is at or above the D.C. breakdown fields which, in general, are higher than used in most triggered large pulsed power applications. The Editor had extended their data a few orders of magnitude to cover this area.

He also found that with a gap operating in this region, i.e. where the initial avalanche phase had occurred in a few places but the other phases were taking place, the introduction of a substantial burst of u.v. caused the gap to break down in a few ns. This was reasonable on the above picture, because fairly large numbers of electrons would be produced in the body of the gas and these would avalanche in a fraction of an ns and the plasma streamers (which move comparatively slowly) could then link together in line without having to traverse the whole gap separation. This mode of gap triggering is completely distinct from the weak u.v. irradation used to reduce statistical delays caused by lack of an initiating electron, and from the very much higher levels of energy deposition involved in a laser triggered spark. The range of applicability of this triggering mode is not very large but as the triggering action is extremely quick, under certain conditions the jitter should be tenths of an ns or less.

EDGE PLANE TIME DEPENDENCY

A rather crude relation is given in NPT and Reference (16) for the relation between point plane breakdown fields and time. For air at atmospheric pressure, the data extends out to 5 metres or so and other measurements have carried it down to a few cm for times from tens of ns to a few microseconds. In general it is a relationship which gives answers to 20% for the breakdown fields and is a useful compilation of data rather than an exact physical relation. A differential relation can be derived from this, but the accuracy of it must be expected to be poor. However, this has been used in a number of cases such as in Reference (16), and for the case of partially completed plasma channels reaching out from a transmission line with a short high voltage pulse on it, and answers obtained which are not ludicrously wrong.

Reference (17) gives some data recently obtained for a range of gases at different pressures for small gaps.

PRESSURISED SF_6 BREAKDOWN

This is covered in Reference (18) and also dealt with in Lecture Note No. 2. SF_6 makes a very good spark gap medium and can withstand very large gradients when under pressure, and used with pulsed conditions supporting gradients of the order of 1 MV/cm at about 100 psig.

TRACKING

This is a very large area, on which a great deal of work has been done, but, as far as we are concerned, very little has been written up, certainly not in a systematic way. As there are three media (gaseous, liquid and solid) of interest here, there are three simple interface combinations. In addition, there are at least three cases which can be readily distinguished - DC, pulse, and DC plus pulse - and it is important to note that some remedies which work with DC may actually make the last case worse. Most work has been done on air/solid interfaces, next on liquid/solid, and least on air/liquid. In most cases of tracking the streamer starts from a metal surface, usually at a triple media point; however, this is not always the case. A proper parametric study would cover all the cases for a range of voltages, pressures, and times in the case of pulse tracking; however, life is too short and these notes overlong already. Thus I will outline a few useful remedies, after mentioning some experiments on a tracking set up which were illuminating, or at least were to me.

Tracking is usually considered to be a very variable quantity and as it is affected by moisture, dirt, and time, this is not unreasonable. However, even when these are nominally controlled, it is still pretty variable in the case of DC or pulse charged set ups working in air. If this variability is "intrinsic" then tests with sections of, say, a transmission line, will give optimistic answers compared with the full length line when it is operated; and indeed this can be so. Also when the system is operated a large number of times there may still be a small percentage chance of a track per shot, even when the set up has been modestly overtested for a few shots. The traditional way around this is to provide a really healthy margin against tracking by using overvolt testing. Some of the overtest factors used by us are outlined in the next section.

In the case of a 30 kV megajoule bank, where a track can have dramatic consequences, the feeds to the condensers and the condenser face were overtested by a factor of nearly 3 i.e. they withstood DC tests near 90 kV. However, in some cases upper limits can be put to the tracking length by doing experiments in which it is made as bad as possible, and this was the aim of the short sets of experiments to be described. The set up was a square of thin copper stuck down to the centre of a large 2 thou. sheet of melinex, which in turn was placed in close contact with an extensive earth plane. This central square electrode was pulse charged from a low inductance capacitor of a few microfarads capacity. This was used because as the corona moves out over the surface of the mylar, the capacity of the hot electrode increases greatly and if the DC capacitor is too small, or the circuit linking it to the hot electrode too inductive, the voltage on the root of the streamers drops and this is a good way to slow them up or stop them.

With this set up, the resulting streamers were quite visible for some way out from the hot electrode at even 10 kV, but were erratic in length and shape. It was guessed that surface charges deposited on the mylar, by rubbing during assembly or by previous shots, might be affecting the streamers and it was resolved to discharge the surface. This was more easily said than done and a couple of hours was spent in devising a means of doing this simply and quickly. When the surface of the mylar was charge-free, the streamers around the edge of the copper became regular in length and were spaced about equally apart at a spacing of the order of their visible length. As the voltage was increased, the length grew rapidly, but the streamers were still of equal length and spaced about "this far apart". The explanation for this is that shielding was taking place and streamers that were slightly longer at any one stage outgrew and killed off their neighbours; the same thing would occur in a forest of trees, densely planted at the time as small plants. The relationship for streamer length was approximately $l_m = 4\ V^{3/2}$ where l_m is in cm and V is in units of 10 kV. Thus at 40 kV distances of about 30 cm could be tracked. Thus for this worst case of a system rapidly charged from a low impedance voltage source, with no previous charge deposition, very long tracks are possible. In a DC (i.e. slowly) charged set up, little tracks run out a small way along the surface, depositing patches of surface charge and relieving the stress on the edge. If, however, this charge is removed or reduced substantially, tracking can occur over lengths up to the one given by the above. This happens when a strip transmission line tracks over, because the track removes a lot of the charge locally. Thus, if a system which has tracked is recharged rapidly, it is likely to track again in the same place. To avoid this, the voltage should be worked up again fairly slowly in a number of shots, re-establishing the charge distributions.

Also, in DC charged systems where previous charge distribution has occurred, the tracking distance is much less than that given by the above relation and fairly erratic. However, as the voltage is raised, the micro-tracks get more enthusiastic and may run out into uncharged mylar. Thus DC tracking lengths in normal set ups are much less than l_m but climb to meet it as the voltage is raised to 100 kV or so, and hence show a higher power voltage dependency than l_m. In addition, moist conditions can discharge the surface charge patterns and usually, but not always, increase the chance of tracking.

Consideration of the phenomena outlined above led to some of the following anti-tracking recommendations.

If possible, end both conductors together where the insulation sticks out and avoid the case where an extensive earth plane overlaps the hot strip. This helps DC and pulse tracking.

Use charge deposition intentionally to reduce the field at sharp points or edges. Light weight power packs can be built by providing a perspex cover over the high voltage points. Charge is sprayed on to this and in reasonably dry conditions does not leak away. This charge automatically distributes itself to reduce the field on the conducting components below their corona level. Such covers should be "seamless", ie stuck together with simplex (see Lecture 5), not screwed together. However, charge deposition techniques are actually harmful under DC plus pulse conditions, as when the polarity is reversed quickly, the deposited charge adds to the applied field.

A layer of coarse paper (blotting or filter) around the sharp edge of the metal in, say, a copper strip transmission line, carries charge out to its edge, because its surface resistivity is much lower than that of, say, mylar. When a streamer starts out from the paper edge, the potential at its root drops rapidly because it is not attached to metal, and this tends to snuff it out. For lowish voltages this works quite well on its own, but for higher voltages the metal/blotting paper combination should be sandwiched between mylar surfaces, so that the root of the streamer has to move back through the fibres and cannot flash across the surface of the paper. An example of such grading is given in Reference (19). In this case another trick is also used and this is to make the mylar insulation out of several sheets and separate these at the edge, folding the outside pair back on themselves and separating the others. The streamers then have to run away from each other, reducing their capacity and the energy available to drive them. This trick can be very effective when used in pulse charged systems (see the section in Lecture No 3 on cheap pulse charged 1 MV capacitors). The gas in the interstices between the mylar sheets as they separate breaks down but each streamer only has a fraction of the full voltage on it and also the internal ones are not connected to metal.

For pulse work, barriers can be erected at right angles to an interface penetrated by two conductors, say. This increases the tracking length greatly for streamers moving across the shortest route at the interface. Such barriers can be simplexed on to the interface, if this is perspex and, properly done, the breakdown can be made to occur between the two metal conductors clear of the interface plus its barriers, before the streamer moving along its tortuous path can complete. This works for pulse voltages up to a million volts or more. For lower voltages, mylar sheets can be twinstuck (a double sided sticky tape) across the interface at right angles to the flash over paths.

Combinations of many of the above tricks can be employed, in most instances, and while the improvement factors they give cannot be multiplied together, at least they are in part additive.

Another class of techniques is to reduce the field on a sharp metal electrode by burying grading conductors in the insulator, just under its surface, and holding these at intermediate potentials with a resistive divider. This is a bit cumbersome but quite effective.

Yet another class of solution is to provide a compliant seal onto the surface so that a zero or extremely thin air space results along part of the insulator surface. This has been done quite a lot recently by German plasma physics groups, using square section silicon "O ring" compressible gaskets. This approach needs good engineering and a degree of cleanliness, and some care in design, otherwise a solid dielectric switch may be created unintentionally when a pulse passes the joint.

Another class of palliatives is to use a high molecular gas, such as freon, to flood the region. SF_6 can be used, but is considerably more expensive. As freon is much heavier than air, it can be poured into a bag surrounding the equipment. However, making gas tight seals through the bag is rather difficult and for a permanent set up something better is usually worth while. The level of the freon can be simply found by floating an air-filled balloon on it, or by lowering a lighted match into it, when a white cloud of smoke will float at some level in the mixed air/freon layer. By introducing the gas at the bottom of the bag via a spreader, freon can be conserved and the interface made pretty sharp.

There are a couple of points about freon worth making. Used in large quantities it can be a hazard, as in a laboratory it may accumulate on the floor, and crawling about in labs or large vessels where freon is in extensive use is not to be recommended. Smoking in air/freon mixture can produce phosgene, and should not be done. On the other hand, a little sparking or corona in freon produces only minute amounts of noxious products and in the normal laboratory is no hazard. Freon is no good in spark gaps, as carbon is produced which deposits on the walls of the gap; SF_6 is much better.

Reference (20) is a brief collection of some data on the pulse breakdown of freon and freon/air mixtures. Also given is the tracking gradient of a Marx in air and a lower limit to the degree of improvement achieved by putting it under freon. In general, in complicated situations such as Marx generators, strip lines, etc., an improvement of about 1.6 is usual. In situations which approximate more nearly to uniform field conditions, higher factors can result, and a peaking capacitor improved by a factor of nearly 2. In genuine uniform field conditions, the factor is more like 3 for pulsed voltages.

Freon being electro-negative, is very good at suppressing DC corona from sharp points, etc.

LIQUID/SOLID INTERFACES

The tracking problems of these are serious, especially as pulsed voltages of several million volts may be involved. Techniques have been developed for coping with most situations, but to cover them would be a lengthy business and not likely to be of immediate interest to Hull University. In the case of high voltage air cored pulse transformers, the tracking problem can be greatly eased by impregnating the transformer with a dilute copper sulphate solution. As the pulse volts rise on the thin metal sheets forming the windings, current flows along the thin liquid sheet resistors, charging the capacity of the insulant between it and the turn beneath. A diffusion type solution applies to this set up and the very high fields which would result at the edge of the copper are graded over a few millimetres during the pulse rise time. High dielectric constant layers can also be used to reduce the high fields at the edges of thin metallic sheets, in contact with plastic sheets.

The outline of the lecture calls for a discussion of a tentative picture of some of the phases of solid plastic breakdown: however, on thinking about it further, it seems the points I was going to make were either obvious or very contentious. I will be pleased to discuss the naive ideas with anyone really consumed with interest after the lecture.

"How to break a 6 inch square of polythene in 200 places". When I put down this topic, I had forgotten it had been referred to in any of the notes and I was relying on my memory of what was done. Consequently I am afraid I got it wrong. The sheet was broken in 600 places, not 200. The experiment is briefly outlined towards the end of Reference (1).

TENTATIVE PICTURE OF STREAMER BREAKDOWN

Early work which led to the light source described briefly below gave rise to one model of the air breakdown process. In these experiments a surface discharge was used. Figure 5a-1 shows a schematic cross-section of the set up. The top sheet of mylar is thin (a couple of thou. or so). The potential of the trigger is initially at V during the charging of the condenser. It is then shorted down to earth by a switch and a very large field develops at the HT copper electrode, which is typically 5 thou. thick. Field emission provides copious initiating electrons and a very energetic avalanche discharge moves across the top of the mylar.

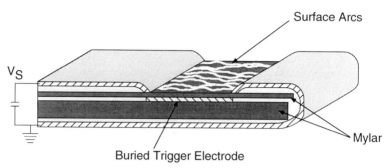

```
Width of Electrode ~ 10 cms
Spacing ~ 1.8 cm for 10 KV
For light source condenser = 1 µF
V~10KV number of channels ~ 8
Energy in channel ~ 30 Joules
```

Fig. 5a-1   Surface Discharge Arrangement

These avalanches lead to a number of plasma channels (~ 8) forming across the gap between the electrodes and then most of the energy (~ 60%) of the condenser dumps into these in about 200 ns. The point of interest here is the stages of formation of the plasma channel.

The very high gradient driven Townsend avalanche crosses in a few ns and has dimensions of a few mm. 20 ns Kerr cell photographs showed that, after a period, relatively low luminosity channels formed which were uniform end to end. These grew brighter but not very quickly. At this time the current (which was much less than that which flowed in the main discharge) would take hundreds of nanoseconds to beat the volume of air to plasma temperatures. During this phase the temperature was a few thousand °K. However, at this stage a very thin, very highly luminous core developed along the axis of the glowing hemicylindrical worms linking the electrodes. These were not completely straight in general but had gentle bands and kinks in them. However, the plasma channel thread followed these perfectly. What was calculated to have taken place was that the warmed half worm-like regions had expanded slightly, sending a rarefaction wave towards their axes. This could have lowered the density here by a factor of 10 to 100. The rate of heating of this central low density core was then an order of magnitude and a half higher than the main mass of ionised fairly weakly conducting worm. The central core thread then reached plasma temperatures quickly and its resistance ceased falling, and the only way the resistance could then fall further, was by expanding the cross sectional area of the plasma column in a hydrodynamic-radiation shock, and the resistive phase was started, during which the real energy was deposited. Thus in the surface discharge channels a hydrodynamic phase (the rarefaction wave) reduces the density and then the limited rate of delivery of energy can heat a small low density volume quickly to plasma temperatures, by which time the resistivity essentially stabilises at a few milliohm/cm. In this case the transition happens simultaneously along the whole length of the ionised region, because the degree of ionisation and intermediate declining resistance is uniform along the channel.

In the case of breakdown in, say, uniform fields, an electron and is emitted from the cathode and a Townsend avalanche sets out. After about 20 generations, during which the electrons drift with a velocity characteristic of their mobility in the applied field, the space charge dominates the applied external field and the avalanche speeds up, the degree of initial ionisation becomes much greater and the gap is quickly crossed. The current starts to rise through the region of amps; however, most of the voltage is across the initially weakly ionised region near the cathode. The current down the ionised region has to be constant, so the heating rate is rather bigger close to the electrode. This region also has the smallest cross-section. This region quickly expands and the root of the plasma starts at the site of the original initiating electron. The highly luminous plasma streamer then grows out across the gap. This sequence is deduced from streak image intensifier photographs for uniform breakdown over longish gaps in nitrogen at 1/3rd atmosphere. These records were shown to me at Strathclyde University, but they are not responsible for my interpretation of their very nice records. Thus, once again, a hydrodynamic phase would seem to enter into the breakdown process, before a plasma channel is established to move across the gap. This tentative picture is far from being well established. Moreover it sidesteps the physics of the propagation of the plasma channel streamer tip and whether this, too, involves a very rapid hydrodynamic phase, or whether, when the plasma channel is established as a highly conducting sharp projection, a new high field rapid ionisation process sets in at its tip.

It should be stressed that the above is a summary of my personal views about what is going on. They should not be taken as much more than opinions or, at best, a working model with which to play.

A USEFUL LIGHT SOURCE

The above surface sparks provide a useful light source. From the 20 ns Kerr cell photographs when the plasma channel was expanding through the resistive phase, the total volume was known. The current voltage records gave the energy deposited and confirmed that the energy density corresponded to a temperature of about 3 eV. The velocity of expansion of the channel (typically about 3 x 10^5 cm/sec in the case) confirmed this energy density. In addition, the luminosity of the channels measured in the photographic region gave a temperature of about 2 eV with modest accuracy. The superficial area of the 8 channels was about 5 cm^2. After the completion of the condenser energy deposition phase, the shock continues to move outwards and the plasma cools. By the time the temperature has fallen to about 1 eV, the Rossland mean free path becomes larger than the radius of the hot channel. The luminosity then falls quickly, partly because of T^4 term in the Stefan-Boltz-mann equation, but also because the emissivity is rapidly decreasing from unity. The measured light pulse had a duration at half luminosity of 0.4 μsec, in reasonable agreement with the calculations. This short duration was a desirable feature from the application for which the light source was designed. The light energy radiated was calculated to be about 1 joule of a black body spectra, with a peak around 1000 angstroms. Subsequently a 200 cm^2 area source was built and used to take reflected light 0.4 μsecond exposure photographs at 100 feet, as I recall.

Surface guided sparks were also made using pulsed high voltage sources, when the trigger strap was unnecessary. The letters "SSWA" were made in guided sparks about 1 foot long, except that the bar in the letter "A" was omitted: I told you we couldn't write.

Reference (21) is the only one I have been able to find dealing with these light sources and in it these are only briefly covered, I am afraid.

HOW TO TRACK 30 cm WITH 10 kV WITH THE AID OF INSTANT VACUUM

Figure 5a-2 shows the set up. Aluminised mylar links the electrodes and an external switch closes the hot electrode to a low inductance capacitor. A brief low level current flows as the aluminium layer melts and then partially vapourises. Its density at

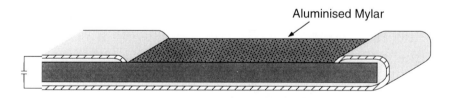

Fig. 5a-2   Long distance tracking at low voltages.

this time is like unity and it expands to 10^{-3} gm/cc when it is roughly in pressure equilibrium with the air, but it overshoots by about the same density factor, producing a mm thick layer of instant vacuum. The voltage across the electrodes is still essentially 10 kV and this low pressure gas layer breaks down for distances up to 30 cm at 10 kV. At slightly smaller distances, multichannel discharge arcs occur, and for considerably shorter lengths an essentially uniform sheet of plasma is generated. This starts below the aluminium, but bursts through this because of Taylor instability. The plasma front can move at 3 cm/μsec, providing a large bank is driving the set up, when few thou aluminium foils can be used as the primary layer. Low energy flash radiographs proved that the aluminium remnants were left way behind the luminous front, travelling at rather less than 1 cm per μsec and in a sadly disorganised state.

REFERENCES

1. "Volume Effect of the Pulse Breakdown Voltage of Plastics", J.C. Martin, SSWA/JCM/6511A (reprinted as AFWL Dielectric Strength Note No 3). (Section 8a)
2. Breakdown Strength of 'Tedlar' Polyvinyl Fluoride Film as a Function of Volume Stressed", I.D. Smith, AFWL Dielectric Strength Note No 6.
3. "DC and Pulse Breakdown of Thin Plastic Films", I.D. Smith, AFWL Dielectric Strength No 8.
4. "Revision of Breakdown Data Concerning Mylar", I.D. Smith, AFWL Dielectric Strength Note No 5.
5. "Pulsed Breakdown of Large Volumes of Mylar in Thin Sheets", J.C. Martin, SSWA/JSM/673/27 (reprinted as AFWL Dielectric Strength Note No 14). (Section 8c)
6. "Pulse Life of Mylar", J.C. Martin, SSWA/JCM/6611/106 (reprinted as AFWL Dielectric Strength Note No 11). (Section 8b)
7. "Comparison of Breakdown Voltages for Various Liquids under One Set of Conditions", J.C. Martin, SSWA/JCM/1164/8 (reprinted as AFWL Dielectric Strength Note No 1). (Section 7a)
8. "Impulse Breakdown of Deionised Water", I.D. Smith, SSWA/JCM/6511/8 (reprinted as AFWL Dielectric Strength Note No 4).
9. "Impulse Breakdown of Deionised Water with Asymmetric Fields", P.D.A. Champney, AFWL Dielectric Strength Note No. 7.
10. "Further Breakdown Data Concerning Water", I.D. Smith, AFWL Dielectric Strength Note No 13.
11. "Pulse Breakdown of Transformer Oil", I.D. Smith, AFWL Dielectric Strength Note No 2.
12. "Breakdown of Transformer Oil", I.D. Smith, AFWL Dielectric Strength Note No 12.

13. "Velocity of Propagation of High Voltage Streamers in Several Liquids", H.G. Herbert, SSWA/HGH/6610/104 (reprinted as AFWL Dielectric Strength Note No 10). (Section 7b Appendix)
14. "Nanosecond Pulse Breakdown in Gases", D, Felsenthal and J.M. Proud, Tech Report No RADC-TR-65-142, Rome Air Development Center, Griffins Air Force Base, New York, June 1965.
15. "High Speed Pulse Breakdown of Pressurised Uniform Gaps". J.C. Martin, SSWA/JCM/708/107. (Section 6c)
16. "Pressure Dependency of the Pulse Breakdown of Gases", J.C. Martin, SSWA/JCM/679/71. (Section 6a)
17. "Results from Two Pressurised Edge Plane Gaps", J.C. Martin and I. Grimshaw, SSWA/JCM/729/319.
18. "High Speed Breakdown of Pressurised Sulphur Hexafluoride and Air in Nearly Uniform Gaps", J.C. Martin, SSWA/JCM/732/380. (Section 10d)
19. "Pulse Charged Line for Laser Pumping", J.C. Martin, SSWA/JCM/732/373. (Section 12d)
20. "Early Data on High Field Gradient Systems", J.C. Martin, SSWA/JCM/HUN/2.
21. "Light Sources and Kerr Cell Modulators", J.C. Martin, SSWA/JCM/HUN/3.

Section 5b

**HULL LECTURE NO 2*,   HIGH CURRENT DIELECTRIC BREAKDOWN
                        SWITCHING**

J.C. Martin

INTRODUCTION

The NPT techniques should be consulted for an explanation of the rise time parameter used that is the e-fold rise time or maximum slope time and its relation or rather lack of it to the 10 to 90% rise time. It is entirely reasonable that no one parameter can adequately specify the full range of pulse rise fronts that nature can provide. You pay your money and take your choice but the resistive phase time is a maximum slope parameter and has to be handled as such, it is not a 10 to 90% time, so confirmed 10 to 90%'ers had better get down to it and derive their own resistive phase relationship, good luck. The inductive component of the rise time is fairly easily calculated to 10% or so, the more novel expression is the resistive phase term. The original note quantifying this is given in Reference (1) and was written some 8 years ago (1967). Numerous experiments in many laboratories have since tended to confirm the relation and I know of no serious disagreement with experimental observations.** Further discussion of the relation and possible variants of it are contained in the multichannel note, Reference (2), which should also be consulted. The NPT note gives typical examples of the rise time of the pulse derived from dielectric breakdown, for circuits with a range of parameters, showing the relative importance of the two terms under various typical conditions.

---

* SSWA/JCM/HUN/5
  June 1975
** Editor's note: (T.H. Martin) Further research shows higher switch energy losses, see Ed. Ref. 12.

The multichannel switching note, Reference (2), was the final outcome of a long series of endeavours to obtain reliable multichannel operation in gas switches. Multichannel operation has been achieved with intrinsic breakdown solid dielectric switches almost as soon as they had been invented. Multichannel operation of triggered liquid gaps had been obtained fairly easily but gas switches had proved obstinate. However, eventually the conditions necessary to obtain multichannel operation had been specified (at least to the author's satisfaction) and Reference (2) was the final outcome. Reference (3) contains some recent experimental results and calculations for two multichannel edge plane gaps which show excellent agreement with theory and also give pulse rise times in very good agreement with the rise time calculations. Some three years ago a DC multichannel switch was operated on a portion of a megajoule bank and switched currents of 4 million amps, the gap operating with a number channels in agreement with theory. Since then Physics International have developed a pulse charged pressurised SF_6 multichannel gap working at 400 kV and switching a 1 ohm line, which works very nicely and whose operation agrees with theory.

It should be mentioned that the theory of Reference (2) applies to the operation of multichannel light sources, again a case where stabilising resistors or transit time effects are not used to increase the isolation of the channels as they form.

TWO ELEMENT GAPS

The DC breakdown of conditioned gaps in pressurised air can be calculated to a couple of percent or so using the treatment given in Reference (4). This treatment is useful up to ten atmospheres or more. But for high enough pressures the breakdown field becomes dependant on the material of which the electrodes are made (see for instance Craggs and Meeks "Electrical Breakdown of Gases"). Also, if large coulombs are carried by the gaps deconditioning can occur, and then the breakdown voltage of the gap will develop a significant jitter and also fall below that at which the same gap carrying much less charge will operate. However, high coulomb carrying gaps have been developed by Mr T. James from Culham Laboratories which are now operating at up to 100 coulombs as start switches. In pressurised SF_6 gaps the electrode material dependancy sets in at a few atmospheres (at fields comparable to those at which the effect appears in pressurised air). However, below this pressure the breakdown voltage is pretty linear with pressure and there is no significant gap length dependancy in well designed gaps as there is inherently in air. DC operated pressurised SF_6 gaps both triggered and untriggered have been extensively studied at Physics International of San Leandro California for high coulomb operation. Also

HULL LECTURE NOTES

Maxwell Laboratory of San Diego have done much work in pressurised spark gaps with both pressurised air and SF_6.

For pulse charged gaps the mechanical strength of the electrodes does not have time to be very involved, so higher gradients can be achieved in two electrode gaps with both pressurised air and SF_6. Reference (5) summarises some experiments on pulse charged nearly uniform gaps. Gradients in pulse charged SF_6 at 6 atmosphere pressure can approach 1 MV/cm and represent one of the best switching media known under these conditions. Also covered in Reference (5) is the external tracking problems of a rail two electrode gap, where a gap with a pressure body diameter of 2" with 4" long fins attached did not track at 800 kV. Reference (6) gives further details of this pulse charged gap and also shows that the velocity of light may contribute to the rise time of the pulse from such a gap when used in a relatively high impedance system. In lower impedance systems two electrode rail gaps can have low inductances. The single channel breakdown is arranged to happen in the central region of the rails and the rest of the electrodes act as pressurised gas insulated parallel-wire feeds to the spark channel. For few hundred kV operation inductances of less than 20 nH are obtainable with 30 cm wide gaps. With pulse charged two element gaps, edge plane arrangements can be used. Reference (3) describes the construction and performance of two such gaps. A modest extrapolation of the pulse sharpening gap data would give a less than 1 nano henry inductance for a meter wide gap under the operating conditions given in the reference.

THREE ELEMENT GAPS

Whilst the summary given in NPT notes is still an adequate summary of the situation, a few additional comments will be given of later work.

<u>Solid Gaps</u>

Triggered solid multichannel gaps have been used in low impedance water generators at voltages of up to 1 1/2 MV with no trouble. The slave triggered gaps were of the more advanced field distortion version where the edges of the trigger strap were sealed with unpolymerised araldite. The trigger pulse that fired these was provided by a master gap which was a version of the older stabbed polythene switch. The reason this switch is still used is that it is rather tolerant of odd tracking etc. effects which lead to pulses on it. The field distortion gap is rather sensitive to any small voltage excursions on the trigger strap. In addition, the stabbed polythene switch can have a wide range of breakdown voltages by simply changing the depth of stab and/or its polarity, thus the firing of the slave switches can be simply controlled by the stabbed master switch over a range of voltages.

Liquid Gaps

Physics International now have multichannel triggered oil switches operating at 10 MV in transformer oil and also have extended the working voltage of multichannel water switching to 4 MV, both of these are of course pulse charged.

Gas Switches

Maxwell Laboratories, Ion Physics of Burlington, Mass. and Sandia Corporation have all extended the use of pulse charged trigatrons to 3 MV or so, and operated several switches in very close proximity in what is essentially a multichannel mode of operation. Mr. I.D. Smith of Physics International has originated a version of the field distortion gap known as the V/n gap. This gap is pulse charged and uses pressurised SF_6 and the field distortion trigger electrode is very close to one electrode, typically at one tenth or one twentieth of the total gap spacing. Thus, a 2 MV gap needs only 100 kV or so of trigger volts to fire it.

Descending from the rarified levels of very high voltage pulse technology to the medium voltage level, the three element field distortion gap operates over a wide range DC charged. Versions have been used over the range 200 kV down to 10 kV with no difficulty. Working at 80% of the self break voltage, the gaps can be triggered in ten nanoseconds or less with a jitter of fractions of a nanosecond. The breakdown time of such gaps can be calculated using point/edge plane breakdown relations given in Lecture No. 1. For the most rapid breakdown, the offset or simultaneous mode of breakdown is best where the triggering electrode is spaced at roughly 2 : 1 in the gap and the two parts of the gap are broken at approximately the same time. A fast rising trigger pulse of amplitude roughly equal to the self break of the gap is needed, but as the capacity it is driving is low, this does not present serious problems. Recently, versions of a triggered rail gap with a self break voltage in air of 15 kV have been triggered down to 6 kV and it is hoped to extend the working range of these gaps downwards significantly further in the near future. Work in the States has shown that similar gaps pressurised can have jitters of ± 0.2 ns and it is considered that by using pressurised hydrogen, this could be reduced further, if it was ever necessary.

On the practical side pulse charged three electrode gaps need careful balancing, if capacitors are used to hold the trigger electrode at its proper potential so that the gap does not operate early because the potential division of this has been disturbed. This difficulty can be obviated by putting a small gap in between the trigger and the trigger pulse generator. Of course, this has to hold off the pulse charging waveform of the trigger electrode but this is not too difficult to arrange. When the trigger pulse arrives it breaks this isolating gap and operates the gap. The isolating gap can be used to pulse sharpen the trigger pulse as well if this is desirable. Some form of auxillary surface discharge u.v. irradiator may be necessary to break the isolating gap at reasonable voltages and so avoid having to provide an unnecessarily large trigger pulse. The spacing of the isolating gap can also be halved by back biasing it with a half voltage DC supply of the appropriate polarity.

Three element field distortion gaps can also be used as clamp gaps i.e. ones that fire at nearly zero voltage but this application is a wide area where much work has been done in the plasma physics research establishments and the reports of these should be consulted.

As an example of the triggering range and fast firing possible with field distortion gaps, Figure 5b-1 shows the self breakdown voltage of a gap with spherical 1" diameter ball bearing electrodes operated with a 0.5 cm gap for a range of SF_6 pressures. The figure gives the breakdown voltage and also the voltage at which the gap closes after a delay of 40 ns. The curve labelled amplitude of trigger pulse is the value at which triggering starts in the gap. In a real system a trigger pulse more than this voltage would be needed, especially as it is desirable to fire on a fairly rapidly rising portion of the trigger waveform.

A mention should be made that with atmospheric SF_6 in a gap with fairly small spacing, breakdown voltages above that deduced from the values in the literature can be observed. These occur after fairly frequent low energy shots and are believed to be due to the formation of insulating layers (possibly sulphur) on the surface of the electrodes. These layers, if they exist, can hold up to ten kV or so in extreme cases. Filling with air and sparking removes most of the effect until the layer reforms. Also shown in Figure 5b-1 are the calculated curves as well as the observed data and the agreement is quite reasonable. The calculations are made using the edge plane breakdown relations. Figure 5b-2 shows the triggering delay for the gap at 3 atmospheres, the calculated values are not shown in this case but have the general form of the observed curve.

The optimum trigger spacing ratio in the case was 0.3 : 0.7 and essentially independant of pressure over the pressure range investigated. Jitter measurements were not made for this gap (they

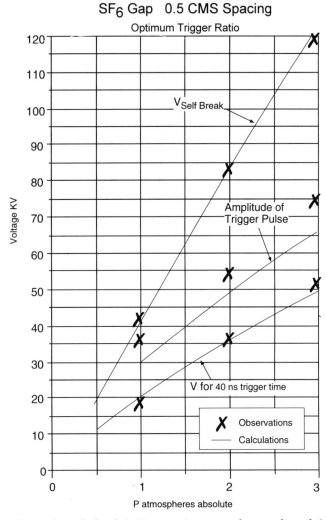

Fig. 5b-1   Example of field distortion spark gap breakdown.

are quite tricky in the sub ns range) but would be expected to be ~ ± 5% of triggering time.

If a 20 cm wide rail version of this gap were made and operated in a single channel mode it would have an inductance of a little under 10 nanohenries.

With a sufficiently fast trigger pulse (involving a pulse sharpening gap) and a good feed system to apply the pulse to both electrodes, multichannel operation should be possible, but to date we have not had the time or the necessity to do this with a pulse charged gap.

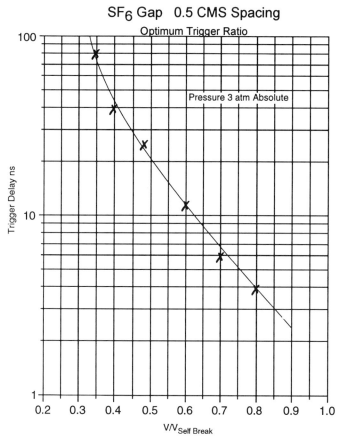

Fig. 5b-2 Triggering delay for field distortion spark gap vs gap voltage.

Inductance

As a rough guide sphere gaps can have inductances around 50 nH without going to great lengths to reduce this, rail gaps (single channel) around 20 nH and edge plane pulse charged multichannel gaps a few nH. With multichannel operation rail gaps can get down to the same range but have the advantage of a considerable smaller resistive phase, and the disadvantage that it is significantly harder to make a uniform field gap multichannel, than it is a pulsed charged edge plane gap. The above rough guides, apply for voltages between about 10 kV and 200 kV. Above this limit low inductances can still be achieved but multichannel operation becomes more desirable. Liquid gaps (triggered multichannel) have inductances in the 20 to 30 nH range up to several megavolts while solids can have 10 nH or less fairly easily up to a couple of megavolts. For DC use at 30 kV solid gaps, such as are used in the high current banks at AWRE, have inductances of a few hundredths of a nanohenry when used in total widths of twenty feet or so.

Quick Constructional Techniques

Using ball bearings (mounted on 0BA screws if about 1 mm pitch per turn is required) and perspex plates simplexed together as spacers, an unpressurised air triggered gap can be made up in about 1 hour. Such a gap can be used up to 70 kV. Using perspex tube and simplexing semipermanently end plates, a pressurised gap can be knocked up in about 2 hours. If access to the gap is required an end plate can be sawn off and resimplexed after the alterations have been made. To make unpressurised triggerable rail gaps takes a little longer as the brass rods have to be contoured at their ends (see Reference (3) for details). Pressurised triggerable rail gaps with anti tracking guards take longer again, the time being of the order of 4 to 5 hours for gaps working up to 150 kV DC. Much of this time is in the pressure sealing of the several feed through points along the length of the rail electrodes. This requires fairly accurate drilling of the holes, something I cannot usually manage and I finish up using a file invariably. These seals can be demountable but as it is rarely necessary to remove the main electrodes, simplex over the feed throughs, suitably applied, can provide a quick final seal.

Up to 3 inch OD perspex tubing of 1/2 cm wall thickness is regularly tested to 100 psig (wrapped in rags in case it blows up) and used up to 60 psig. We have never had any gap disintegrate in this pressure range but the possibility should be borne in mind. The only occasion a gap did physically break was in a 2 MV pulse charged system working under oil at 300 psig or thereabouts.

A flashover occurred (probably across the outside of the gap in the oil) the shock from this cracked the perspex pressure vessel and the expanding SF_6 gas took out a face of the perspex oil filled box, with the obvious results.

Care is always taken in simplexing up pressurised gap bodies to clean and roughen the mating surfaces and to provide an adequate length of joint. The vapour from setting simplex can cause vapour crazing of some versions of perspex tube which have not been completely stress relieved. Either a flow of air or moistening the surfaces near the joint can alleviate this surface crazing. The ultimate solution is to stress relieve the tube by warming to about 80°C for an hour or two. Quite bad surface crazing has not been observed to greatly reduce the strength of perspex tube, but it seems desirable to avoid it.

SF_6 as a Gap Medium

Examples have been given of the very desirable properties of pressurised SF_6. For large coulomb operation brass electrodes are as good as more exotic materials, as far as data exists, and are much more easily worked. The SF_6 should be dried and the gas changed every few shots, if full DC hold off voltage is required. As the hot decomposed gas rises, the gas should preferably be introduced at the bottom of the gap and extracted at the top. We have never experienced any trouble with the nasty decomposition products of SF_6, the volume produced per shot is minute compared with that of the average lab. For high rep. rate and for large coulomb use, a hose pipe stuck out of the window might be necessary. For pulse charged gaps, the removal of the decomposition products is much less necessary and flow rates can be considerably lower. Pulse sharpening gaps are described in Reference (3) which covers the issue pretty thoroughly from the point of view of achieving high di/dt outputs. They can also be used in other circuits, but where the pulse being sharpened is not of great amplitude a small sphere gap can be employed. Such a gap will work at large gradients on a fast rising pulse but will also display a jitter in its closure voltage. If this is undesirable the gap should be UV irradiated, this will remove the jitter in breakdown voltage at the cost of reducing this significantly. A very quick way to make a UV irradiator gap is by placing a sharp metal edge on a few thin mylar sheets which is in its turn stuck down to an earth plane. This sub gap is fed by a high resistance (~ 10 kΩ) and coronas across the mylar produce UV which irradiates the main gap. Another way is to include a small insulated pin drilled through one of the spherical electrodes. The trigger pulse is fed to this pin and causes an arc between this and the main electrode, irradiating the gap as it overvolts it.

THE CORONA GAP

These are extremely useful little gaps, which very few, apart from ourselves, have learned to love and appreciate. Basically they are gaps operating at low voltages (typically 5 to 10 kV) which can be triggered by a 300 volt pulse via a 50 ohm cable fed to the trigger strap via a step up transformer. They normally are not used to handle more than a few joules but can provide several hundred amps of input current rising in 20 to 30 ns.

Used with a reasonable trigger pulse they can have jitters of less than 2% and firing times of 40 ns or less. We have used them to provide pick-up-immune combined delay and output pulse units which can be quickly and cheaply constructed. For instance in one application a seven gap unit (size 1 1/2 feet x 1 foot x 1/4 foot) provided six delayed outputs (delay 0.1 to 100 μseconds) three of which were used to trigger surface discharge, light sources and three of which directly produced the 8 kV (via a double Blumlein circuit) pulse to operate a 20 ns Kerr cell.

With this system reflected light 20 ns multiframe photographs were obtained, which is quite an achievement in view of the fact that the optics were operating at something like f/150. While the units themselves are pick-up-immune they may have to be boxed in order to prevent them firing other low level systems, but in fact all the normal functions of delay, trigger pulse production, and even the scopes themselves can be done using these gaps. They are simple, cheap, robust and reliable and can be made in a few minutes, if the components are stockpiled, or an hour a piece starting from scratch.

Unfortunately we do not have extensive write up of them and Tommy Storr has kindly produced a brief description of them as we make and use them (Reference (7)).

A very old write up of a system using an early version was excavated from the filing system (the early Cambrian strata) and is included as Reference (8). There is also a patent referred to in the NPT notes, but I could not understand it. A sealed off version has been produced at AWRE by another section (Reference (9)) and works well, but to my mind one of the great advantages of the gap is its ease of construction and this, of course, is lost once you start encapsulating in glass. There is a further write up in an AWRE report (Reference (10)). If this report is obtained there is a statement on page 4, paragraph five with which I do not agree in general, although in the particular set up the author used, it could have applied because of the way the pulse transformer was wound.

Improved versions of this gap have recently been developed and it is hoped to make considerably better performance versions shortly. The SSWA group should be consulted if anyone becomes seriously interested.

A further reference has been unearthed which deals with an early version of the Kerr cell, light source arrangement. A copy of this is included as Reference (11).

REFERENCES

1. "Duration of the Resistive Phase and Inductance of Spark Channels", J.C. Martin, SSWA/JCM/1065/25, reprinted as AFWL Switching Note No. 9. (Section 10b)
2. "Multichannel Gaps", J.C. Martin, SSWA/JCM/703/27. (Section 10c)

3. "Pulse Charged Line for Laser Pumping", J.C. Martin, SSWA/JCM/732/373. (Section 12d)
4. "DC Breakdown Voltage of Non Uniform Gaps in Air", J.C. Martin, SSWA/JCM/706/67 (reprinted as AFWL Dielectric Strength Note No 16). (Section 6b)
5. "High Speed Breakdown of Pressurised Sulphur Hexafluoride and Air in Nearly Uniform Gaps", J.C. Martin, SSWA/JCM/732/380. (Section 10d)
6. Notes for a Report on the Generator TOM, J.C. Martin, SSWA/JCM/735/407. (Section 12e)
7. "Corona Switch", T H Storr, SSWA/THS/5/73.
8. "Brief Outline of Operation of Trigger Unit", J.C, Martin, HUN 1. (Section 5a)
9. "Triggered spark Gaps for Image Tube Pulsing", R.J. Rout, Journal of Sci Instr, 1969, Series 2, Vol 2, Page 739.
10. "The performance of the Miniature High Voltage Spark Gaps Used in the E14 Camera", B.R, Thomas, SPA2/E14/BRT/71-5 AWRE Report
11. "Light Sources and their Cell Modulators" J C Martin HUN 3.
12. Editor addition, "Energy Losses in Switches", T.H. Martin, J.F. Seaman, D.O. Jobe, 1993 IEEE International Pulsed Power Conference Proceedings.

Section 5c

HULL LECTURE NO. 3*   MARX-LIKE GENERATORS AND CIRCUITS

J.C. Martin

MARX GENERATORS

Two useful versions of Marx generators will be briefly described in these notes, untriggered and triggered.

The untriggered Marx works by overvolting the higher gaps by linking an impedance across a number of gaps(n). All impedances directly across one gap tend to prevent the voltage on the gap rising and hence inhibit triggering. However if these impedances are counteracted by ones across more than one gap this aids triggering. Figure 1 shows a schematic with impedances linked across every third gap, a n = 3 case. The parameter n of course can have any number greater than one and when the higher order impedances swamp the n = 1 impedances a total of nV appears across the gaps. A minimum of n + 1 gaps have to be triggered at the base of the Marx to start the propagating erection wave. If capacity coupling is used, the needed linking capacities can be built in as strays up the Marx, but then it is not usually possible to make these greatly bigger than those across the individual gaps and less than nV appears across the nth gap up, after the ones below have fired. However this voltage will appear quickly, the rise time of the triggering pulse being determined by the breakdown time of a gap, and the inductance and capacity of the circuit involved in feeding the pulse up.

---
*   SSWA/JCM/HUN/5
    June 1975

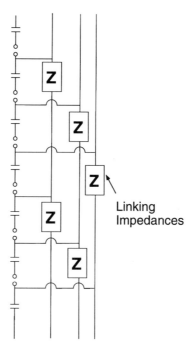

Fig. 5c-1 Untriggered n = 3 Marx.

If resistive coupling is used this will provide the full nV after a time characterised by the discharge time of the gap strays via the multistage linking resistors. Thus the mode tends to have the largest triggering range but a slightly slower erection. In a Marx built of small value capacitors, this mode of coupling may not be of use, as the lower stages of the Marx may discharge while the erection is taking place.

The two modes can be combined and Figure 5c-2 shows an example of a n = 3, R and C-coupled Marx.

The minimum number of charging resistor chains needed is n + 1, where one high impedance chain links across all the gaps and n much lower impedance chains link each across n gaps. However if helpful stray capacities are built into the Marx, by stacking it in n columns of capacitors, it is frequently more convenient to use 2n charging chains, especially if plus and minus charging is being employed.

A n = 2 Marx will have a triggering range down to about 60% of self break voltage, provided the set of gaps in the Marx has a low standard deviation of breakdown, and gets UV irradiation from the lower gaps. However, in general n = 3 is the best number for a

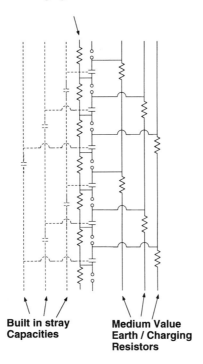

Fig. 5c-2 Untriggered R = 2, R C Coupled Marx

reasonably wide range Marx, a good version of which will trigger down to about 40% of self break.

The triggered Marx obviously has the added complication of a third electrode in each gap, but has three significant advantages which may more than justify this. Firstly it can operate with arbitrarily large capacities across each gap. Such capacities are an essential consequence of building low inductance Marx generators because of the strip line feeds from the gaps to the capacitors. Secondly, the gaps can be in separate bodies without paying a penalty in triggering range as UV irradiation between gaps is unnecessary. Thirdly, much smaller transient voltages can be made to appear across each gap during the erection and this can considerably ease internal flashover problems in high gradient generators.

Figure 5c-3 shows a schematic for a triggered Marx with impedances across three gaps (m = 3). This provides a total voltage excursion of 3V on the trigger electrode when the added impedances swamp the strays. However the voltage applied to the second part of the gap after the first part has fired, will not be much bigger

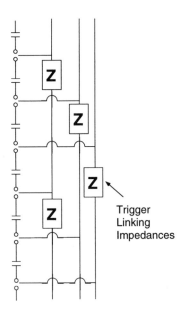

Fig. 5c-3   Triggered Marx, m = 3.

than V unless steps are taken to ensure this. Figure 5c-4 shows a schematic of an m = 3 Marx with added resistive coupling. In this case plus and minus charging is employed using high impedance chains and the m coupling is obtained by use of 3 earth resistance chains attached to the mid point of the two capacitors in each stage.

For very low inductance Marx generators with big strays across each gap, this technique may not be of too much use as it may take too long to operate, but better operation can still be obtained by off-setting the trigger electrode so the gaps tend towards "simultaneous" operation. Considerable expertise is required to make the Marx operate in this mode; however it is considered to be the best one to achieve multichannel operation with and offers the lowest possible inductance in a single Marx. Such a system was designed at AWRE in 1969 but has not been built and operated to date.

Reference (1) gives details of an n = 2, RC coupled Marx, while Reference (2) covers the case of an m = 2, C coupled Marx.

In general a very good range of operation can be obtained with an m = 2 triggered Marx providing this is not of the very lowest inductance type, and triggering down to 30% of self break can be

HULL LECTURE NOTES

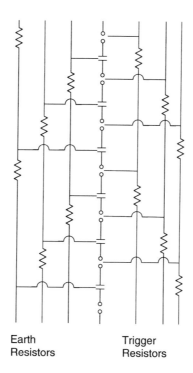

Earth Resistors   Trigger Resistors

High Impedence Charging Chains Omitted

Fig. 5c-4   An m = 3 Marx with added resistive coupling.

obtained with R coupling added. For the lowest inductance systems m = 3 with offset triggers appear to be necessary.

ERECTION TIME

Both of the above notes give the erection time for the particular Marxes and show that (a) fast erection times can be easily obtained and (b) there are wide ranges of operating conditions where these do not change significantly.

Physics International on one occasion built several Marx generators of the same design, each being a large energy system with m = 3 and containing 20 stages with an output voltage of 2 MV. The erection time for a given Marx was about 210 ns with a shot to shot jitter of ± 3 ns and a set of four of the Marx systems had mean erection times within a span of 15 ns under the same set of gap pressures etc.

The erection time can be calculated roughly. Firstly the collapse time of the voltage across a gap is calculated, then the integration effects of the feed up the trigger resistors is calculated. A guess at the equally spaced time of operation of the preceding m - 1 gaps is made and the resulting trigger waveform on the mth gap in the set is obtained; the edge plane streamer relation is used to obtain the firing time of the first part of the gap, and then that of the second. These times are used to check that the total time to fire the set of m gaps is consistent with the value assumed earlier in the calculation. This total firing gap time is divided by m to give the effective single gap time, on the assumption of a stable propagating erection phase moving up an infinitely long Marx. However real Marx generators have a finite number of stages and as the wave propagates up the Marx, the voltage on the higher gaps increases. The way this happens depends on whether the output of the Marx is effectively earthed (as in a peaking capacitor case) or is psuedo unearthed (only having its stray capacity to earth grounding it) thus the erection process speeds up as it proceeds and may indeed start from the top and fire downwards in some cases towards the end of the erection phase. Thus the erection time is not the above single gap firing time multiplied by the number of stages in the Marx, but something smaller. For earthed Marxes a fudge that is usually good is to multiply by 0.4 of the number of stages, while for a pseudo open circuit case a factor of 0.7 is frequently applicable.

In general the resulting erection times so calculated are in agreement with the observed times to a factor of 1 1/2 or so and have a pressure dependency similar to the observed one. Their real value is in deciding what changes can be usefully made to the many circuit parameters in order to speed up the process and which changes are unlikely to be worthwhile. Of course the first m - 1 gaps have to be properly fired, and that this is the case should be checked by monitoring the time of closure of these gaps in a separate experiment.

MARX INDUCTANCE

This can be crudely calculated using the methods outlined in NPT notes to obtain the inductance of each bit of the circuit; an example is given in Reference (2). Again the real value of such calculations, apart from deriving an approximate value of the inductance of the system before it is built, is that if the inductance of some portion of the system is an unduly large fraction of the total system inductance, steps may be taken to reduce this and so make a worthwhile improvement. If the system has several bits, all of which have roughly the same inductance, improving one will have little effect. Equally if one is increased by a factor of 2 for high voltage or mechanical assembly reasons, it will have little effect on the overall inductance. In such a balanced set up, if an improved inductance is needed a completely new approach is usually necessary, such as a change from untriggered to triggered systems. Of course the overall inductance is not just the sum of the component capacitors, feeds, and spark gaps but also involves the inductance viewed as a conductor rising the height of the Marx as well as that of the output feeds joining it to the load.

PEAKING CAPACITORS

This topic is dealt with in Reference (2) and will not be further covered here as is the topic of cheap high voltage pulse charged low inductance capacitors, except to point out that as the peaking capacitor becomes comparable in value to the Marx capacity it becomes what is called a transfer capacitor. This is simply a pulse charged high voltage capacitor of superior performance, to which the energy of the Marx is transferred so that it can be more quickly applied to the load. As such, most of the systems mentioned in the fourth lecture can be described rather inadequately as transfer capacity systems.

L/R MARX OUTPUT CIRCUIT

In some cases the essentially exponential tail of the output waveform of Marx feeding a constant resistance load is undesirable. This is the case for some e-beam generators where a few microsecond roughly flat top waveforms can be desirable. There are a number of approaches to this, the first is to use a Marx with a large output capacity but this is wasteful of energy. Another approach is to build a lumped constant delay line out of high voltage components. However a third rather crude approach is to put an inductance L bypassed by a resistance R in the output of the Marx. Figure 5c-5 gives the circuit.

Initially the resistance R is in series with the constant load resistance Z but as the voltage of the Marx (viewed as a lumped circuit capacitor) falls, the current transfers to the inductance. The circuit can be arranged so that the voltage across Z rises after the switch-on front, peaks and then falls back to the same value at the end of the phase under consideration. Thus an output pulse across Z has a variation of say ± 5% or ± 10% up to some time at which the voltage begins to decay fairly rapidly in a roughly exponential fashion. If a dump or divertor gap is then fired, an approximately flat top pulse is obtained. The energy transfer to the load can be quite large (> 50%) at the expense of having an output voltage of about 3/4 of the open circuit value of the Marx. Figure 5c-5 gives the results of some calculations made a couple of

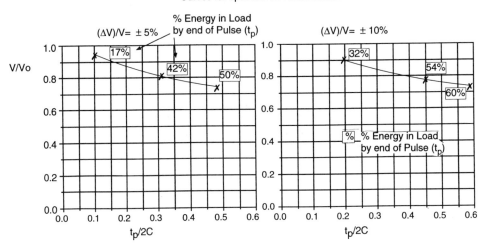

Fig. 5c-5   "Flat-top" L/R output Marx generator.

years ago for the circuit and uses the optimum values for L and R. The calculated values have not been checked by myself since doing them but are within a few per cent of the correct ones.

The inductor and resistor are very simple components to make from a high voltage point of view and can be easily altered to cover any desired range of parameters. Probably the best way is to wind the inductance along a copper sulphate resistor where the best grading is achieved. 20 kV/cm gradients can be supported on such an arrangement in air, probably twice this in freon and of course the components can be oil immersed or, less desirably, potted.

For further details of the calculations the author could be consulted but the circuit analysis is pretty simple.

If a fast rise time to the pulse is required, (and razor blade cathodes need a fairly fast switch-on pulse) a peaking capacitor can be included in between the Marx and the L/R arrangement. Some small oscillations on the output waveform would be expected to result from this combination, but if these cannot be accepted damping resistors can be used to reduce their amplitude to low levels.

HULL LECTURE NOTES

AIR CORED TRANSFORMERS

The section in the NPT note describes the techniques for use in high voltage versions of these systems. However for lower voltage operation, copper sulphate solution impregnation is not necessary. The air cored transformer can have either a single sided output referenced to earth or can have a plus and minus output of twice the total output. The rough guides given below to the output pulse amplitude are for the single sided output; the values have to be doubled for the other case. For voltages up to 100 kV plus the winding need not be impregnated although above 100 kV some loading due to corona in the windings is to be expected. Up to 250 kV plus transformer oil impregnation can be used, which is an advantage as it does not evaporate. Using solution resistive grading, voltages up to 2 MV can be generated in transformers a couple of cubic feet in volume. The above guides for output voltage are only rough and depend a bit on size of the transformer.

Reference (1) gives a complete calculation for a trigger transformer and the approximate relations used. These relations are summarised in Figure 5c-6 and can be used not only to calculate the open circuit gain but also that to be expected from addition resistive or capacitive loading of the transformer. Care must be taken to keep the impedance of the capacitor driving circuit low and the relations given show how the inductance and resistance of this effect the output obtained. The relationships and expressions are only approximate and give answers to about 5% in general. However they are simple enough to allow an optimisation of a transformer to be done. In general there is a turns ratio which gives a calculated maximum output, however it is desirable to work a little below this, to keep the output impedance lower than would be the case with the calculated optimum. Also of course the high voltage tracking lengths have to be borne in mind otherwise these may get too small, and these dictate some of the dimensions. Reference (3) gives a detailed description of a ± 1 MV pulse transformer built at AWRE by Dr Dave Hammer, however in general for this voltage range a Marx is a better bet especially if it has to transfer many kilojoules.

It is not necessary to go the system outlined in the above notes for many trigger applications and well insulated wire, intelligently wound on a few turn primary can be used to give gains of ten or so and output voltages up to 100 kV or more. However for high gains and fast rise times it is usually worth the effort to construct them as described.

CHAPTER 5

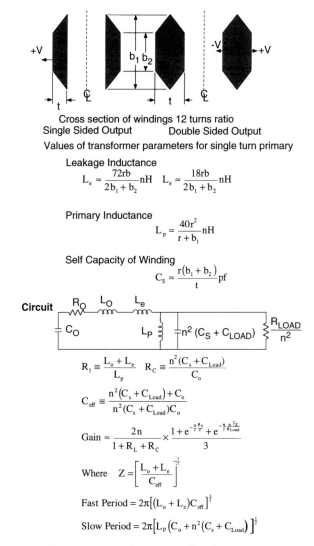

Fig. 5c-6   Approximate transformer relations.

Cross section of windings 12 turns ratio
Single Sided Output   Double Sided Output
Values of transformer parameters for single turn primary

Leakage Inductance
$$L_e \approx \frac{72rb}{2b_1 + b_2}\,\text{nH} \qquad L_e \approx \frac{18rb}{2b_1 + b_2}\,\text{nH}$$

Primary Inductance
$$L_p \approx \frac{40r^2}{r + b_1}\,\text{nH}$$

Self Capacity of Winding
$$C_s \approx \frac{r(b_1 + b_2)}{t}\,\text{pf}$$

Circuit

$$R_L \equiv \frac{L_o + L_e}{L_p} \qquad R_C \equiv \frac{n^2(C_s + C_{Load})}{C_o}$$

$$C_{eff} \equiv \frac{n^2(C_s + C_{Load}) + C_o}{n^2(C_s + C_{Load})C_o}$$

$$\text{Gain} \approx \frac{2n}{1 + R_L + R_C} \times \frac{1 + e^{-\frac{\pi}{2}\frac{R_o}{Z}} + e^{-\frac{\pi}{2}\frac{n^2 Z}{R_{Load}}}}{3}$$

Where $Z = \left[\dfrac{L_o + L_e}{C_{eff}}\right]^{\frac{1}{2}}$

Fast Period $= 2\pi\left[(L_o + L_e)C_{eff}\right]^{\frac{1}{2}}$

Slow Period $= 2\pi\left[L_p\left(C_o + n^2(C_s + C_{Load})\right)\right]^{\frac{1}{2}}$

REFERENCES

1. "Mini B Design and Operation" T.H. Storr SSWA/THS/692/9.
2. "Notes for a Report on the Generator TOM" J.C. Martin SSWA/JCM/735/407. (Section 12e)
3. "Notes on the Construction of a High Voltage Pulse Transformer" J.C. Martin, P.D. Champney, D.A. Hammer, Cornell University CU-NRL/2.

Section 5d

HULL LECTURE NO. 4*   FAST CIRCUITS, DIODES AND CATHODES FOR e-BEAMS

J.C. Martin

FAST CIRCUITS

The topics which should be covered in these notes represents the majority of the SSWA group activity over the past eight years or so. The material of the other four lectures represents peripheral or side issues which were necessary to tackle in order to work effectively in the desired area. Thus ideally these notes should and could be many times the length of the others combined. However, apart from e-beam triggered discharges Hull University is not too likely to have any need for data from this area, hence both the lecture and these notes will only briefly cover the topics involved. This is so that a properly balanced overall picture can be given and in the event that some new area does assume importance (such as direct pumping with e-beams) an appreciation can be made of what has been done and what sort of effort is required. For details of what is still a rapidly evolving field, we should be approached for the latest position as far as we know it. As ever the NPT note should be consulted for a general review of the high speed output lines of use and Reference (1) (now sadly outdated) gives a popular resume of some of the systems we built years ago. Reference (2) gives a detail of a medium sized system of some five years ago. Reference (3) gives a detailed description of a pulse charge low impedance line using gas switching and also gives details of the real life performance of a strip line version of the Blumlein circuit in some of its variants.

---
*   SSWA/JCM/HUN/5
    June 1975

The trend of recent years has been towards lower impedance systems with output voltages in the 1 MV range. Two sorts of approaches have been employed to achieve megamp outputs. In one, solid mylar dielectric sheets are used to make Blumlein circuits and series-parallel arrangements of these are used to provide the desired output voltage and current. These Blumlein circuits are pulse charged and graded by resistive films at the edges of the conductors and were initially switched with solid dielectric switches. Later these were replaced by multichannel pressurised SF_6 rail gaps which have performed well. Physics International of San Leandro California, have raised the state of the art of such systems to a high level. A number of groups in the States (including Physics International) have followed the route of a pulse charge water filled coaxial line. This of course needs to be charged to 2 MV to deliver 1 MV into its matching load but this is now a routinely used modest voltage. Those using this route are Naval Research Laboratories, Washington DC, Maxwell Laboratories, San Diego, Sandia Corporation, Albuquerque and ourselves. It is probably fair to say that the later route has proved the most useful and least complicated for the working range mentioned above. Generally but not always high speed pulse lines are used to accelerate large currents of relativistic electron beams which are then used directly or are in their turn used to produce X-rays or sometimes microwave radiation. All of these and other applications require the high voltage pulse to be fed from the line to a vacuum filled diode. This involves two portions of the system, the output feeds and a vacuum envelope or diaphragm. The output feed is designed on transmission line concepts and can have a closer spacing than the original pulse charged lines, as the pulse duration it sees is much shorter than that applied to the main lines. Also in a matched coaxial line system it sees only half the voltage that the pulse charged line has to withstand. Thus in general the design problems in this area are not too severe.

THE DIODE VACUUM INTERFACE

This represents an area of considerable concern to the high voltage pulse designer and in particular in low impedance systems, the design must be very good in order to keep the inductance of this portion of the system acceptable. Reference (2) describes a typical high voltage tube insulator of the multisection type, where the pulse voltage is applied along a carefully graded structure made out of perspex. Fortunately modern high current cathodes do not need a very high vacuum in order to operate and pressures of 0.1 to 1 micron are acceptable. Reference (4) summarises the available data and references on vacuum flash over of various materials for short pulses. Typically tubes to withstand high voltage pulses of 100 ns duration can work at gradients up to 2 MV per foot for pulses of a few million volts.

There is another version of the diode interface which was also developed at AWRE and that is the diaphragm tube. In this case a large disc of plastic is used, usually without intervening rings, and the adjacent metal surfaces are contoured so the electrons leaving the plastic surface do not re-enter it, and thus the field lines have to have the optimum angle to the surface on the vacuum side.

After the vacuum interface, a stalk carries the cathode which is located close to the anode. From the cathode a large current of electrons is extracted and directed at the anode or passed through a window for use in the region beyond.

Reference (5) describes a large X-ray machine built at Sandia Corporation some years ago while Reference (6) covers some of the work at NRL on high current systems. Physics International have built a very wide range of systems, including the largest system to date, a 5 megajoule stored energy one. In addition they have done much work on beam propagation as have the NRL group, Sandia Laboratories and a group at Cornell University. A considerable body of published literature exists on the machines and beam propagation from these and other groups working in the field. More complete references can be supplied if desired.

CATHODES

As will be described in the talk an enormous range of cathodes have been used at one time or another. Apart from very low current systems all of these have in common that they either intentionally or unintentionally get covered by plasma blobs early in the high voltage pulse. These plasma blobs originate from whiskers which vapourise by $\int j^2 dt$ heating of the field emitted current and then the resulting debris gets heated to a few eV temperature. Provided they are not affected by the self magnetic fields of the current they are emitting, these blobs expand hydrodynamically at a velocity of about 2 cm per microsecond. Near the front of these plasma blobs at a density of about 10^{18} atoms per cc the electrons run away and are emitted at a surface of nearly zero work function.

The trick in making a good cathode is to provide many such blobs distributed over the area from which it is required to carry current. Reference (7) gives a description of some earlyish cathodes. This note also covers focusing and drifting of moderate level (~ 50 kilo amp) electron beams.

The plasma blobs, as was mentioned earlier, expand hydrodynamically unless the magnetic pressure becomes comparable with the particle pressure. When this happens the placid and usually acceptable velocity of expansion of 2 cm/sec increases by an order of magnitude or more in what we call the bulging instability. The plasma can then close across the anode cathode gap (typically a 1 cm or so for 1 MV) during the pulse and usually causes impedance collapse.

For some time now we have been using arrays of razor blades as cathodes and find we can produce hundreds of blobs, each of which carries a few kilo amps of current and hence does not go unstable. Such cathodes have a slowly decreasing impedance as the blobs expand but even 1 cm like spacings can be used for 300 to 400 ns.

The current carried from the plasma covered cathode obeys the Child Langmuir space charge limited relation when allowance is made for the motion of the effective cathode surface, providing the total current in the diode is not too high.

However many systems produce a prepulse before the main pulse arrives, unless considerable care is taken to suppress this and this prepulse can cause the plasma blobs to form well before the onset of the main pulse, Reference (8) covers some of the effects of this prepulse. The prepulse while usually a disadvantage can be turned to advantage and, intentionally tailored, can give rise to a class of cathodes (the "plasma filled" ones) which can provide very large current densities of 1 MA/cm^2.

As was stated above, space charge limited current flows in between the virtual cathode and the anode providing the self-magnetic field of this current is not too large. As the current drawn from the cathode is increased, eventually the radius of curvature of the electrons on the outside of the beam in the self field is so small that a simple treatment says that they cannot reach the cathode. However, nature is subtler than simplistic 2D calculations and indeed increasing current can still be drawn but now the current pinches into a region on the axis of the cathode and current densities up to 100 $kiloamps/cm^2$ result. Reference (9) gives a treatment of the conditions necessary to avoid the diode pinch. If one wishes to distribute the electron beam over a large area then it is necessary to arrange the cathode-anode set up so that diode pinching does not occur, or alternatively to introduce an axial magnetic field whose strength is comparable to the self field of the electron beam in the anode-cathode space. On the other hand the diode pinch (first observed by Mr I.D. Smith of Physics International) provides simply large current densities. Various theories exist as to what is actually going on in a diode pinch, of which a model based on parapotential flow, is probably one of the best. In any event the current in the pinched anode is no longer proportional to the area of the cathode but goes as its radius, an observation in agreement with the parapotential theory predictions.

The resulting very large electron beams can be transported over several meters without too much loss and can be focused to some extent. Transport can be carried out in low pressure un-ionised gases (p ~ 1 torr) when a return current establishes in the background gas, which largely cancels the self magnetic field of the primary electron beam. Alternatively the electron beams can be controlled by B_z or B_θ fields and transported at high efficiencies. The whole field of transport of large current relativistic electron beams is quite complex and much very good work has been done in this area mainly in the States. To even summarise it would be a time consuming business, but suffice it to say that there is a plethora of interesting plasma physics involved and a number of instabilities have been noted and, in the case of plasma heating experiments, exploited.

Large pulse microwave generators have been built at NRL by modulating a drifting relativistic beam with a rippled magnetic field. Also by making potential wells with focused electron beams, collective accelerations of ions was achieved, producing ion currents of energy considerably larger than that of the original electron beam. Just recently current densities of several megamps per cm^2 have been observed and while the total currents in such beams is quite small and only exist for 10 or so ns, it is to be expected that rapid advances will be made in high density high current beams in the very near future.

The transport and focusing of high current electron beams is closely related to their "temperature". This is the measure of the perpendicular momentum of the electrons. The cooler the beam the more paraxial are the electrons and the easier it is to transport or the more it can be focused. For very high current density beams, all materials placed in their path vaporise and this makes many methods of measuring the temperature of the beam difficult. The method which is still applicable is the analysis of the X-rays produced by the beam, both spacially, temporarily, and spectrally when it hits a target. Reference (10) summarises some of the techniques and also contains illustrative data on the behaviour of the beams from razor blade cathodes. For e-beam laser work it is unnecessary to go to such methods as the current densities are very low and the total currents modest and essentially unaffected by their self fields. However, as the reference shows, razor blades backed by non emitting metal surfaces can produce beams whose mean angle of divergence is only a few degrees, also we have found X-ray techniques useful even in this case. Arrays of razor blades have been used to produce current densities down to a few amps per cm^2 for times essentially limited by the hydrodynamic velocity of expansion of the plasma blobs at 2 cm per μsecond. However it is probably necessary to have a reasonably fast rise to the high voltage pulse in order to light up the cathode edge in a large number of places. When this is done, good uniformity is achieved. Such beams can be drifted 10 cms or more in vacuum with out any magnetic field, because the low current density involves little space charge blow up, and the initial trajectories can be made only weakly diverging. The great advantage of such cathodes is their robustness, simplicity and the fact that they can operate in relatively poor vacuo. However I do not know of their operation at current densities of fraction of an amp per cm^2. With sharp enough edges this should be possible but of course the velocity of expansion of the plasma puts a practical limit of some microseconds to the pulse length usable.

Not really relevant to this section are two further notes which are included for interest. These are a description of a cheap 1 megajoule low voltage (30 kV) bank using solid dielectric start switches (Ref. (11)). Also included is some very early work with a much smaller bank, on which pulsed magnetic fields of 2.5 megajoules were achieved with simple coils (Ref. (12)).

These notes describe low inductance banks which are typical of those built in SSWA and give some measure as to what is involved in low voltage high current megajoule systems. Of course in such banks the energy is delivered in a few microseconds, not tens of nanoseconds.

REFERENCES

1. High Speed Pulse Power Technology at Aldermaston, M.J. Goodman. (Chapter 1)
2. "Mini B Design and Operation", T.H. Storr, SSWA/THS/692/9.
3. "Pulse Charged Lines for Laser Pumping", J.C. Martin, SSWA/JCM/732/373.
4. "Fast Pulse Vacuum Flash Over", J.C. Martin, SSWA/JCM/713/157, reprinted as AFWL High Voltage Notes No. 2. (Chapter 12d)
5. "Summary of the Hermes Flash X-ray Program", T.H. Martin, K.R. Prestwich, D.L. Johnson, SC-RR-69-421, Sandia Laboratories Albuquerque.
6. "Pulsed Power Technology for Controlled Thermonuclear Fusion", L.S. Levine and I. M, Vitkovitsky, IEEE Transactions on Nuclear Science, NS-18 $\underline{4}$, August 1971.
7. The Production Transport and Focusing of Electron Beams with $\nu/\gamma < 1$, M.J. Goodman, SSWA/MJG/6812/106, reprinted as AFWL Radiation Production Note No. 6.
8. Prepulse and other Factors Affecting Fine Self Focusing of Large Currents, D.W. Forster SSWA/DWF/6812/109 reprinted as AFWL Radiation Production Notes No. 7.
9. "Magnetic Pinching in Vacuum Diodes at Relativistic Electron Energies", D.W. Forster, AFWL Radiation Production Note No. 9.
10. "Electron Beam Diagnostic Using X-rays", D.W. Forster, M. Goodman, G. Herbert, J.C. Martin, T. Storr, SSWA/JCM/714/162.

11. "A Cheap Megajoule Bank", J.C. Martin and A. MacAulay, SSWA/JCM/686/36, reprinted as AFWL Circuit and Electromagnetic System Design Note. 3.
12. "2.5 Megagauss From a Capacitor Discharge", D.W. Forster and J.C. Martin, AFWL Circuit and Electromagnetic System Design Note 2.

Section 5e

HULL LECTURE NO. 5*   ODDS AND SODS

J.C. Martin

VOLTAGE MONITORING

The section in NPT notes should be consulted. A flexible copper sulphate solution high voltage divider is described in Reference (1). This monitor can be curved in the way described in NPT notes so as to reduce the integrating effect of its stray capacity to a minimum.

The alternative approach of capacitor dividers or E-field probes as they are sometimes known, has not been used in the SSWA division much but has been employed in a number of situations by other groups. In the event that references to these are required, a request to us should be made. The major disadvantage as far as we are concerned with this approach is the output signal tends to be at a low level and a rather pick-up free recording system is an essential.

CURRENT MONITORING

Circuit shunts have been the method most used by us. For large current low voltage banks, thin brass sheet low inductance resistors have been used to date, but the tendency is now to move over to Rogowski loops, integrating or otherwise. For more modest currents, 100 kamps and below, we have used rings or strips of many

---
* SSWA/JCM/HUN/5
 June 1975

10 ohm resistors in parallel. Where the current is possibly asymmetric a number of outputs (usually 4 suffice) are taken off and mixed. This averages the current across the shunt and can produce answers to a couple of per cent even when the current distribution across the shunt varies by up to a factor of two. The advantage of using current shunts is that again the signal is produced at a level of a few kV. Integrating Rogowski coils tend to produce output waveforms a couple of orders of magnitude lower. Both approaches have various effects which reduce the accuracy of the current measurement and purity of waveform. In the current shunt these are primarily the inductance of the resistor array and to a lesser extent internal capacity effects. With the Rogowski coil approach the internal resistance of the coil and its self capacity enters and the effects of those and other limitations are rather subtler and less readily appreciated perhaps.

SAFETY

NPT notes should be consulted and it should be re-emphasised that it is to be taken seriously. The major real risks are the DC charge on the capacitors and high current power packs. The nominal lower lethal limit is 50 joules for a worst case but discharges of lower than this can be quite unpleasant when they flow through the experimenter. There are a number of cases in the literature of people being involved in accidents with many impulse generators of up to 25 kilojoules stored energy at several million volts and in all cases that I know of the unfortunate recipient survived. This is possibly because the current flows on the outside of the body but more importantly because of its short duration. Thus empirically the pulse aspects of high voltage systems are probably less dangerous than those connected with the DC charged condensers which drive them. However, accidents involving the pulsed voltage output can be painful and potentially dangerous.

Many of the systems we now build are so heavily insulated that it is not possible to find points on which to hang an earthing probe, as is suggested in the NPT note. A slightly modified approach is employed in such systems, which are obviously safer against accidentally touching a capacitor terminal with a few kV residual charge on it.

As before, the first line of defence is the charging voltage meter or meters. These record the volts on the capacitors and power pack and have to operate each time the system is charged. If they become inoperable the system must not be used (it is of course not very sensible to try and use it in this state, safety consideration apart). On firing the system it is checked that they have dropped to near zero. The entry to the caged area is via a door or other barrier and when this is opened a gravity operated ball shorts out both the power pack and the condensers in the bank or Marx. At this stage it is checked that the meters do in fact go to zero. In addition the mains volts to the power pack is routed via the door or barrier so that in order to open it, the mains supply to the power pack is broken and this cannot then be operated by accident. As there is now only one dump in the system, the integrity of the lead from the power pack to the condensers has to be insured, and in particular it has to be arranged so that it cannot be pulled out at either end by accident. Even if this were to happen, the power pack meters would show a rapid charging rate or discontinuous charging behaviour with sparking at the partially opened plug and warn the experimenter, as he charges the capacitors. Only copper sulphate resistors are included (usually in the Marx charging columns) in the charging leads, as these can be easily seen to be intact. Solid resistors are not used in these links because they can go open circuit without any obvious external signs. In a heavily insulated Marx, as has been mentioned, there may be no access to the high voltage conductors on which to hang a shorting stick. However, when the system is being repaired or taken apart, a shorting stick can be used to short across the condenser terminals. In addition a high impedance bleed is always present to discharge the low value leakage currents which recharge the condensers after they have been used (the voltage monitor chain can frequently fulfill this requirement).

When a pulse charged capacitor is being disassembled, the mylar sheets out of which it is made can have significant local deposited charges on them, making the disassembly process mildly painful but not directly dangerous. A procedure for combating this is outlined in Reference (2). The same difficulty applies to any system where localised charges are deposited on high resistivity surfaces to prevent tracking, a last resort to avoiding small but unpleasant shocks is some sort of earthed metal gloves or their equivalent. The real danger from such discharges is that they may cause someone to drop a heavy weight on himself or to jump or fall over and damage himself mechanically.

MODELLING

Most single dielectric breakdown fields scale (area/volume effects apart), consequently scaled models can give quickly and cheaply any required data on breakdown voltage. In general however the breakdown fields can be calculated to an accuracy adequate for system design purposes. However, sometimes portions of the systems with complex geometry may not be easily treatable and a scale model may be useful in designing these sections so that they are adequately safe. In addition it is of course much quicker to work out ad hoc solutions on a small arrangement, than on a large fully engineered system. Surface tracking may not scale but to a first order linear scaling applies; however it would be possibly dangerous to scale by factors of more than 2 or 3. However, there is one sense in which tracking problems are worth looking at on a scaled model and that is the case when the model does track. This almost invariably means that on the full scale tracking will occur and that remedial action is essential. In general it is not necessary to make slavishly exact scale models and indeed the models can be intentionally distorted to emphasise the effect feared or to be studied. Frequently scale models will show that cruder solutions than those initially envisioned may be perfectly adequate and then the time and effort spent in modelling can be recouped manyfold.

In electromagnetic wave propagation problems, scaling is exact and the only limit to the scale used is the response time of the pulse monitoring equipment, the bandwidth of which has to improve as the scale factor. However, frequently it is details of the top and tail of the waveform which are of importance and the rise time of the pulse in the model need not then be exactly scaled.

In order to build small models of high voltage pulse systems, water with its low velocity of light can be used to reduce the length, compared with that of, say, transformer oil or air systems. Of course in this case the model would not necessarily be a linear scale but could be physically distorted to scale various features of the full low dielectric filled generator.

The effect of transitions, mixed dielectric media, bends in finite width strip transmission lines etc. can all be studied in small scale systems operating in water, whose length can be quite modest without requiring the use of wide band width sampling scopes.

While the inductance of a Marx can be estimated to 20% or so quickly, to get a more accurate evaluation or to compare two arrangements, scaled models of the conductor layout can be made in half an hour or so using plastics and cardboard, covered by aluminum foil then stuck down. The inductance of these scaled models is most easily measured by ringing a 10 kV low inductance capacitor into them, when any poor joints will spark over. The capacity should be chosen so the ringing frequency is scaled from the real systems fundamental frequency, in which case the current paths should be the same in the model as in the full system.

Modelling, when intelligently carried out, can be a very powerful method of solving problems in a few hours which would otherwise require weeks of calculations, calculations moreover which are frequently only approximations to what is really going on, sometimes fatally inadequate ones. Intelligent modelling enables nature to do the hard work.

COPPER SULPHATE RESISTORS

When these are used as dump or charging resistors they should be made leak proof and also be readily visible so that it can be checked that they are properly full. If they are constructed of flexible tube they will take rapid temperature rises of 25 to 30°C and not come to any harm. If they are in rigid walled tubes (typically perspex tubes) the rapid deposition of energy leads to a pressure pulse when the time of deposition is less than the velocity of sound transits across the radius of the tube, since the liquid cannot expand. This can lead to cracking and, in extreme cases, shattering of the containing tube. Typically with 5 mm wall tube up to 75 mm OD we restrict the temperature rise to less than 2°C. For larger tubes or thinner walls 1°C rise may be all that can be tolerated. When copper sulphate solution resistors are employed as charging columns for high energy systems, ionic transfer occurs. This leads to a layer of sulphuric acid being produced at one end of a simple tube with copper end electrodes while distilled water accumulates at the other electrode. If the resistor is vertical this state of affairs is unstable when the positive electrode is uppermost and the layers remix. However when the charging polarity is the other way round a stable situation may result. Sparking can then occur at the top, water isolated, electrode and the copper is deposited in a black tree like growth. The situation can be eased by mounting the resistor horizontally, or by making the electrodes out of cylinders. In the case of a chain of resistors, these can be conveniently made out of a single tube of perspex or flexible PVC, and the intermediate electrodes are arranged so as not to block the tube completely. The individual separated layers then easily mix, except for the electrode at the top of the column and a reservoir of liquid is provided above this electrode. This above difficulty is usually only encountered in very large energy systems, or those with a fairly rapid rate of firing, but can be met where very small diameter liquid resistors are used.

With regard to measuring the resistance of a copper sulphate solution resistor the high voltage pulse value is well defined and constant apart from a temperature coefficient of resistivity. However, a low voltage calibration measurement can be well out because of polarisation effects and electrolytically deposited insulating films on the metal electrodes. Reference (3) covers these effects in part.

CHEAP GRADING STRUCTURES

An outline of these is given in Reference (4). No very great accuracy is required in producing the contours, in general a pleasingly smooth surface based upon an intuitive solution of Laplaces equation is quite satisfactory. Playing around with an electrolytic tank for an hour or so can rapidly give one the necessary intuition as to the sort of surfaces needed and how sensitive the folds on them are to departures from the optimum.

QUICK CONSTRUCTIONAL TECHNIQUES

In general we stick glue or use cold setting simplex for nearly all of our constructional work in the high voltage field. Simplex can mechanically join perspex tube and sheet as strongly as the parent material can stand. Thickish pellets of set simplex have a good hold off voltage and while electrically weaker than perspex of the same thickness, are not grossly so.

While evostick or other impact adhesives can be used to stick mylar sheets together, etc. (see Reference (5)), Twin Stik (a double sided tape) is much quicker, although more expensive by a considerable factor. (See Reference (6)).

We also use quite a lot of a cloth based tropicalised black tape to hold things together, and it is worth learning how to attach such tape so that it does not slide off under tension over a period of time, a little thought can repay itself many fold. Properly applied such tape can take a lot of tension over an indefinite time and has the advantage of distributing the force over a large area rather than concentrating it in a few places as would screws or bolts.

Occasionally PVC tape is used, this has a significant degree of stretch and many turns of it can apply a large compressional force, which is not released as the compressed member contracts a per cent or so.

The great advantage of these quick and dirty assembly techniques is that modest systems can be built in a few days, and do not need to call on sophisticated engineering support. Frequently perspex sheet can be scored and quickly cracked in a sheet metal bender into the required shapes. Perspex, in reasonably thin sheets, can also be warmed in a gas flame and bent or moulded into more complex shapes before assembly by simplexing. These shapes can be made quickly which would be prohibitively expensive to make out of solids. In addition not infrequently the result of the quick and dirty techniques, can be better than could have been achieved by a more standard engineering approach (see the details of the peaking capacitor in Reference (2)). Also the use of metal bolts or screws in high voltage systems can give rise to breakdown or tracking problems which may be very difficult to circumvent and while nylon and other plastic screws are available and sometimes used by us, the necessary hole through a perspex sheet can likewise give a lot of trouble.

RUNNING UP HIGH VOLTAGE GENERATORS

In general we follow over-test procedures wherever possible, rather than firing a large number of test shots at the nominal maximum rating of the system. Thus in testing capacitors we do not fire 4000 shots to weed-out the few per cent of weak components but overtest in peak current i and impulse ($\int i^2 dt$) by substantial factors for say ten shots. These tests are followed by the re-application of the maximum over voltage test that the manufacturers use to show up an incipient fault growing because of the above tests.

Because surface track-over is a rather variable phenomenon (σ is frequently 10% to 15%) it is very desirable to have a healthy safety margin against it. Typically we would aim for a safety factor of at least a factor of 2 in systems up to 100 kV, 1 1/2 for systems up to 2 MV and as much as can be got without too much expense above this (see Reference (2)). For the output face of a 50 kV capacitor, this obviously cannot be over-volted to 100 kV but if a simple model of the output face is made, tracking tests can be carried up to the desired levels. It is necessary to put a small high voltage capacitor (several hundred pf's) where the capacity would be, otherwise a surface track starting across the face of the mock-up, would drop the HT terminal volts and choke itself off, something that would not happen in the proper arrangement. The mode of working is to find out where and at what level corona or tracking occurs and then try ad hoc experiments to raise these levels, experiments guided by physical pictures of the processes going on. By means of extra insulation, rounding metal surfaces, etc. the tracking voltage is raised until either it reaches the desired safety factor, or until so many different tracking phenomena show up that they cannot all be beaten. In the later case the system is certainly safer than if it had just been built from scratch, and it is usually possible to make the weak region easily repairable and limit the effect of a track occurring at the now known weak point in the design. In one recent case one stage of a spark gap column designed to work at 80 kV was built and a couple of simple tricks enabled the tracking voltage to be raised to greater than 145 kV, the construction and tests taking about 4 hours to perform.

If the portion of the system has to work in a dirty atmosphere or in high humidity, then it is very desirable to get some representative dirt (swept off the floor of the laboratory) or to boil a kettle near by. Such tests will sometimes show that a tracking solution which works well in clean conditions is no good for dirty and/or damp real life working. In case of wind borne fine dust, face powder makes a good scaled substitute and is usually available in modest quantities from a charming nearby source of supply.

In testing a full scale system a relatively small number of shots at 20 or 30% above the maximum rated output of the system will disclose rapidly any weaknesses, which then can be corrected. While repairing a brand new system that has just tracked under an overall test it is worth bearing in mind that the fault would almost certainly have shown itself during the many shots of the systems use and that then it might cause more damage and lead to secondary faults elsewhere. Again the system can be fired at modest volts into short circuit to test the current capabilities of the connector, gaps, etc., typically a current two to three times the maximum expected being a useful overtest. The system is then examined for signs of burning at vulnerable plugs, joints, etc. Overtesting requires judgment and not inconsiderable willpower and courage but is a vital factor in obtaining reasonably trouble free use afterwards. Indeed some complex systems have never operated properly because the individual components and sub-systems failed more often than the number of sub-systems.

FAULT MODES

All systems malfunction at sometime if they are used at all extensively and it behoves the system designer to consider these occurrences. Usually components can be designed to take the fault mode energy deposition in charging resistors, etc, without too much extra cost and effort. If not the component concerned can be made easily replaceable (consider the normal fuse link). Extra impedances can usually be added at a cost of a modest energy loss in normal operation, which however absorbs much greater fractions of the energy in fault mode operation. We normally add some resistance in series with the Marx when this is feeding a pulsed high speed section to damp out the Marx ringing after the fast section switch has fired. This limits the current sloshing backwards and forwards in the Marx and makes life much easier for the spark gaps and condenser. This is not strictly a fault mode operation but is a more general precaution which incidentally helps greatly in limiting fault mode damage. The system should ideally be operated in its likely fault modes and indeed during overvolt testing this sometimes happens incidentally. However not all faults are equally likely and this is impracticable in general. However certain faults are either inevitable or very likely and it should be ensured that the systems survives these either with no, or with easily repairable, damage.

The larger the system the more imperative it is to consider fault mode behaviour and, a sad fact but true, the less likely the system is to be intentionally tested in these modes. I know of one case where a 5 Megajoule system was intentionally fired in several fault modes, something which required real guts to do but which I am certain has paid off handsomely in trouble free usage.

HULL LECTURE NOTES

Largely for amusement, Reference (7), is included as an example of a possible system designed to work continually in at least one fault mode. I trust Hull University will soon be wanting to build systems of this size.

REFERENCES

1. "Mini-B Design and Operation", T.H. Storr, SSWA/THS/692/9.
2. "Notes for a report on the Generator TOM", J.C. Martin, SSWA/JCM/735/407. (Section 12e)
3. "Measurement of the Conductivity of Copper Sulphate Solutions", J.C. Martin, SSWA/JCM/667/67 reprinted as AFWL Energy Storage and Dissipation Note 4. (Section 12a)
4. "Electrostatic Grading Structures", J.C. Martin, SSWA/JCM/706/66, reprinted as AFWL High Voltage Note 1. (Section 12c)
5. "Pulse Life of Mylar", J.C. Martin, SSWA/JCM/6611/106, reprinted as AFWL Dielectric Note 11. (Section 8b)
6. "Pulse Charged Line for Laser Pumping", J.C. Martin, SSWA/JCM/732/373. (Section 12d)
7. "The Self Healing Marx - a Possible Way to 100 Megajoule Systems", J.C. Martin, SSWA/JCM/704/38.

Chapter 6

GAS BREAKDOWN

Section 6a

PRESSURE DEPENDENCY OF THE PULSE BREAKDOWN OF GASES*

J.C. Martin

The following measurements were rather preliminary, being undertaken in haste, and were performed by Dave Forster and Phil Champney. Following on some very preliminary measurements made by J.C. Martin, it was assumed that the positive point yielded a lower breakdown field for a given pulse length and so measurements were initially confined to this mode, since it was the minimum breakdown field that was required. Also following this early work, the power of the time on the breakdown pulse (t in microseconds) was taken as 1/6th. The pulse length applied by the test transformer was more or less of a constant duration and lay between 0.1 and 0.2 μsecond effective value.

In the early work (which was all at 15 p.s.i. absolute) the values given in Table 6a-I were obtained for k' where k' = $Ft^{1/6}d^{1/6}$ and t is in microseconds, d is in cms and is the distance between the point or small sphere and the plane and F is the breakdown voltage divided by this distance d in kV per cm.

---

* AFWL Dielectric Strength Notes
  Note 15
  SSWA/JCM/679/71
  26 Sep 1967

Table 6a-I

Values of $k' = Ft^{1/6}d^{1/6}$

	Point -ve	+ve
AIR	28	24
FREON	65	37

Pressure 1 atmosphere

Analysis of the new data showed that a better fit for the power of d was one tenth. Thus the new data is expressed in terms of

$$k = F\, t^{1/6} d^{1/10}$$

The experiments were conducted in a perspex cylinder of I.D. 5 1/2 inches and the large radius electrode was a portion of a sphere of 8" O.D. which was cut so as to just fit in the bottom of the cylinder. The cylinder, which could be pressurised up to about two atmospheres, was placed under oil and was driven by a pulse transformer. The needle point or small radius ball was located on the axis of the pressure-containing cylinder and could be set at various distances. The facts that the "plane" electrode was of distinctly limited extent and the presence of the oil do not appear to have affected the results obtained at one atmosphere, when these are compared with those obtained previously in a setup of much better geometrical conditions. However, some effect of the experimental set-up cannot be excluded and it should again be emphasised that the variation of k with time was not investigated to any significant extent.

Table 6a-II gives the results of the point to plane experiment and also gives values of k where the point was replaced by a sphere of 1 inch diameter. As was found previously, in the case of air, this actually reduces the breakdown voltage. The values given are the average of k for distances ranging from 2 cms to 15 cms and the individual values of k were usually within 10% of the mean. Small systematic variations are apparent in the original data but for approximate predictions of the pulse breakdown voltage of a given set-up, answers good to 10% should be obtained.

GAS BREAKDOWN

Table 6a-II

Values of $k = Ft^{1/6}d^{1/10}$ for positive point or small sphere

	p absolute P.S.I	15	25	35
AIR	Point	24	33	37
	Small Sphere	20	28	33
FREON	Point	40	43	46
	Small Sphere	45	51	55
SF_6	Point	48	55	59
	Small Sphere	60	66	69

As an example of the meaning of the data, a pair of points or a point plane to which a positive pulse of 1 million volts of 1 microsecond is applied will spark over for air, freon and sulphur hexafluoride at atmospheric pressure for approximate distances of 63, 36 and 29 cms respectively.

Subsequent to these measurements with positive points, another set was performed which repeated them with negative points. The point results fitted the new relation very well but the fit for the data for small spheres was considerably less satisfactory, the data suggesting a slightly stronger dependency on d. However the data for the small spheres are included in Table 6a-III but the fit of the data over the full range of d (2 to 15 cms) was not better than 20% for spheres.

At one pressure of Freon, the time dependency was looked at and fitted the one-sixth power well. One set of data for the highest pressure longest gap in Freon did not agree with the relation, but the pressure vessel after these measurements showed tracks down its inner wall, so this set of data was ignored.

Table 6a-III gives the values for k- and it shows that now in all cases the small sphere breakdown voltage is lower than that for a sharp needle. In addition, for pressures over one atmosphere absolute the negative point breakdown voltage was higher than that for positive points, in some cases considerably so. For a negative pulse of 1 million volts of duration 1 microsecond, the breakdown distance in the gases at atmospheric pressure would be approximately 60, 20 and 16 cms.

Table 6a-III

Values for k- for negative point and small sphere

	p absolute p.s.i	15	25	35
AIR	Point Small Sphere	25 17	38 -	49 28
FREON	Point Small Sphere	67 47	84 -	100 77
SF_6	Point Small Sphere	79 49	- -	116 93

Section 6b

D.C. BREAKDOWN VOLTAGES OF NON-UNIFORM GAPS IN AIR*

J.C. Martin

One of the minor irritations of my life has been the fact that while approximate calculations of the breakdown volts of practical sphere/sphere and cylinder/cylinder gaps gives reasonable agreement to 10-20 per cent, when more accurate attempts are made to compare experiment with crude theory, reality seems strangely perverse. In addition to this, for pressurised gaps the plot of breakdown voltage against pressure is significantly non-linear. This non-linearity is distinct from the one that develops about 10 to 15 atmospheres, which, amongst other things, depends on the nature of the electrodes.

As a result of a succession of pressurized gaps which invariably just failed to meet their designed operating range, I was driven to derive a relatively simple picture of what was going on and as a result of this performed a series of measurements covering the range of parameters of practical interest to myself. When analysed these appeared to support the relatively simple picture proposed with satisfying accuracy. However, in analysing the effects of a fairly small correction, it became apparent that reality must be significantly more complex than the theory allowed it to be. However, in the last analysis a simple empirical approach has been derived which enables the breakdown voltage of a large range of pressurised air gaps to be calculated to some 2 or 3 per cent.

---

* AFWL Dielectric Strength Notes
  Note 16
  30 Jun 1970

The starting point is the relationship for the DC breakdown field of a uniform field gap which can take the form

$$E = 24.5\, p + 6.7\, p^{1/2}/d^{1/2} \text{ kV/cm}$$

where p is the pressure in atmospheres and d is the spacing in centimetres. An approximate derivation of the form of this relation can be obtained by considering a Townsend avalanche which after 20 or so e foldings creates sufficient charge to start seriously modifying the field imposed by the electrodes. This then leads to the start of a modified discharge pattern such as the formation of a streamer. If the distance (obtained from the net ionisation coefficient $(\alpha - \beta)\, p$) is less than the electrode spacing, then a breakdown occurs. If, however, the electrodes are closer together than this distance, a breakdown does not take place. It should be stressed that using published data, the numerical agreement is not by any means perfect but it is certainly in the correct street. In this picture breakdown in diverging fields is modified because the avalanche is moving out into an ever-lower field and if the twentieth generation is not reached before the field has fallen too low, a breakdown will not occur. Consideration of the net ionisation curves suggested that if the field falls to about 80 per cent of the field on the electrode within the twentieth generation distance the avalanche peters out. For non-uniform gaps this picture suggests that the breakdown field at the surface should be taken to be

$$E = 24.5\, p + 6.7\, p^{1/2}\, \beta\, /\, r_{eff}^{1/2} \text{ kV/cm}$$

where
$$r_{eff} = 0.115\, r \text{ spheres}$$

$$r_{eff} = 0.23\, r \text{ cylinders}.$$

The coefficients were determined from experimental data with spheres and cylinders separated by rather more than their radius and agrees well with the expectations of the theory outlined above. The coefficient β was included to allow for the fact that as the spheres or cylinders approached each other the field between becomes more uniform and eventually does not fall below 80 per cent. Thus the factor β tends to 1 as the spacing (d cms) becomes large compared to the radius of the electrodes and one of the aims of the experiments was to evaluate it for small gaps and to check it against theory. In order to complete the calculation of the breakdown voltage of a gap, the published expression for the field enhancement on the axis of the electrodes is used. For the maximum field on the electrodes, the factor f is given in Fig's. 6b-1 and 6b-2 for spheres and cylinders, respectively, where f is defined as

$$f = E_{max}/\, (V/d).$$

GAS BREAKDOWN

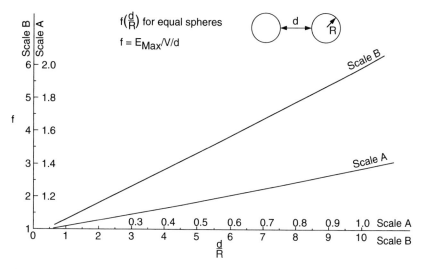

Fig. 6b-1 Field enhancement factor for equal spheres.

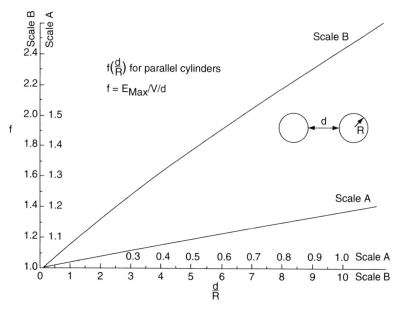

Fig. 6b-2 Field enhancement factor for parallel cylinders.

As an example of the mode of use of the expression, the case of a sphere/sphere gap is considered where the radius of the spheres is 1 cm and the separation is 2 cm. The expression for the breakdown field is

$$E = 24.5\ p + 19.8\ p^{1/2}\ \text{kV/cm},$$

the factor β being 1 at this separation. For 1 atmosphere air $E = 44.3$ kV/cm and using Fig 6b-1 $f = 1.77$, giving an effective uniform gap separation of 1.13 cm and a breakdown voltage of 50 kV. As can be seen the breakdown voltage is significantly non-linear with pressure: in the example treated above, the breakdown voltage at 4 atmospheres is calculated to be 155 kV.

A fairly extensive experimental programme was undertaken to check the proposed treatment and to obtain the value of β for small gaps. Three radii of spheres were used, namely 0.635, 1.14, and 1.91 cm and spacings of 0.5, 1.0, 1.5, 2.0 and 2.5 cm. The sphere/sphere gap could be pressurized and data was obtained at 1, 2, and 3 atmospheres absolute, subject to a voltage limitation, and no data were taken for breakdown voltages over 110 kV. For the experiments with cylinder/cylinder gaps (the actual experimental set up was cylinder/plane) three radii were again used, namely 0.64, 0.1285 and 2.035 cm, with separations of 1, 2, 3 and 4 cm. However, these gaps could not be pressurised, so all the measurements were at atmospheric pressure. However, a small series of experiments was performed with one rod/rod gap which could be pressurised and these results agreed well with the theory.

The voltage measurements were made with a calibrated voltage divider and had an error of the order of 1 per cent.

Including errors in the density of the air in the gap, the distance between the electrodes, and errors in calculating the large numbers of measurements, the error in the experimental data is estimated to be ± 1.8 per cent.

Using the β correction curve given in Fig. 6b-3, the mean difference between the theory and the experimental results is given in Table 6b-I, as well as the standard deviations between the two. One comment which should be made with regard to the cylindrical data is that a polarity effect was observed. The measurements were made with a cylinder over a plane and it was found that the positive breakdown voltages were some 3 per cent higher than those for the negative voltage. Both the theory and the data quoted are for the negative polarity which of course would be that applying to a cylinder/cylinder gap.

GAS BREAKDOWN

Table 6b-I
Difference between theory and measurements of β factor

	Radius (cms)	Difference Measurement to Theory (%)	Standard Deviation (%)
Spheres	0.635	+ 0.3	2.5
	1.14	+ 1.5	1.8
	1.91	+ 0.5	2.3
Cylinders	0.64	+ 1.0	2.1
	1.285	- 0.9	2.0
	2.035	+ 1.1	1.0

As is shown by Table 6b-I, the weighted mean difference between measurement and theory is only some 0.6 per cent and the standard deviation not much greater than the estimated measurement errors.

With regard to the correction factor β given in Fig. 6b-3, the dotted curve shown is for a calculation for spheres at 1 atmosphere pressure, and once again agreement is very good considering that errors in experimental values correspond to something like ± 6 per cent in β on this plot. However, the Fig. 6b-3 conceals two fairly serious failures. The first of these is that for the data for spheres, a pressure dependency of β would be expected but the

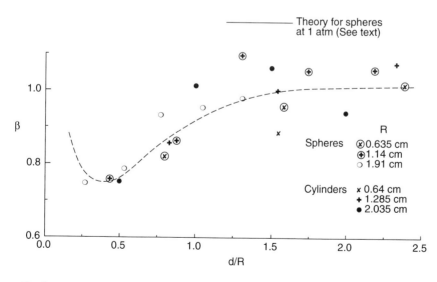

Fig. 6b-3 β correction vs d/R.

data shows no such effect and hence the points plotted are the average of the three pressures used in the experiments. The second failure is that the detailed application of the theory would suggest that the correction curve would be shifted to the right by a factor of two for cylinders compared with the curve for spheres. Once again there is no real evidence for such a shift apart from the one point for r = 0.64. The other points for this radius are off to the right of the graph and average out to $\beta = 1$ as would be expected. Unfortunately the errors are greatest for this point and probably no significance can be attached to it being low.

Thus after a very satisfactory degree of agreement between experiment and theory, the details of the correction for only mildly non-uniform gaps fails to display detailed agreement. While there are arguments which can account for some of the differences, it is probably simpler to regard the β correction as more or less purely experimental. Again it is probably worth stating that substantial errors in β reflect only weakly in the final calculated breakdown voltage.

In conclusion I would like to express my appreciation of Mr. Brian Harle's help in obtaining the experimental data. Not only did he assist in a rather lengthy and boring series of measurements, but he also served to keep my scientific standards from wilting under the tedium.

Section 6c

HIGH SPEED PULSE BREAKDOWN OF PRESSURISED UNIFORM GAPS*

J.C. Martin

INTRODUCTION

Mr. Laird Bradley (1) has performed some nice experiments on pressurised nitrogen gaps, measuring their breakdown voltage for fast rising pulses. The results show breakdown fields considerably above the uniform field DC values. For instance, for a gap of 1/2 cm at 2 atmospheres pressure, the breakdown field is some 6 times the DC value. His note shows that this behaviour can be accounted for by assuming that a streamer transit phase occurs after a rather rapid Townsend avalanche phase and that the streamer time predominates over the latter. He fits a relation of the form $Fd^{1/6} t^{1/6} = \ell$ to the data and derives the pressure dependency of ℓ. The values of ℓ fitting the experiments are some 3 times that applying to point plane data. The aim of this note is to estimate these values using the point plane streamer data and while it is to be expected that the fit to the data obtained will only be rather approximate, it may then be used to deduce trends.

APPROXIMATE THEORY

Following Mr. Bradley, the picture assumed is of a Townsend avalanche happening quickly and producing a cone of ionisation which after about 20 generations has produced enough electrons to affect the originally pulse applied field. At this stage a streamer starts out from what is effectively a small point sticking

---
* 24 August 1970
  SSWA/JCM/708/107

out from the cathode. This streamer then grows out from the backing electrode and crosses the gap.

The approximate integral relation for point plane streamer takes the form, for air,

$$F\, d^{1/6} t^{1/6} = k$$

from which a very approximate differential velocity expression can be obtained

$$\frac{dx}{dt} = \frac{1}{5k^6} \frac{V^6}{x^4}. \qquad (6c-1)$$

To the level of accuracy that this expression applies, the effective voltage on a tip sticking out of a plate a distance x can be taken as

$$V(x) = \frac{V_o}{d} \cdot x$$

where V_o is the voltage on the uniform field gap with spacing d.

Integrating (6c-1) gives

$$F t^{1/6} = k\, 5^{1/6}\, \left(\frac{1}{\delta} - \frac{1}{d}\right)^{1/6}$$

where $F = V_o/d$ and δ is the initial protruberence of the point from the cathode and is assumed to be equal to the length of the Townsend avalanche at the time it begins to control the applied field. To an adequate accuracy this can be approximated by

$$F t^{1/6} = k\, 5^{1/6}\, (\delta)^{-1/6}.$$

For air (and very probably nitrogen, too) $k = 23\, (P/p_o)^{0.6}$ for negative points and where F is in kV/cm and t in microseconds. The power of the pressure dependency is slightly lower than that given in the streamer velocity note where the value of the power is 0.7. However those experiments were for a limited pressure range (1 to 2 1/2 atmospheres) and other data suggests that for the large pressure range considered here the power will be significantly lower. Thus for nitrogen at p = 640 mm Hg (assumed Albuquerque atmospheric pressure)*

$$F t^{1/6} = 27.5\, (\delta)^{-1/6}\, \left(\frac{P}{P_o}\right)^{0.6}. \qquad (6c-2)$$

Editor's (T.H. Martin) note: Use 640 mm Hg for this note, 620 mm Hg is actual Albuquerque pressure for the original data.

GAS BREAKDOWN

EVALUATION OF δ

The effective first Townsend coefficient (α') has been determined by a number of workers for many gases including nitrogen. The measurements are frequently reported to three figures, hence it is somewhat unfortunate that two of the most recent determinations (Harrison Phys. Rev. 105 366, 1957 and Heylen Nature 183 1545, 1959) differ by a factor of up to six. After a judicious survey of the data, I have selected a curve quite close to that of Heylen (shown in Fig. 6c-1 by the Editor).* The normal mode of use of α' is to assume that after approximately 20 generations enough space charge has accumulated to dominate the applied field. I have used a probably totally unnecessary refinement to estimate the number 20 a little more accurately. If n_e is the number of electrons resulting from one electron in a cascade, then $n_e = e^{\alpha' d}$ and n, the number of generations, is given by $n = \alpha' d$, let $d = \delta$ then $n = \alpha' \delta$. Using the approximation for α' (in Fig. 6c-1) provides**

$$\delta = \frac{(n)\ e^{320/(E/p)}}{15\ p}.$$

COMPARISON WITH EXPERIMENTAL DATA

The approximate theory can be compared with the experimental data in a least two ways. In the first, the smoothed experimental data are used to generate values of δ and then the values of ℓ so obtained are compared with the experimentally determined ones. The second method is to calculate the breakdown fields ab initio and compare these directly with the smoothed ones measured by Mr. Bradley. Both approaches will be covered below, the first showing the theoretical pressure dependence of ℓ; the second, while being the more rigorous comparison, suffers from the fact that small errors are amplified in the final answer.

Table 6c-I gives the smoothed experimental breakdown fields for the uniform field gap working in nitrogen at various pressures in kV/cm. The deduced value of δ in cms is in the second column and the calculated values of ℓ in the third column.

---

*   Editor's note (T.H. Martin): Notice that the approximation shown of $\alpha'/p = 15\ e^{-320/(E/p)}$ is reasonably accurate for air and N_2.

** Editor's Note (T.H. Martin): Fig. 6c-2 shows the original calculations by J.C. Martin along with the editing aapproximation generated for 20 generations. Fig. 6c-2 gives δ at 640 mm Hg as a function of E/p. This paragraph varies from the original. The result is the same but a slight different derivation was used.

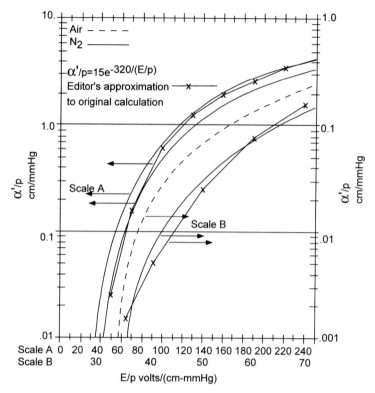

Fig. 6c-1 Editor's approximation to JCM's approximation for the Heylen data.

Table 6c-I
Breakdown Fields for Uniform Field Gap in Nitrogen

d	1/2 cm			1 cm			2 cm		
p/p_o	F	δ	ℓ	F	δ	ℓ	F	δ	ℓ
2	250	.006	83	180	.011	88	102	.07	72.5
4	282	.016	113	206	.033	111	127	.33	87
6	289	.030	128	217	.11	115	148	1.2	71
8	294	.077	130	226	.37	114	169	~10	
10	302	.2	128	242	1.1	110	195		

Note: Expression (6c-2) gives $Ft^{1/6}$ and has to be multiplied by $d^{1/6}$ to obtain ℓ.
Also that for p/p_o, p_o = 640 mm Hg.

GAS BREAKDOWN

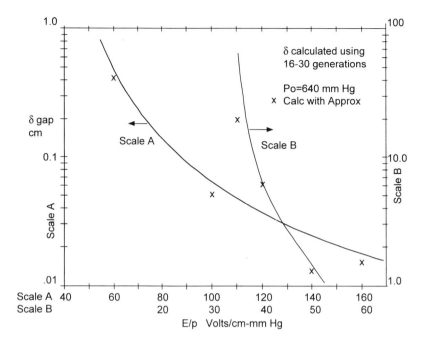

Fig. 6c-2 JCM origional calculations for δ with Editor's overlay which uses 20 generations.

Figure 6c-3 shows the experimental and calculated values of $Fd^{1/6}t^{1/6}$. As δ approaches one, the calculated values of ℓ fall and the dotted lines give an extrapolation, very crudely, involving the Townsend avalanche phase. The agreement between crude theory and experiment is not very impressive but the general order of the values of ℓ has been obtained.*

---

* Editor's note: Table 6c-I provides the calculation curve for Fig. 6c-3 with $\ell = Fd^{1/6}t^{1/6}$.)
Figure 6c-1, where JCM summarized the Heylen data, was added from a letter written to L. Bradley. To allow calculation the approximation of $\alpha'/p = 15\ e^{-320/(E/p)}$ was added by the Editors and reasonably approximates the added graph.

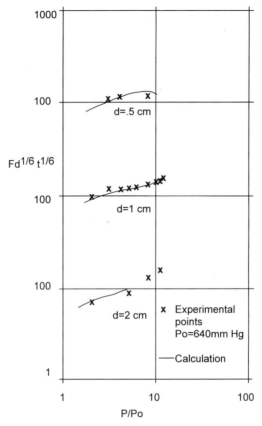

Fig. 6c-3 Experimental and theoretical values of $Fd^{1/6} \; t^{1/6}$ vs p/p_o.

Calculation of Breakdown Fields*

The more useful comparison between experiment and theory is to calculate the breakdown fields from scratch and this is now done.

A series of values for E/p are taken and for various values of p the field is calculated. From the values of E/p the values of δ at various pressures are obtained and using expression (6c-2) the values of $F' t^{1/6}$* are then found. Figure 6c-4 shows the values obtained for pressure ratios of 2, 4, 6, and 10. To the left hand side of the curves δ tends to one and the Townsend avalanche phase begins to dominate. However, for the experimental values used by Mr. Bradley, a fortunate coincidence enables a limit for the value

---

* Editor's (T.H. Martin) addition.

GAS BREAKDOWN

of $F't^{1/6}$ to be obtained. Using his values of $(F_{townsend}(t_{townsend})^{1/6})$ for a given pressure which are nearly independent of d (Ref 1). Thus for $p/p_0 = 2$ and $d = 1/2, 1, 2$ cms, the values are 34, 32, and 31. Thus limits to $Ft^{1/6}$ for the particular experimental rise times used by Mr. Bradley exist and these are shown by the dotted horizontal lines in Fig. 6c-4. These are smoothly joined to the streamer $Ft^{1/6}$ curves. This coincidence is not a completely unreasonable one, since, as is shown by Felsenthal and Proud, the slope of log $t_{townsend}$ against log E/p is about 1/6 in the range of values applying in the experiments.

The calculated values of F may now be obtained from Fig. 6c-4 for each smoothed experimental value of $t_{eff}^{1/6}$ by finding the value at which F' and $F_{townsend}$ are equal. This may be simply done by making a piece of paper with the value of $1/t_{eff}^{1/6}$ and sliding it along the curve until the desired condition is obtained. It can of course be done analytically but the above crude measuring technique has the advantage of showing that quite small errors in the streamer curve give rise to substantial errors in the calculated breakdown field.

Table 6c-II lists the values so obtained and compares these with the smoothed experimental values.

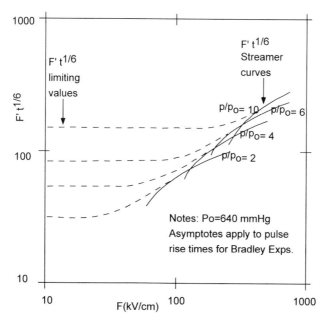

Fig. 6c-4 $F't^{1/6}$ vs F at various pressures.

Table 6c-II
Breakdown Fields Calculated for the Experimental Pulse Times

p/p_o	2		4		6		10	
d cms	"obs"	calc.	"obs"	calc.	"obs"	calc.	"obs"	calc.
1/2	250	195	282	280	289	340	302	430
1	180	145	206	185	217	220	242	270
2	102	120	127	135	148	160	195	220

Again the agreement is not outstanding but in view of the numerous uncertainties it is probably acceptable. The departure between experiment and crude theory for small gaps and high pressures is not unlikely since the basic point plane streamer relation is probably poorest in this area. Again experience in other calculations has suggested that while agreement has been obtained for gaps of separation of about 1 cm on the Townsend avalanche plus streamer theory, large changes in gap separation are not likely to lead to good agreement, and that while the general picture used leads to reasonable agreement in other calculations, such as the uniform and non uniform field DC breakdown, as soon as the theory is extrapolated, divergences begin to appear. However, an adequate degree of agreement has been achieved to justify some conclusions to be drawn.

DEDUCTIONS FROM THE THEORETICAL CALCULATIONS

If the theory dealt with in this note applies, then the high fields measured for small gaps should be independent of gap length, providing no new electrode damage mechanism enters the picture. This is because most of the time is spent in the streamer climbing out from the small initiating Townsend avalanche. However, in order to achieve these fields the pulse charging should occur at the same rate as in the small gap experiments. For the 1/2 cm gap, the t_{eff} was 11 ns and firing half way up a ringing wave form (as in an output gap for a peaking capacitor arrangement). The half period to which this corresponds is a fairly long 250 ns or so. Thus for peaking capacitor output gaps fields of 300 to 400 kV cm might be available at 8 atmospheres sea level.

A disadvantage of the uniform field gap is that any dc/dt loading would be enhanced because of the relatively large stressed area. However, a pseudo uniform gap with a rather flat bump in its middle would reduce these effects while still retaining the other very considerable advantages of the uniform gap. As the voltage on such a gap is raised above 1 million volts or so, the possibility of $\int j^2 dt$ blow up in the electrode material begins to be a serious one because of the very large di/dt experienced by the root of the channel. Thus copper might have some advantages and conditioning with relatively low voltage pulses before a main pulse might have beneficial results. All these remarks apply with more force as the value of t_{eff} is reduced. In addition pressurised SF_6 can be taken towards 1 MV/cm if conditions are maintained correctly. However, as the pulse field is raised the value of δ decreases and any surface whiskers or damage becomes more critical. Hence for really high field pulse operation conditioning or controlled chemical attack on the electrodes may be all-important.

With regard to the IB (intentionally buggered) gap, the calculations given indicate the order of the height of edge necessary to degrade the breakdown field slightly and stick out heights are very small for short pulses. Thus some form of recessed edge would be necessary, it seems.

The Townsend avalanche plus streamer theory used in this note also enables the fast operations of trigatrons to be calculated, insofar that δ is small compared to the gap between the trigger disc or pin and the electrode in which it is placed. If this is the case streamer bushes move out across this gap and partly stick into the main gap. If the length of stick out into the main gap can be estimated, the closure time of the trigatrons can then be calculated. It is expected that the higher the voltage on the gap, the higher its pressure and the larger the spacing between trigger and the trigger electrode, the faster the trigatron will close. This progression means larger and larger trigger pulses, but this is not a serious matter these days. In addition, the trigger electrode must be placed in the electrode so that it does not significantly degrade the pulse breakdown of the gap when it is tied to its adjacent electrode. However, too much recessing of the pin will lead to the streamer bush not entering the main field of the gap. With certain configurations I believe it is possible to make the main discharge close between the main electrodes without going via the trigger pin and even close before the trigger pin has discharged itself.

CONCLUSIONS

A fair measure of agreement has been obtained (or forced?) between calculations using a streamer transit relation based on the stick out of a partial Townsend avalanche. Based on this and agreement obtained in similar calculations for the IB gap, it is felt that the differential streamer velocity expression can be used (with considerable care and discretion) to calculate other pulse charged systems where the pressure is atmospheric and above. However, such calculations (as with the above) should be treated with caution and be mainly used for the first design stage of gaps and should be checked with model or full scale experiments at the earliest moment.

I would like to thank Mr. Laird Bradley of Sandia Corporation for his kindness in making available to me the results of his experiments and his analysis of them.

REFERENCES

1. L. Bradley, "Preionization Control of Streamer Propagation", JAP, Vol. 43, No. 3, March 1972, p 886 (written after this note).

Section 6d

PULSED SURFACE TRACKING IN AIR AND VARIOUS GASES*

J.C. Martin

INTRODUCTION

This short study of surface tracking was partly a return to some simple experiments performed several years ago and partly in the hope of providing a zero work function surface covering a large area. The potential application for this is in molecular gas lasers, where a uniform background of electrons is required before a large amplitude multiplying voltage pulse is applied. Such a background of electrons can be provided by u.v. pre-ionisation, but an alternative approach is to drift the electrons from an array of hopefully zero work function surface sparks through the gauze cathode of the laser discharge cell and across this. It is this objective (yet to be tested) which provided the necessary impetus to investigate surface tracking and the characteristics of the subsequently produced surface arcs. As these investigations were rather developmental in nature, the accuracy of the measurements is not very high and also, as usual, the range of parameters studied rather limited. However, even with the above limitations, the data obtained in three weeks or so may be worth recording.

---
* J.C. Martin
   SSWA/JCM/745/735
   May 1974

HANDWAVING PHYSICS BACKGROUND

In the work done several years ago, it was found that under certain conditions pulsed surface tracking became quite regular. These conditions were that the impedance of the source had to be low, so that as the surface sparks moved out, the voltage at their roots did not drop. More importantly, the surface of the insulator (typically mylar -UK- melinex -USA-) had to be properly discharged between the applied pulses. In addition, the longest tracks were obtained with the thinnest mylar backed closely by an earthed plane. Under these conditions the tracks originating along a sharp metal electrode resting on the top of the mylar were quite regular in length and spaced about their length apart. As the applied voltage pulse was increased, the length of the tracks increased fairly rapidly and some 30 cm was tracked with an applied pulse of about 40 kV. The normal very irregular tracking lengths usually observed experimentally were shown to be due to patches of charge laid down by previous discharges, in the case of pulsed voltages, or small incomplete discharges where an edge in contact with a surface is DC charged. Indeed, in the case of DC, it is only because small length discharges occur as the voltage is raised, so depositing charge, that tracking lengths much smaller than the one mentioned above can be used. However, without precautions the process is erratic and humidity-dependent, leading to the normal experience of erratic behaviour. As examples of what can be achieved either way, the following cases may be of interest. For a thin metal strip transmission line with the insulator sticking out 2 1/2 cm beyond the two edges, using every trick known, 100 kV DC was held between the electrodes in air. However, with a metal edge and an extensive metallic backing, a 10 cm length can be tracked, using a power pack with a reversible 6 kV voltage, and a condenser and switch.

Considering the possibilities for the initial deposition of surface charge of either sign before the pulse is applied, the following five conditions can be delineated, the order being in the direction of increasing tracking length for a given voltage.

(a) Surface charge deposited of the same sign as the voltage applied to the sharp metal edge. This charge can either be intentionally deposited by stroking the surface with a flexible, thin metal brush, while this is attached to a power pack, so that the mylar is charged up with reference to the metal backing; or the other way is to pulse the edge a number of times with increasing voltage, each pulse depositing more charge by small surface discharges. As the aim of this investigation was to produce long tracks, this condition was not investigated.

GAS BREAKDOWN 157

(b) When a surface track has been closed over the mylar surface, a highly conducting arc results and the surface potential rises to that of the spark channel. However, as the condenser feeding the arc discharges, the voltage falls and at some point the arc extinguishes. This then leaves the surface charged up more or less linearly from the earthy end to the pulsed voltage electrode at a voltage which is a function of the resistance in series with the arc, but typically is a few kV to over 10 kV. If it is desired that another surface track result when a second voltage pulse is applied, then the applied pulse has to overcome this residual charge. However, this initial charge density is quite reproducible and the threshold tracking voltage is, too.

(c) In this case the surface is discharged carefully between each firing and in a sense this is the most fundamental tracking mode. It should be mentioned that discharging a surface is not all that easy but, as was mentioned above, stroking the surface softly with an earthed metallised brush was found to be simple and effective.

(d) If the polarity is reversed between firings, the surface charge deposited at arc extinction now aids the propagation of the streamer and a lower tracking voltage will be found than for a completely discharged surface.

(e) The surface may be charged intentionally with charge of the opposite sign to that of the voltage pulse applied to the sharp electrode. Very long tracking distances can then result However, where the mylar sheet has been chosen to be as thin as possible, this initial DC charge cannot usefully exceed the applied opposite polarity pulse in magnitude and the useful minimum ± voltage to track a distance is obtained.

In all the above it is assumed that the insulator film is dry and does not carbonise. In practice, mylar seems to be very good in this respect for the gases studied. In addition, its surface resistivity is very high, providing the relative humidity does not approach 100%. Those who have disassembled pulse or DC charged systems, days after their last use, and have collected a shock off the mylar will be able to attest to the very long self discharge times that can occur.

In terms of the experiments to be described, a fair number of experiments were performed in air, using condition (c) and the functional dependency of the track length on mylar thickness, applied voltage pulse amplitude, and duration, was determined. A small amount of work was done with the surface track operating in the condition (b) but the main work was done using the surface track operating in the condition (d) which gave the lowest voltage for a given track length under conditions which could be operated in a sealed system. While condition (e) would have given even longer tracks, calculations suggested that it would be very difficult to charge a large area of mylar a few thou (mil) thick by non-contact methods. The alternative, of a charged little mechanical bug gently stroking the surface while it was moving about, was also daunting under the space constraints imposed by the possible application. Thus condition (d) was selected to investigate the dependency of the tracking length as a function of voltage and pulse duration, for a range of gases and mixtures of interest.

To conclude this section, a few observations and speculations about the nature of the surface discharges will be given.

As the pulse voltage is raised (with the surface discharged between shots), at first a very dim glow is observed, reaching out from the edge. At a somewhat higher voltage this glowing region moves out from the sharp electrode and when it becomes a couple of cm long, bright channels begin to form, linking it to the electrodes. At a pulse large enough to track, say, 10 cm, a number of quite bright channels link the 2 cm faintly glowing uniform region back across the 8 cm to the sharp metal electrode. These observations are obviously made when the surface discharge does not link all the way across the surface. The width of the leading faintly glowing region given above is approximate, but is typical for air, nitrogen, or carbon dioxide. However, for helium, the glowing region can be up to 6 cm long and if this is made to link across to an earthy electrode, the subsequent current is carried in a uniform sheath producing a lovely pinkish uniform glow. This lasts for several hundred nanoseconds, eventually going over into rather diffuse channels.

A very tentative picture of the above observations is that in the faintly glowing uniform region, impact and u.v. ionisation cause the gas to become poorly conducting. This layer is probably very thin: visually it is less than 1 mm thick and may be very much less. The energy deposited in this thin layer causes it to heat and expand. A region of pressure very much less than atmospheric results at the plastic surface and the current flowing through this to the advancing front causes the low density gas to heat rapidly and a plasma channel can now be formed. Because of very small surface irregularities, residual surface charge, etc., the formation of the low density zone is not very regular and hence the plasma channel is locally curved and zigzags along the surface. When the discharge links the electrodes, the main current then flows along the preformed zigzag channel. On this picture it would be expected that there would be some constant voltage (that dropped across the faintly glowing head region) which would have to be subtracted from the applied voltage to determine the voltage down the main brightly glowing channel, so as to obtain the relation between the length of

GAS BREAKDOWN

this channel and the voltage driving this. As is explained below, the observations agree with this deduction.

SOME EXPERIMENTAL ASPECTS OF THE TESTS

When using mylar film of a few thou thickness, the earthy backing electrode must be in very close contact with the underside of the mylar. Even a fraction of a thou of air decreases the capacity in this area and leads to more irregular tracking. Initially aquadag was used to make intimate backing between the earth electrode and the mylar. As the resistance was quite high, this aquadag layer had to be backed by thin aluminium foil. However, a better solution is to use silver paint. The conducting paint we use is that supplied for making electrodes on conducting paper for solving two-dimensional electrostatic problems. Two coats of this provide a layer about 2 thou thick, with a resistivity of about a tenth of an ohm a square. The layer is quite tough and can be flexed without changing its resistance. It can take significant currents for reasonable times, without blowing up. The time integral of the current density squared for blow up is about 10^{-4} of that of silver.

The sharp metal electrodes have been made out of 3 thou copper lightly sprung against the top of the mylar layer. Figure 6d-1 shows a cross-section of a typical set-up.

The capacity feeding the channels is 0.043 microfarads and the switch is a simple mechanically operated one. The inductance in the circuit is about 150 nH. The duration of the applied pulse is controlled by a series of resistor chains (R) which are placed in parallel with the surface track, the circuit being given in Figure 6d-1. The current in the surface arc, if closure results, is controlled by a series resistor (S).

In some experiments a single channel was used, but in most of them an array of 4 or 6 parallel earthy straps were provided, so as to produce data on the ability to light a series of parallel channels and also to increase the statistics of channel formation.

Apart from fairly short channels, it was found that there was little polarity effect in all the gases, except helium. Using the reversing condition (d), sometimes the negative edge required more volts than the positive (nitrogen), but for other gases or mixtures the reverse was true. In general, the difference was less than 10% and seemed to decrease with increasing voltage and track length. For helium, the negative edge gave completed arcs much more easily than for the positive edges. Since for the envisioned application it was very desirable that the applied pulse be negative, the results are quoted for this polarity. However, the positive edge results would be very similar for track lengths over about 5 cm.

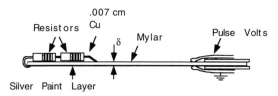

Cross Section of Experimental Arrangement

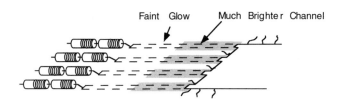

Oblique View of Setup With 4 Understraps Showing Incomplete Closure Tracking

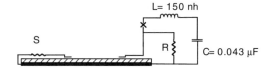

Circuit

Fig. 6d-1 Surface tracking experimental arrangement.

It was noticed that there was some hysteresis. That is, when an arc had been struck and the voltage reduced, the voltage at which the complete tracking stopped was a bit lower than when the voltage was raised until restrike occurred. This was despite every effort to discharge the surface (condition (c)). Either the discharging was not complete or there were free ions or excited molecules around which helped the surface discharge slightly. The effect of the latter would be most likely to be felt on the copper electrode, perhaps in helping this to emit electrons to start the discharge. In any event, for reasonable tracking distances the effect was small, again being less than 10% and mostly less than 5%. Because the track is zigzagged (for the pulse durations of most interest) there is a significant standard deviation in the closure voltages, even when everything else is kept nominally

GAS BREAKDOWN

constant. Ideally a lengthy series of shots should have been done under every condition and the mean and standard deviation obtained. However, life is short and the criterion used was that the voltage applied should cause closure in about 80% of the shots. Because of hysteresis, care had to be exercised to allow for this and the effect of this and other uncertainties is that the error in any figure quoted can be ± 5% or more. The standard deviation of the observations is probably more like ± 3%. For the application under consideration this was more than adequate.

SURFACE TRACKING RESULTS

Air

Condition (c) Tracking

With discharging the surface between each shot, the following relation was found to give answers with about ± 5% of the observed ones:

$$ l = 1.6 \ (V - 5.5 \ \delta^{1/3}) \ t^{1/4} $$

where l is the distance tracked in cm

   V is the peak applied voltage in kV

   δ is the thickness of mylar in thou (10^{-3} inch)

   t is a time in μsec equal to discharge time constant of the condenser circuit divided by 6, i.e. the time width of the pulse at 87%.

I must apologise for the weird mixture of units, but they happen to be convenient. The range over which the parameters were varied was as follows:

   l was varied in general from 4 cm to 12 cm.

   δ was varied from 1 thou to 30 thou.

While δ is specifically for mylar, a short series with polythene suggested that for this parameter the capacity was the important factor and that hence 1 thou polythene was equal to 1.12 thou of mylar: however, this was not established with any great accuracy.

The time parameter (≡ RC/6) was varied between 0.1 and 1.6 microseconds. I must particularly apologise for this parameter. The early work had suggested a 1/6th time dependency; thus the

initial choice of RC/6. When this section of the work was done I
did not have a pick-up-suppressed 'scope to hand, so could not
measure the delay between time of application of the pulse volts
and the time the track closed. Later on a Tektronix 'scope became
available, but because it is rather pick-up-sensitive and I now
have some doubts about its time of triggering, the results from it
were not completely reliable. However, they indicated that the
closure time did roughly obey the 1/3 power law but were a factor
of 2 to 3 <u>shorter</u> than the time parameter as defined above, instead
of about twice as long, as the 1/3 power would imply. Thus the
absolute time of closure of the surface track is not very certain
but is of the order of a half of the time parameter and there is no
sense in re-analysing the records using RC/3. I am very sorry
about this and can only suggest the t be considered a time-like
parameter whose significance is not exactly clear.

The relation was also very roughly checked against some long
distance tracking data at 160 kV and gave answers roughly in line
with these results. While the results of these (and some later
work referred to below) favoured a $t^{1/3}$ dependency, a power significantly above or below this value would have given quite a reasonable fit, with a change in the threshhold voltage value of 5.5 $\delta^{1/3}$
kV. Thus the time power is not determined with great accuracy.
Moreover, for times longer than about 5 μsec, the time dependency
vanished.

Some study of the effect of the width of the understrap was
made. Where this was wide and the time parameter around 1 μsec or
less, a number of channels formed, the spacing being dependent on
the mylar thickness. For 1 thou mylar these were on average about
1/2 cm apart, and about 1 1/4 cm apart for 10 thou mylar. Using 3
thou mylar, the effect of reducing the width of the understrap was
roughly investigated. Down to 0.5 cm width of understrap there was
no change in required voltage; below this there was a very slow
increase in theshhold voltage, so that at 1 3/4 mm width the constant 5.5 had risen to maybe 6.0, but, again, this change was not
investigated for a wide range of parameters.

Other than Condition (c) Tracking

A rather shorter series of tests was performed with conditions
(b) and (d). In condition (b) the track was started by discharging
the surface and the voltage determined which allowed at least 4
subsequent tracks to occur. In these experiments 4 or 6 understraps were used, each 1/2 cm wide and either 2.5 or 1.25 cm between edges. In air, where a track failed to form, it strayed out
on all subsequent shots.

GAS BREAKDOWN

In condition (d), the voltage on the capacitor was reversed between each shot. Again the length was varied between 4 and 12 cm and the time parameter varied between 0.3 and 5 microseconds. Only 2 thicknesses of mylar were used, 2.8 and 5 thou. Again the results could be adequately fitted by a relationship

$$l = k (V - 5.5\, \delta^{1/3})\, t^{1/4}$$

where k = 0.85 condition (b)
         = 1.6   condition (c)
         = 2.2   condition (d)

OTHER GASES

The gases used were carbon dioxide, nitrogen, and helium, and some mixtures of these. In these mixtures the notation used is the volume percentages of the gases in the order - carbon dioxide, nitrogen, helium. Thus pure carbon dioxide is 100/0/0 while pure helium is 0/0/100.

The data was obtained for only one thickness of mylar - 5 thou. The length was again varied between 4 and 12 cm, while the time parameter was varied between 0.3 and 5 μsec*.

Apart from pure helium, nearly all the data was reasonably fitted by a $t^{1/3}$ dependency, with the reservation that above or around 5 microseconds the tracked length became independent of time. The date for a 5 thou mylar thickness and with voltage reversal between each shot (condition (d)) could be fitted by relations of the form

$$l = k (V - V_{th})\, t^{1/4}$$

where the values of k and V_{th} are given below in Table 6d-I.

---

* Editor's note: original copy has " - - 0.3 and 5 thou."

The values in Table 6d-I are those obtained from fitting the individual sets of data points. However, because of the hysteresis and other effects, the scatter of experimental values would have allowed a range of nearly equally good fits. Because of this, a simplified set of values is almost as good. These are given in Table 6d-II.

Table 6d-I
k, V_{th} Values for 5 Thou Mylar, Negative Edges

	V_{th}	k
Air	10.3	2.7
100/0/0	11.1	1.0
0/100/0	10.6	1.6
0/0/100 *	10.5	2.8
10/10/80	12.6	1.8
25/25/50	12.4	2.0
50/0/50	12.2	1.55

* The value for helium was obtained only for l = 8 cm. For 4 cm, the values of k were considerably higher, but the definition of when a sharp channel formed from the glow phase very uncertain. Values for l = 12 cm could not be obtained because of very long length tracking taking place at the pulse feedthrough into the gas cell. Thus the helium values are of very limited use.

Table 6d-II
Simplified k, V_{th} Values

	V_{th}	k
Air	)	2.8
100/0/0	)	1.0
0/100/0	) 10.5	1.6
0/0/100	)	2.8
10/10/80	)	1.75
25/25/50	) 12.5	2.0
50/0/50	)	1.6

These values were used in calculating the tracking voltages and the results compared with the original 72 sets of data for gases other than helium and the 8 cm data for this gas. The overall average error for each gas was within ± 2% and the average standard deviation for each result ± 5%. Of this, some ± 3% would be expected in the experimental scatter of each point, the rest being due to errors in the simplified relationships.

With regard to the earlier data quoted for air and independently measured, these give for 5 thou V_{th} = 9.7 kV and k = 2.2. The larger value of k = 2.7 quoted in Table 6d-I arises because a value of V_{th} = 10.3 was found from the second set of data. Over the range of lengths studied, the two different sets of parameters give results within ± 2 1/2% of each other and therefore the differences fall within the experimental errors. A wider range of parameters would have shown which fitted better, but time was not available to do these experiments.

With regard to the behaviour of the mixtures, a few points can be seen by plotting the data on a triangle for, say, 10 cm track length. These show that the carbon dioxide tends to dominate the mixture. This was further shown by some data at 10 cm for a 50/50/0 mixture which gives a value for k much closer to that of carbon dioxide than that for nitrogen.

A second and somewhat surprising result was that for the n/n/(100-2n) mixtures the trend of k is not uniformly upwards as n decreases. Indeed, the values for the 10/10/80 mixtures are consistently slightly below 25/25/50 instead of being above it as would be expected.

The third result is that, helium apart, all the gases and mixtures need more volts to track a given length than air does. The reason this is a bit surprising is that the attachment coefficient in air (due to the oxygen) is much higher than for pure carbon dioxide and very much higher than for the other gases. However, nature, as usual, resolutely refuses to be simplistic.

This completes the section on tracking voltages. It should perhaps again be repeated that the relationships are only good to some ± 5%. In addition, if the voltage is raised 10% above that given by the relations, 6 tracks will result essentially every time for a 6 strap array with both polarities.

VOLTAGE DROP DOWN ARC AFTER TRACK CLOSURE

After the track closes the 2 metal electrodes, a current flows, determined by the value of the series resistor S. This resistor is added for 2 reasons. 1) It keeps the current flowing for many microseconds with reasonable values of the main capacitor. 2) It reduces the requirement for simultaneity in closure of the surface tracks. This is because a track closing does not rapidly discharge the capacitor but ensures the voltage stays up on the pulse charged electrode. The only thing that then causes failure to close of a track of different length is cross-tracking from one channel to another. Where the understraps are separated by 2.5 cm and the length of the track is 10 cm, adjacent channels can be different in physical length by at least 3% and still not affect the formation of both channels on each shot.

Because spark channels can be formed with gradients of 1 kV/cm or even less down them, their behaviour is substantially different from ordinary arcs. In addition, because typically several kilohms are in series with the spark channel, the resistive phase formula is inapplicable, even during the the tracking phase.

By measurement it was found that the voltage across S differed from the voltage applied to the electrode fed by the capacitor by a voltage ΔV (kV). As the capacitor discharges, this difference of voltage remained more or less constant, rising only some 20% by the time of spark extinction. Thus over the portion of the waveform of interest to me, from the point of view of practical application, it is possible to assume a constant voltage drop from end to end of the weak spark channel. This voltage difference was found to be a function of arc length and current flowing (i.e. value of resistor S).

A short series of tests in air gave quite a good fit to the relation

$$\Delta V = 0.25 \ S^{0.46} \ (\ell - 1.5)$$

where ΔV is in kV
S is in kilohms
and ℓ is in cm.

In the experiments ℓ was varied from 5 to 15 cm and S from 2 to 16 k-ohm, where this was possible. For the longer lengths and the higher resistances, the arc channel either did not form or went out discontinuously while the source voltage was falling.

A rather more limited series of measurements was made for the value of ΔV for various gases and mixtures, for a track length of 10 cm. Within the experimental error of about ± 10%, the results can be collected into two sets and the average values given in Table 6d-III. No data was taken for pure helium. The applied voltage pulse was some 10% above the reversing voltage criteria given in the previous section. In a short separate series of

Table 6d-III
Average Values for ΔV for Various Gases and Mixtures

S kΩ	Gas/Mixtures	
	100/0/0 0/100/0 50/50/0	10/10/80 25/25/50 50/0/50
2 4.2 8.2	3.9 5.1 6.3	3.0 4.4 6.1

tests, it was shown that ΔV was independent of the applied pulse over the range of 25% or so studied.

The power dependency for the helium-containing mixtures is close to that of air, while that for the carbon dioxide, nitrogen and equal ratio mixture is lower, being about 0.36.

The above measurements were made with only one time constant of discharge (an RC of about 20 microseconds). It is possible that the approximate constancy of the voltage drop down the arc channel during the pulse is only true for times of this order. The rate of delivery of energy to the channel is typically 1 to 2 kilowatts per cm (assuming a uniform voltage gradient down the channel) at the start of the pulse, dropping by about a factor of 5 shortly before the time of channel extinction. Crude calculations of the various cooling mechanisms of the channel suggest that diffusion of cold molecules into the channels is the main cooling mechanism, with radiation and thermal conduction being rather smaller but still significant terms. Using visually determined estimates of the channel radius, the theoretical estimates of the cooling rates give values around 1/2 kW/cm, in reasonable but possibly fortuitous agreement with the observations. However, this model implies that for rather different time scales from those used experimentally, the channel voltage drop may not remain roughly constant.

CONCLUSION

Approximate relationships are provided for the voltage pulse required to track mylar surfaces for various conditions of initial surface charge. After surface track closure the behaviour of the arc can be pretty closely approximated to by a constant voltage difference between the electrodes. This voltage difference is given as a function of track length for air and current limiting resistor value for various gases and mixtures. No great accuracy is claimed for the relations, but they provide a good basis for estimating the conditions necessary to track a surface in multi-channels and to sustain the current in the channels for many microseconds, with a known voltage drop down the channels.

Section 6e

HIGH SPEED BREAKDOWN OF SMALL AIR GAPS IN BOTH UNIFORM FIELD AND SURFACE TRACKING GEOMETRIES*

J. C. Martin

INTRODUCTION

This note briefly summarises some data obtained on the high speed breakdown of air for small gaps in various geometries. The data in all cases was obtained incidentally during other work and while reasonable care was taken to calibrate the voltage monitors and to allow for 'scope response, etc., the accuracy of the data is not outstanding, being no better than ± 3%.

UNIFORM FIELD, UNIRRADIATED, 8 mm GAP

This 'gap' was actually part of a possible laser pumping cell and had electrodes 50 cm long and was connected to a 1.5 nF capacitor by inductance of about 0.7 nH when uniform current filled the length of the electrodes. The 1.5 nF capacitor was charged with a $\sqrt{LC}$ time of about 3.5 ns from a larger pulse charged capacitor via a pair of multichannel edge plane gaps of total length about 120 cm. The gap operated in pressurised nitrogen and gave a total of 10 to 15 channels and operated closely in accord with the data given in "Results from Two Pressurised Edge Plane Gaps," (J. C. Martin, F. Grimson; AFWL Switching Note 20, September 1972) without any significant signs of fizzle at voltages up to 190 kV.

---

* AFWL Switching Notes
  Note 24
  April 1977
  Originally published as
  SSWA/JCM/774/423

In the air breakdown experiments the effective time the voltage was applied to the gap (full pulse width at 89% of peak volts) varied from 2 to 13 ns, with most of the times between 2 and 5 ns. These times were corrected to a time of 2.5 ns, assuming $Ft^{1/6}$ was constant. Such a correction brought the data into good agreement. The air pressure was varied between 0 to 67 psig and the 'scope response corrected breakdown voltage varied between 71 and 180 kV.

The air results from "High Speed Breakdown of Pressurised Sulphur Hexafluoride and Air in Nearly Uniform Gaps" (J.C. Martin, AFWL Switching Note 21, February 1973) suggest that for the very limited range of data studied, $Ft^{1/6} d^{-1/6}$ is constant. The data from the last graph in that note, when scaled to a gap of 8 mm and a time of 2.5 ns, gave the solid line in Figure 6e-1. Some other data obtained with a fast charged fairly uniform weakly irradiated pressurisable gap operating at up to 600 kV agreed well with the data of Switching Note 21, except near one atmosphere, where the breakdown field was some 15% lower, suggesting, as is speculated in the note, that there was little overdrive because of lack of initiating electrons, even though the original gaps were unirradiated. Thus the curve in Figure 6e-1 is expected to be also very close to that applying to an irradiated gap.

The experimental breakdown field data, scaled to a time of 2.5 ns, is also shown in Figure 6e-1 and the agreement at the high pressure end is within the expected experimental error. At low pressures the present data lies a little above the curve scaled from Switching Note 21. This is probably due to a delay in the

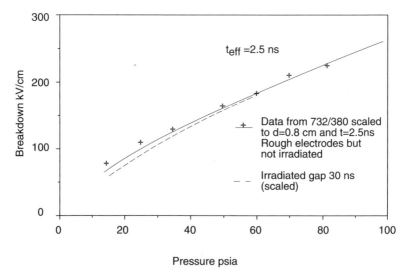

Fig. 6e-1 Air breakdown field for uniform gap.

GAS BREAKDOWN

production of initiating electrons and if the gap had been irradiated I would expect the low pressure data to drop to the curve, or slightly below it, to the dotted curve.

Thus the new data supports the results of the previous high speed breakdown note and extends the data to 2.5 ns. Less certainly (because of the limited range of the parameters), the air breakdown in this region of spacing and time scales as $Ft^{1/6}$ $d^{-1/6}$. The time range over which the expression has been tested is from 40 ns to 2 ns and d has ranged from 0.8 to 3 cm.

HIGH SPEED ATMOSPHERIC BREAKDOWN OF SMALL, NEARLY UNIFORM GAPS

During work on a four element low voltage irradiated gap some data for the high speed well-irradiated breakdown fields in air at one atmosphere was obtained. Again the data is not of exquisite accuracy, largely due in this case to possible errors in measuring the spacings. The gap was between relatively large diameter long rods, i.e., it was a near uniform rail gap.

Figure 6e-2 gives the breakdown voltages observed when the gap was very well irradiated from a nearby surface flashover irradiator. When this was removed the breakdown fields could easily treble, as well as developing very large jitters. The voltage jitter of well irradiated gaps was difficult to measure directly, but as multichannel operation was observed regularly it was plausibly a fraction of a percent.

In Figure 6e-2 the times are again full pulse width at 89% of peak volts. The curves for 10 and 2 1/2 ns approximately obey a $t^{1/6}$ dependency and this has been used to indicate the breakdown voltage required for 0.6 ns, which was beyond the 'scope's measuring capability.

The high speed breakdown field data over the range 0.2 mm to about 2 mm obey a relationship of the form $Fd^{1/6}$ = constant, where d now has a positive power rather than the negative one indicated by the larger gap data discussed in the previous section. The reason for the larger gaps having a breakdown field which weakly increases with gap distance is that for very fast pulses the final plasma channel streamer takes a significant time to cross the gap, and hence the larger gaps need a little more volts. For small gaps the plasma channel closure time is still very short, even at a couple of ns, and hence this field dependency disappears and is overcome by other effects.

The DC breakdown for very small gaps eventually goes as $d^{-1/2}$ and while the high speed breakdown does not seem to have as strong a dependency as this, it still exhibits some of this effect. Figure 6e-3 sketches the breakdown field dependency for 2 1/2 ns at

172 CHAPTER 6

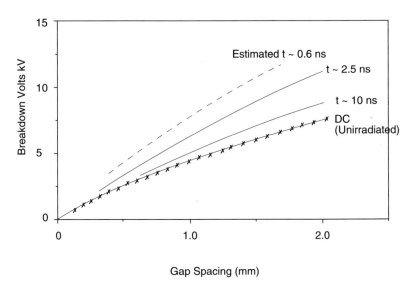

Fig. 6e-2 Irradiated pulse and DC breakdown voltage of atmosphere air gap.

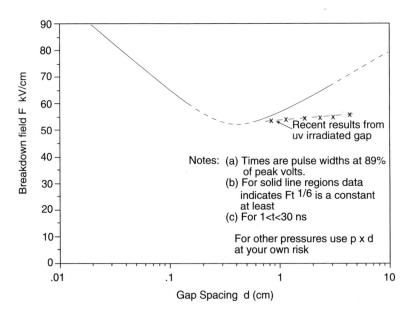

Fig. 6e-3 2.5 ns breakdown field for air at one atmosphere (irradiated).

GAS BREAKDOWN

atmospheric pressure as a function of gap spacing indicated by the data of this and previous notes. The curve may well be changed in detail for other times and for other pressures. Indeed one would expect that Figure 6e-3 might be more generally applicable if plotted as a function of p x d rather than as d. If this speculation is right, then pressurised small gaps would continue to obey $Ft^{1/6}$ $d^{-1/6}$ constant down to very small gaps.

As a brief aside for those unfamiliar with previous notes, for mildly non-uniform gaps the breakdown field used above is that applying to the electrodes, that is the breakdown voltage increased by the field enhancement factor divided by the gap spacing. This is most conveniently done by using an effective gap spacing, the real distance divided by the FEF. However, in the field dependency relations d is still the real gap separation in mildly non-uniform gas as this is the distance the plasma channel has to close over.

The DC breakdown voltage shown in Figure 6e-2 is the one obtained without pulse irradiation. With sufficiently intense uv pulse irradiation small gaps can, of course, be broken down at voltages below their DC levels.

Before closing this section it should be repeated that functional dependences suggested in this note have only been tested over very limited ranges and are known not to be true, say, for longer times and may well not be those that apply for large gaps. However, fortunately there are unlikely to be too many requirements for many atmosphere, metre spaced, few ns, breakdown air gaps, but if anyone has such a need, I would strongly advise them not to use the relations given, unless utterly desperate.

SMALL GAP PULSED SURFACE FLASHOVER DATA

In a previous note, "Pulse Surface Tracking in Air and Various Gases," (J.C. Martin, AFWL High Voltage Note 4, May 1974), the length tracked (1) was found to be a function of the state of DC charge on the surface, as would be expected. Also the length was observed to obey a relation of the form

$$\ell = k (V - 5.5 \delta^{1/3}) t^{1/4}$$

where V is the voltage in kV, δ is the mylar insulation thickness (in thou., i.e., 10^{-3} inch) when ℓ was more than a couple of cm or so. For the case where the surface was discharged between pulses and for t in nanoseconds, this relation becomes

$$\ell = 0.3 (V - 5.5 \delta^{1/3}) t^{1/4}$$

In the original note the definition of t is rather peculiar but corresponds to the pulse width at about 87% of maximum volts. To adequate accuracy this is the same as the pulse width at 89% of maximum volts, which is the definition used in this note.

In those experiments it was noted that ahead of the bright discrete channels there was a faintly glowing uniform region, typically of the order of 1 to 2 cm long, for times of 100 to 1000 ns. It was speculated that the voltage ($5.5\ \delta^{1/3}$) taken away from the applied voltage represented a drop down the uniformly glowing avalanche region prior to the development of current instabilities which lead to the formation of plasma channels. As such, the relationship really applied to the growth conditions of these channels.

For the experiments reported here the gaps were much smaller (0.3 mm to 5 mm) and hence apply to conditions where this uniformly glowing avalanche region bridges the gap before the instabilities have developed much and plasma channels formed. As might be expected, the relationship for the length tracked was very different. Firstly there was much less effect of previous non-sparking pulses on the breakdown voltage, especially for the smaller gaps. This means that there was not much hysteresis in the breakdown field for the smaller gaps, the same voltage roughly being obtained whether the voltage was slowly raised until sparking occurred, or lowered until sparking ceased. Secondly, for small gaps this sparking voltage was proportional to gap length and one could use a breakdown field relationship to adequate (± 10%) accuracy. For a range of mylar thicknesses from 0.25 to 3 thou. (mil. for USA readers) the following rough relationship applied reasonably well in air:

$$F\ t^{1/6} = 36\ \delta^{1/6} \qquad F \text{ in kV/cm}$$

For one set of conditions (δ = 1 thou., t ~ 30 ns) it was shown that the tracking relationship changed from the new small gap one to the old long track one, the changeover distance being as calculated, around 3 mm for these conditions.

In the experimental gaps 2 thou. steel sharply cut edges were used and some care taken to ensure that this and the mylar were tightly and permanently pulled down onto the backing metal plane.

The range of times studied were from about 2 ns to 30 ns in these experiments and once again the reader is warned that no great accuracy is claimed for the relationship given.

ACKNOWLEDGMENTS

It is a pleasure to thank Tommy Storr for his help in obtaining the results given for the 8 mm gap. Equally, however, he is entirely blameless for what I have done with them. My thanks are also due to Mrs. Vikki Horne, who I defeated with one horribly written word. However this victory doesn't really count, as I couldn't read it either.

Chapter 7

LIQUID BREAKDOWN

Section 7a

COMPARISON OF BREAKDOWN VOLTAGES FOR VARIOUS

LIQUIDS UNDER ONE SET OF CONDITIONS*

J.C. Martin

The conditions used in these experiments were parallel plates of the order of 16 square inches in area which had carefully radiused edges so that there was no field enhancement at them. The gaps used varied between 6 mm and 13 mm usually, and the effective times of charging between .15 and .3 microseconds. The relation to which the breakdown voltages are fitted is $F^{3/2}t$ where F is in MV/cm and t is the duration of the pulse between 63% and 100% of the peak voltage reached. In general the results gave a reasonably constant value of this function (k) and in the case of alcohol the standard deviation of the results was very good, being about 12% for a single result. However, for ethylene glycol there was evidence that the fit was not so good and for water the relationship is significantly different. A value of k is quoted for water but *this is estimated for these particular conditions*; for high voltages and bigger gaps water gets progressively better.

The figure of merit you employ will vary with the application you have in mind. What I have selected to tabulate is $\epsilon k^{4/3}$ which is proportional to the energy per cc at the breakdown point for a given constant charging time.

---

* AFWL Dielectric Strength Notes
  NOTE 1
  SSWA/JCM/1168

Some work was also done with crossed cylinders and these gave values in reasonable agreement with the parallel plate values. The plates used became slightly dented as the experiments proceeded but care was taken that there was no tendency to break consistently at one point or at the edges. In the case of glycerine, the liquid had to be stirred after each breakdown and it was very pretty to see the discharge trees at points other than that which got across first. These patterns were thin white threads frozen in place after the voltage had been removed by the breakdown and were very heavily branched. There were usually a few (like 3 or 4) fairly big ones in any one shot, all coming from the positive plate. There was nothing at all growing from the negative plate. All the liquids were stirred reasonably well but glycerine required a lot of stirring. The plates in these tests were vertical, so bubbles could escape upwards.

Not given in this list is data obtained with smooth ball bearings in a small volume test. This data was in line with these tests but in general gave bigger values for k, perhaps implying an area term. One substance that give quite remarkable results with the small area test was acetic acid. However, I could not bring myself to test it in the larger area test, in case it continued to be good. (Imagine several tons of the stuff). Likewise several other materials were checked in the ball bearing test but excluded in the tests reported here, which were of materials it was possible to conceive of working with.

The conductivity of the alcohols was a bit embarrassing, but with care the resistance of the cell was kept in the kilohm region. Ian also did some work on de-ionizing them with resin columns, which seemed to work quite O.K.

Table 7a-I
Values of $F^{3/2} t_{eff} = k$

Area of plates: 16 square inches

The dielectric constant was taken from standard tables not measured.

Material	k	ϵ	$\epsilon k^{4/3}$
Ethyl Alcohol	.057	24	
" " + 1% water	.059	24.6	0.56
" " + 10% water	.052	29.6	
Methyl Alcohol	.052	33	0.65
Ethylene Glycol	.03	38	0.34
Glycerine	.010	44	0.10
Castor Oil	.11	4.7	0.25
Transformer Oil	.08	2.4	0.09
Water	.03	80	0.72

Note: Values of k are good to perhaps 10 to 15%

Section 7b

A POSSIBLE HIGH VOLTAGE WATER STREAMER VELOCITY RELATION*

J.C. Martin

INTRODUCTION

In a companion note the data backing the oil streamer velocity relationship for voltages above 2 MV or so is discussed. This relation becomes approximately independent of polarity for sufficiently high voltages and I have long speculated that the same is true of water. In this very tentative (and hence hand written note) some very crude estimates of what this relationship might be are presented. These are based on the data provided by Pace Van Devender and Tom Martin's measurements reported in "Untriggered Water Switching" (IEEE TRANS. NS-22, 979, JUNE 1975**). Needless to say my estimates could and should be checked more directly with some of the high voltage water generators now in existence. However, in the absence of these, what follows may be better than nothing.

NEGATIVE STREAMERS

The negative data given in George Herbert's "Velocity of Propagation of High Voltage Streamers in Several Liquids"** are plotted as a function of $\bar{u} \, t^{1/3}$ (where $\bar{u}$ is the mean velocity in cm/μsec and t is in μsec) as a function of V. Most of the data was obtained for t_{eff} between 0.2 and 0.6 μsec. To convert George's curve to $\bar{u} \, t^{1/2}$ as a function of voltage, his best fit curve has to

---

* Hand written note, no date
** Editor's Note: This reference is an appendix to this section. Reading is recommended.

be dropped by about $(0.4)^{1/6} = 0.85$. In addition the slope should be decreased slightly as the higher voltage points would have slightly longer times. This gives a negative streamer curve close to $\bar{u}\ t^{1/2} \sim 11\ V*$ with, of course, a slightly poorer fit to the original data. Considering Pace and Tom's data, the original negative relation gave results too low for the longer pulses but I suspect that the experimental data here is too high. Taking the example given above Figure 4 in their paper, a corrected velocity will be calculated as an example.

For the negative case $d = 5.7$ cm, $\bar{u} = 32$ cm/μsec, $t = 0.18$ μsec and from Equation 4 of the paper by Pace and Tom, $V = 1.75$ MV. Also from the FEF = 12, I calculate the radius of the edge was of the order of 0.03 cms giving a highly stressed area of about 5 cm^2. Using the negative water breakdown relations $Ft^{1/3} \simeq 0.55$ MV/cm and trying a starting voltage of 50 kV at which streamers first start, $t_{eff} \sim 0.06$ μsec giving a field on the negative electrode at which streamers start out at 1.4 MV/cm and, hence, at a voltage of 670 kV $\frac{Ess}{FEF} \cdot d = \frac{(1.4\ MC/cm)\ (5.7\ cm)}{12} = 665$ kV.

Using the differential velocity relation $dx/dt \sim 5\ t^{-1/2}\ V$, it can be shown that for a linearly rising voltage ramp $d/dpp \simeq 1 - (Vs/Vo)^{3/2}$ where d is distance the streamers move when they first start out at V_s compared with the distance (dpp) which they would have gone had they started out at a smaller voltage. For the above example $V_s/V_o = .38$ and dpp = $5.7/.76 = 7.5$. Thus if the edge had been substantially sharper it is calculated that for the conditions given above that the streamer would have gone 7.5 cm or thereabouts. This gives $\bar{u}\ t^{1/2} = 17.8$ for a voltage of 1.75 MV i.e. a relationship of $\bar{u}\ t^{1/2} \sim 10V$. A similar sort of correction is estimated to apply to the short pulse data where the applied voltage was a little lower.

Thus I take the negative streamer data relation up to and presumably beyond 2 MV to fit reasonably well with $\bar{u}\ t^{1/2} = 10V$.

<u>Positive Streamer Data</u>

The reason George's old relation fails to apply above 1 MV or so is that the bush on bush phenomena sets in. This has been seen by us and several others (including Sandia) and the assumption is that the low voltage streamer velocity (with its heavy branching) turns into something similar to the negative streamers and certainly goes faster. If you take the differential + ve velocity relation $dx/dt_+ = 4.4\ t^{-0.5}\ V^{0.6}$ and equate this to the negative one of

---

* This means 11 times V (MV) here and in the following pages.

$dx/dt_- = 5t^{.5} V$ this gives a velocity where these are the same of about 0.7 MV. This is obviously a simplistic picture and could well be in error. However, it is supported by the data given above Figure 4 (in the paper by Pace and Tom) for a particular case. $\bar{u}$ is given as 58 cm/μsec for d = 5.7 and (t ~ .097) giving $(\bar{u}_+ t^{1/2} \approx$ 18.5 for a voltage from Eq. 4 of 1.6 MV (in the paper by Pace and Tom). However the distance travelled is slightly too high because during the positive low voltage phase, the streamer goes further than it would have done if it had been going solely as a negative one. Indeed integrating the two positive streamer relations gives $d+/dpp-Ve \approx 1 + (0.3/V^{0.5})$ where d is the positive streamer closure distance for a voltage V (> 0.7 MeV) and dpp- is the equivalent negative streamer closure distance based on $\bar{u} t^{1/2}$ = 10 V as above. Thus for the case of d+ = 5.7 and V = 1.0, dpp- is about 4.6 cms giving a -Ve mean velocity of 47 cm/μsec and giving $\bar{u} t_{-ve}^{1/2}$ = 14.5. This has to be increased slightly because the + Ve streamers do not start at zero volts for the experimental blades giving $\bar{u} t^{1/2}$ ~ 15.5 for a voltage of 1.6 MV. This is again in agreement with the assumed relation.

Figure 7b-1 shows my best guess at the streamer mean velocity relations for both negative and positive polarities. Also shown on the figure is a corrected data point from John Shipman, who found a closure distance of 13 cms for a 7 MV pulse.

With a t_{eff} of ~ 0.05 μsec, the point was slightly blunt and close to a plane backing plate (stick out divided by the gap ~ 0.1) and from this I estimate that with a sharp needle and no backing the closure distance would have been about 15 cms. This gives a value of $\bar{u} t^{1/2}$ of 68 in probably fortuitous agreement.

It should be repeated that the proposed high voltage streamer velocity relation for water is only an approximation to be used in the absence of better direct experimental values. With regard to these, I would like to suggest that any experiments to obtain these should use fairly sharp points and these should be reasonably removed from any large metal backing.

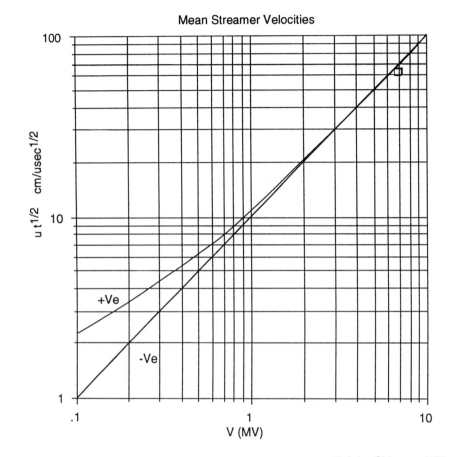

Figure 7b-1   Streamer mean velocity relation vs voltage.

Chapter 7b Appendix

VELOCITY OF PROPAGATION OF HIGH VOLTAGE

STREAMERS IN SEVERAL LIQUIDS

> H.G. Herbert
> Atomic Weapons Research Establishment
> Aldermaston, England

INTRODUCTION

These experiments were carried out using a pulse transformer producing voltages up to one million volts rising in $\simeq 0.5$ μsecs., and the "Dagwood" 1.3 million volt Blumlein generator with a pulse length of $\simeq 45$ nanosecs. The test electrodes consisted of a 2" ball bearing and a sewing needle. The fine needle tip was to minimise any threshold field effects.

Measurements were made of the needle to ball gap over a range of 3-30 mms, peak voltage, and the effective time measured as that for which the voltage is greater than 63% of its maximum. Data was obtained for transformer oil, carbon tetrachloride, glycerine and deionized water.

The results for the first three suggest that the mean velocity $\bar{v} = d/t_{eff} \propto V^n$ (applied voltage) where n lies between 1.2 and 1.8, except for the positive streamers in glycerine which are much less dependent on voltage and n = 0.55.

The effective time of 63% to peak is the calculated time for an equivalent square pulse of peak amplitude as derived from the relationship $\bar{v} \propto \bar{V}^{3/2}$ and the generator's waveforms. The timing errors introduced by the variation of and n are small.

AFWL Dielectric Strength Notes
Note 10
24 Oct 1966

CHAPTER 7

The case of water is an exception: according to Mr. Martin's original positive streamer formulae one expects this to be the case, since

$$\frac{dx}{dt} \propto \frac{V}{x} \quad \text{or} \quad \bar{v}t^{1/2} = kV^{1/2}$$

The results obtained are in close agreement with this, giving $\bar{v}t^{1/2} = 8.8\ V^{0.6}$.

Results for the negative water streamers suggest $\bar{v}t^{1/3} = kV$ as being the best fit for the data obtained.

The water (and the positive glycerine streamers) are also physically different from the others, being more bushy as compared with the spidery nature of the oil, carbon tetrachloride and negative glycerine streamers.

Graphs 7b-(2-9) show the plotted results which have been obtained by two operators using different monitoring systems over a large period of time, making it difficult to estimate the accuracy of V, t, and x. Analysis of the results give the following coefficients of correlation and standard errors of the constants K and n.

		r
Oil, positive	$\bar{v} = (90 \pm 12)\ V^{1.75 \pm 0.12}$	0.96
Oil, negative	$\bar{v} = (31 \pm 5.5)\ V^{1.28 \pm 0.15}$	0.95
Carbon Tetrachloride, positive	$\bar{v} = (168 \pm 28)\ V^{1.63 \pm 0.15}$	0.97
Carbon Tetrachloride, negative	$\bar{v} = (166 \pm 29)\ V^{1.71 \pm 0.19}$	0.95
Glycerine, positive	$\bar{v} = (41 \pm 1.5)\ V^{0.55 \pm 0.03}$	0.98
Glycerine, negative	$\bar{v} = (51 \pm 8)\ V^{1.25 \pm 0.13}$	0.98
Water, positive	$\bar{v}t^{1/2} = (8.8 \pm 0.4)\ V^{0.6 \pm 0.03}$	0.97
Water, negative	$\bar{v}t^{1/3} = (16 \pm 0.25)\ V^{1.09 \pm 0.02}$	0.99

$\bar{v}$ is measured in cms/μsec.
V "   "   " megavolts
t "   "   " microsecs.

An approximate calculation based on the oil positive streamer formulae predicts that for a pair of parallel plates with a small projection on the positive plate $Ft^{0.59}$ = constant and independent of separation, F being the mean field V/d.

For a given area of plates, investigations have shown that $Ft^{1/2} d^{-1/4} = k$ is a better estimate, being only slightly dependent on d.

Velocity of streamers in polythene.

Similar experiments have been carried out with polythene. Sewing needles were pushed into 3/4" and 1" polythene sheets, with needle tip to plane separations ranging from 6-13 mms. Care was taken to avoid errors due to threshold field effects since with such small volumes polythene has a strength > 5 MV/cm.

Graph 7b-10 shows the data obtained and analysis gives:

Polythene, positive needle $\bar{v} = (190 \pm 20) V^{1.00 \pm 0.08}$ r = 0.97

$\bar{v}$ in cms/μsec
V in Megavolts

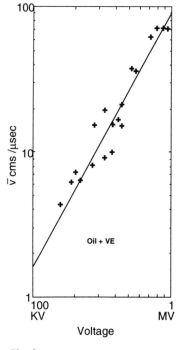

Fig. 7b-2

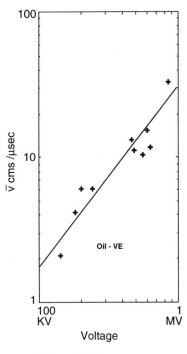

Fig. 7b-3

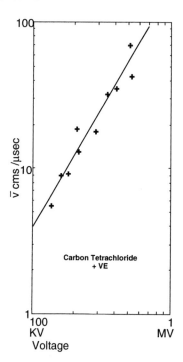

Fig. 7b-4

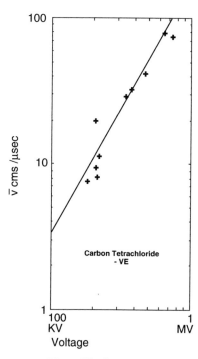

Fig. 7b-5

LIQUID BREAKDOWN

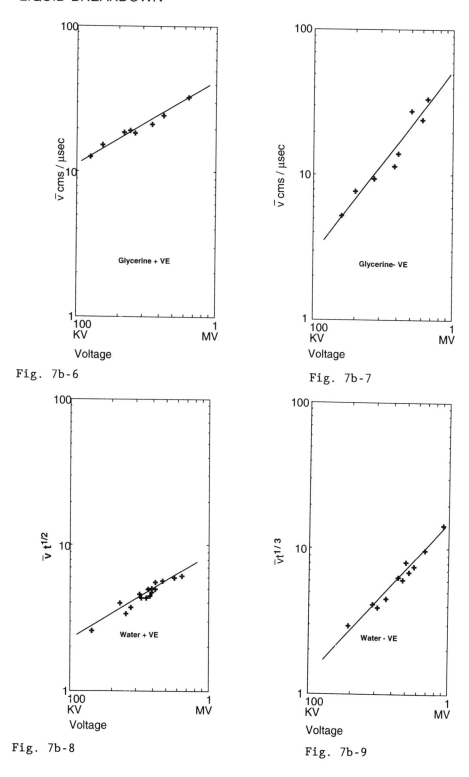

Fig. 7b-6

Fig. 7b-7

Fig. 7b-8

Fig. 7b-9

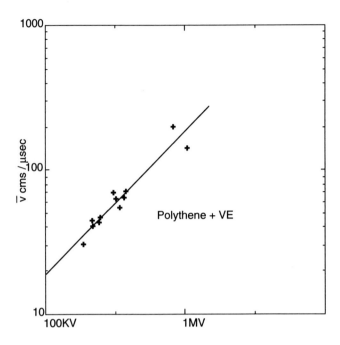

Fig. 7b-10

Section 7c

LARGE AREA WATER BREAKDOWN*

J.C. Martin

INTRODUCTION

It is with considerable reluctance that I attempt to write this note. At various times over the passing years, I have attempted to relate some aspects of uniform water breakdown to the streamer velocity relations originally measured by George Herbert in "Velocity of Propagation of High Voltage Streamers in Various Liquids", SSWA/HGH/6610/104.

I have recently hand written a note "A Possible High Voltage Water Streamer Velocity Relation" which attempts to combine George's measurements for plus and minus points at voltages below 1 MV to give a relation applicable to rather higher voltages. This relation is shown as Figure 7c-1.

This reluctance is partly because I think the integral relations are significantly uncertain and hence the differential ones even more so. In addition any attempt to use the differential relations to calculate the propagation of a streamer from a very small projection, or gas bubble formed from it, has to make a staggering number of uncertain or indeed unlikely assumptions. Hence while being prepared to play around in private, I have been very reluctant to commit the results to paper (even hand written).

However for what its worth (and that's very little) here goes.

---
* Letter
SSWA/JCM/7712/611
16 December 1977

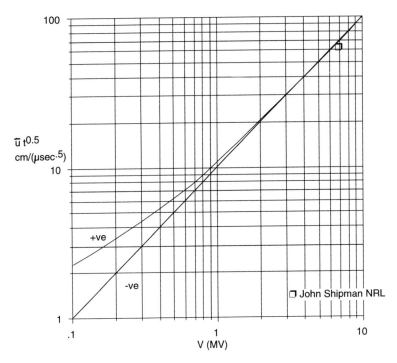

Fig 7c-1 Mean streamer velocities.

DIFFERENTIAL RELATION FOR A LINEAR VOLTAGE RAMP

Let $V = kt$ as shown in Fig. 7c-2.

Put $dx/dt_+ = At^{-0.5} k^{0.6} t^{-0.6}$, $dx/dt_- = B t^{-0.5} kt$

$$d_+ = (A/1.1) t_m^{0.5} V_m^{0.6} = 8.8\ t_{eff}^{1/2}\ V_m^{0.6}$$

$$d_- = (B/1.5) t_m^{0.5} V_m = 10\ t_{eff}^{1/2}\ V_m$$

i.e., $A = 5.8$ $\qquad\qquad\qquad B = 9.0.$

Thus $dx/dt_+ = 5.8\ t^{0.1} k^{0.6}$ and $dx/dt_- = 9\ t^{0.5} k.$

These become equal when $V_s^{0.4} = 0.65$, $V_s \sim 0.35$ MV.

In the "Possible High Voltage Streamer" note the voltage at which the velocities are equal was given as $V_s \sim 0.7$ MV. This applies for square pulses. If the reader wants to work this out he can think about what the old t_{eff} definition has done, it gave me a

LIQUID BREAKDOWN

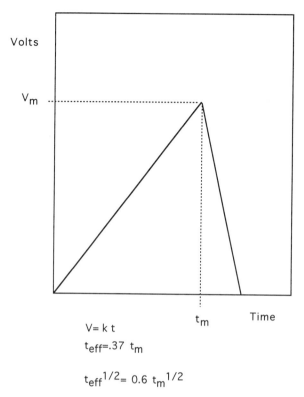

$V = k\,t$
$t_{eff} = .37\, t_m$

$t_{eff}^{1/2} = 0.6\, t_m^{1/2}$

Fig. 7c-2 Linear Voltage Ramp.

headache but I came to the conclusion that a linear ramp was closer to the partial (1-cosωt) waveforms usually used in tests than a square waveform. Even so it is only a partial approximation and different waveforms give different equal velocity voltages.

The first (of many) assumptions is that when the negative streamer is faster than the positive, the swap over occurs. That is

$dx/dt_+ = 5.8\, t^{0.1}\, k^{0.6}$ $\qquad k_+ t = V_+ < 0.35$

$dx/dt_{\pm} = 9\, t^{0.5}\, k$ $\qquad k_+ t > 0.35 \quad k_- t > 0$

Integrating these + ve relations give the + ve streamer curve given in the recent "Streamer" note.

FIELD ENHANCEMENT RELATION NEAR A BACKING PLATE

From "A Method of Estimating the Fields on Backed Razor Blades Cathodes and Their Space Charge Limited Impedances" (that will teach me to write long titles) SSWA/JCM/759/193 (ditto). It can be shown that an edge moved back into a backing plate has a field enhancement factor which is independent of x for $x/d > 0.3$. The geometry is shown as Fig. 7c-3. Below $x/d = 0.3$ the FEF falls roughly linearly with x. This suggests that for the front of the advancing streamer in the differential relation V should be replaced by $(x/0.3d)V$ for $x/d < 0.3$. This may seem a somewhat cavalier assumption but it's not much worse than assuming the voltage drop up a channel is the same whether it is climbing out from a point or from a small protruberence on a plate, which assumption I am about to make (is there no stopping him?).

INTEGRATION OF POSITIVE STREAMER MOVING OUT FROM A BACKING PLATE

(a) For $V_m < 0.35$

$$dx/dt_+ = 5.8\ t^{0.1}\ k^{0.6}\ x^{0.6}/(0.3d)^{0.6} \qquad x/d < 0.3$$

$$dx/dt_+ = 5.8\ t^{0.1}\ k^{0.6} \qquad x/d > 0.3.$$

These give

$$2.5\ x\ .485d\ \{(0.3)^{0.4} - (\delta/d)^{0.4}\} + 0.7d = 5.3\ t_m^{1/2}\ V_m^{0.6}$$
$$= d_{pp+}$$

for $v < .35$ and $x/d = .3$,

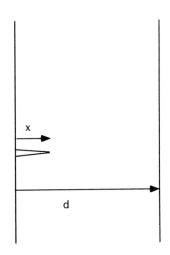

Fig. 7c-3 Sharp edge with backing.

where d_{pp+} is the distance a point plane gap will close over and δ is the initial stick out.

That is $d/d_{pp+} = 1/(1.45 - 1.21\,(\delta/d)^{0.4})$, $\delta/d \ll .3$.

On the above multiple assumptions and where there is no wait for the streamer to start, i.e. no threshold field needed, the closure distance is ~ 0.75 of the positive point closure distance. While the above treatment only holds for $V_s < 0.35$, in fact the same relation holds if the d_{pp+} is calculated from the positive curve given in the recent "streamer" note.

For $V_+ < 0.5$ MV $\qquad d \sim 6.6\, t^{0.5}\, V^{0.5} \qquad$ t μsec, V MV.

For $V_+ > 1.0$ MV $\qquad d \sim 7.5\, t^{0.5}\, V \qquad$ or a bit less V$\gg$1 *.

For the negative streamer a similar integration gives

$$\frac{d}{d_{pp-}} = \frac{1}{0.7 + 0.3\,\mathrm{Ln}\left(\frac{.3d}{\delta}\right)}.$$

Values for this are given in Table 7c-I. For $V \gtrsim 1$ MeV $d_{pp+} \sim d_{pp-}$ and hence a polarity effect of about 2 will be observed for $\delta \sim 3\times10^{-4}$ d i.e., $\delta \sim 10^{-3}$ cm, where the breakdown is streamer transit time dominated. Such a value is reasonable allowing for bubble motion with typical test waveforms.

Table 7c-I
d/d_{pp+} and d/d_{pp-} for various δ/d

δ/d	d/d_{pp+}	$d\backslash d_{pp-}$
10^{-2}	.8	.58
10^{-3}	.73	.42
10^{-4}	.7	.32
0	.69	0

---

* Editor's Note: Interpretation: Read as "or a bit less than this formula for V $\gg$ 1 MV".

COMPARISON WITH EXPERIMENT

The uniform field breakdown of liquids proceeds in at least these main phases, it seems to me. Firstly, some site (probably a whisker) with field enhancement gives rise to a bubble after a time strongly dependant on the field. This bubble then elongates in local field until its dimensions are very roughly of the order of 10^{-3} cms and it is pointed. Then the genuine water streamer starts from this. It has always appeared promising to me that any simple breakdown relation can apply over such a diverse range of phenomena. What the above "treatment" has attempted to calculate is the streamer transit phase. This should, therefore, apply to tests with smallish areas and fast pulses (and hence large fields) with rough electrodes. However as the area of electrodes is made very large it should provide an asymptotic value for the breakdown field, unless the electrodes are specially treated all over.

Table 7c-II lists some water breakdown data mainly taken from AWRE results (see "Further Breakdown Data Concerning Water", I.D. Smith, AFWL Dielectric Strength Note 13), but including a later data point from some stainless steel of area 3×10^3 cm^2. Also included is a lower limit value from MLI tests and also fast breakdown data from "Electron Beam Fusion Progress Report" January-June 1976, Sandia 76-0410 where Pace Vandevender's work is reported. Where it has been possible to obtain it, the standard deviation per shot is also noted. For areas of ~ 1 cm^2 or less σ was of the order of 13% giving the traditional $A^{-0.1}$ dependency. A word of warning is necessary, the values of $\bar{V}$, $\bar{t}$, and $\bar{d}$ are time averages of experimental ranges of the parameters. The mean experimental field quoted is that applying roughly to these conditions and $\underline{is}$ derived from the mean $Ft^{1/3}$ average. Thus don't work F out from $\bar{V}$ and $\bar{d}$, also no great numerical accuracy has been attempted. F_{pp} is the "lower limit" breakdown field calculated for the positive streamer as above.

Considering the results of Table 7c-II there is only one entry number where $F_{pp\delta} > F_{exp}$ and this is only 10% high if 0.70 were used. Entries 4 and 11-14 are fast breakdown ones with no special attention paid to the electrodes and here the agreement is good. The other entries where the field values are close are 5 and 9, both fairly large areas, and entry 6 where the electrodes could have been fairly gritty and damaged from the testing.

Also of possible revelance for results 4 and 11-16 a $t^{0.5}$ dependency was found. These are the points in favour. Those against are No 11-14 and show a significant standard deviation. If the streamer transit calculations are right, the σ should have been small. However there are a number of possible explanations: (a) there is a dependancy on the exact wave form shape, (b) the test plates were horizontal and very small bubbles might have been

LIQUID BREAKDOWN

Table 7c-II
Water Breakdown Data

No.	A cm²	$\bar{V}$ MV	$\bar{d}$ cm	$\bar{t}$ μsec	E_{exp} kV/cm at $\bar{t}$	$F_{pp\delta}$ kV/cm at $\bar{t}$	σ % per shot
1	2200	.3	1.5	.5	200	130	4-5%
2	120	.3	1.0	.3	280	170	
3	400	1.0	4	.4	260	200	
4	50	.4	0.6	.02	700	700	
5	1000	.9	5	.6	180	160	
6	100	1.5	7	.35	230	220	
7	6	.4	1.0	.2	420	230	
9	3000	1.85	10.5	.75	160	150	4-5%
10	55000	2.1	~15	.5	≥150	~175	
11	1000	.63	.6	.0085	1050	1250	~ 9%
12	1000	.53	.6	.010	870	870	~ 9%
13	1000	.98	.95	.016	1030	970	~ 9%
14	1000	1.1	1.27	.027	870	770	

There are a number of additional comments.

No. 1   I reanalyzed the original data and decided $Ft^{1/3} = .16$

No. 10   This was a cylindrical line test and as such the + ve streamer would be speeding up. $F_{pp\delta}$ has been decreased to allow for this approximately.

No. 11-14 Here the pulse length was very short and Table 7c-I indicates there is a weak dependency of $F_{pp\delta}$ on δ. The value 0.75 was used but 0.70 could be a better figure.

trapped for some shots, (c) backfiring from the -ve plate might have occurred, i.e. as the +ve streamer nears the -ve plate a streamer starts from this and links up.

For experiment No. 4, while a $t^{0.5}$ dependancy was found the results fit an F dependancy. Again this might just have been a subtle dependancy on waveform shape but it does not appear too likely. Summarizing, a possible streamer transit dominated limit might have been found for large area water breakdown fields. This has a different time dependancy from the standard one and also for voltages under 1 MV it is not solely dependant on the field. If the plates are very carefully treated values of the breakdown field well above this limit may well be on, but it is of interest that without such treatment large areas may well not obey the present area extrapolation formulae, but will asymptote to a limit.

Any numerical values deduced from the above treatment should be treated with great reserve until they have been checked experimentally. This I hope to do indirectly in the not too distant future. Meanwhile any reasonably clean data on point-plane or uniform field breakdown would be very gratefully received by myself especially it it involves large gaps (point plane or near) or large areas.

INDEPENDANT EMPIRICAL ANALYSIS

I include this only for completeness. I hoped that it would lead to the same answers. Needless to say, it didn't in detail.

The data listed in Table 7c-II falls into a short pulse group ($\bar{t} \sim .02$) and one that averages out at a $\bar{t}$ of about 0.4. The breakdown fields of all the data was converted to that applying to a time of 0.6 using a $t^{1/3}$ time dependency. The resulting data is plotted in Figure 7c-4, the short pulse data being ringed. The numbers besides the points is the mean separation at which the data was obtained. Also included in a mean breakdown field at an area of 1 cm^2, from Ian's note. At this area and lower the standard deviation per shot was about 13% giving rise to the traditional area dependancy of $A^{-0.1}$ which is the slope shown passing through this point. The points labeled 1.5 and 10 both had abut 5% standard deviation and the corresponding area dependancy is shown as dotted tangents. Using the eye of truth and ignoring the ringed points the lines are drawn as possible indications of the trend towards large areas. The mean voltages to which they correspond are very roughly 0.35 MV and 1.4 MV. These values and t = 0.6 μsec are used to calculate the corresponding transit time limited fields and these are the lines on the RHS of the figure. The value for V ~ 0.35 MV is reasonable since the data time corresponds to the mean breakdown field whereas the transit limit should be a little lower.

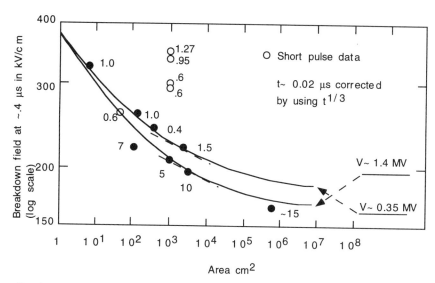

Fig. 7c-4 Breakdown field versus area at t=0.4 μs.

However, the limit for 1.6 MV is above the extrapolated experimental. In view of the approximations used in the derivation at the beginning of the note the disagreement is not too bad. However stretching credibility well beyond the limit, if the negative higher voltage positive streamer voltage dependance were a rather higher power than 1.0 (say the 1.1 which George Herbert originally measured) this disagreement would largely disappear.

SUMMARY. A possible large area limiting breakdown field relation is suggested based on the streamer velocity relations. An extrapolation of the existing water breakdown data is not badly in disagreement with the results of the suggested relation. In addition Pace Vandevender's largish area short pulse breakdown data are in reasonable agreement with the crude theory.

However the relation badly needs checking more directly and should only be used in desperation.

Section 7d

INTERIM NOTES ON WATER BREAKDOWN*

T.H. Storr
and
J.C. Martin

INTRODUCTION

These very preliminary notes have been brought on by the imminence of a trip to the USA by J.C. Martin: they should not be treated as a considered report even of the rather low standard to which our usual notes aspire.

The work started out with a very clear picture of what was needed to be done and what the results would roughly be. The ten year old water streamer breakdown relation of George Herbert gave a relation of the form $d = 8.8 \, V_{mv}^{0.6} \, t^{0.5}$ for the positive streamer.** Deriving a differential relation and using a crude assumption as to how the field enhancement factor varied as a very sharp needle is retracted into a backing plate, suggested that the streamer transit time became very significant for points with small stickouts. Thus the mean breakdown field of a sharp needle would be expected to increase fairly rapidly over that of a point-plane gap as the needle is retracted into the plate. Thus the experimental programme was going to re-investigate George Herbert's old data (taken with 1-cos waveforms), then look at the breakdown of very sharp points as they were retracted into a plate and show that useful fields resulted providing points longer than ~ 0.01 cm were prevented from appearing on the plates. These breakdown fields would then presumably represent limiting breakdown fields for very large area plates. Thus with our hearts hopeful and heads held high we started out. This note records the steps by which we have

---

* Editor's Note: This reference is appendix to Section 7b.
** SSWA/JCM/785/147, May 1987

gone backwards to our present position, with bloodied and bowed heads, sadder but hopefully a little wiser.

Because the work is still in progress, not all the data has been used to derive the approximate integral velocity relations given below and therefore the fits will change slightly when this has been done. In addition, as the work has progressed we learned how to make "sharper" points and some of the earlier data is slightly contaminated by blunt point effects and needs to be repeated. In nearly all that follows only the positive point breakdown has been studied: a small amount of work was done with negative points and the little that is reported was done very early in the programme.

EXPERIMENTAL SET UP

Initially a few-kilojoule Marx capable of working up to over 300 kV was built. This was used in two different ways. It either directly applied a "square" waveform to the point-plane gap with a decay time of about 50 μsec, or by using a water capacitor could give a 1-cos waveform up to about 450 kV with a range of different periods. The energy deposited in the point-plane gap could be quite spectacular with 6 cm gaps and the needles were good for one shot only under the conditions. Later, partly because of this and partly in order to achieve 20X magnification optics, a small 50 kV system was built with an auxiliary gap which (a) crowbarred off the volts before the main gap completed its breakdown; and (b) provided the light to take the picture after the streamer had been frozen in space. This later system, too, could be used to provide square waveforms or 1-cos ones.

The needles used have been Singer hand sewing needles No. 8, which have a tip radius of about .001 cm as new. The tips rounded off after a few shots and it became necessary to re-sharpen them periodically, even in the square waveform tests, and with the shielded or retracted point experiments every time. One of us (THS) did prodigious feats in making ever sharper needles by grinding, finishing up with long tapering needles with a tip radius of about .0005 cm. In the small system, when it was crowbarred before complete breakdown, such tips survived for many shots, but on the larger system they had to be re-ground each time to achieve reproducible results. However, even these tips did not guarantee an essentially "instantaneous" start to the breakdown phenomena when shielded by a close backing plate. To achieve this, it was found that they had to be covered with a thin layer of shellac. Optical photography showed that breakdown phenomena started on such coated needles at very early times in the voltage waveform, and at voltages well below those which could cause complete breakdown of the gaps, even at the smallest spacings. Ideally all the data we have

LIQUID BREAKDOWN

should be repeated with the most recently developed needles, but in any experiment that involves complete breakdown of the gap the needles have to be re-sharpened and re-coated, and life is short. However, it is intended to repeat some of the earlier measurements where it is believed needle bluntness had a small effect on the results, when there is time.

RESULTS WITH SQUARE WAVEFORM

Figure 7d-1 shows the data obtained with the large system with square pulses for point-plane gaps. The rise time of the pulse was a few tens of ns and there was a 25% overshoot on the front which

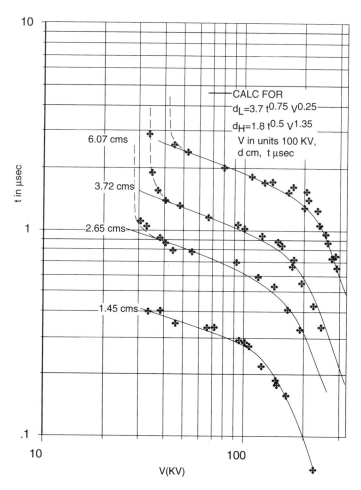

Fig. 7d-1 Breakdown times in water with position point, square waveform.

was rapidly damped. As can be seen, these tests showed two very distinct phases of breakdown; a low voltage (L) phase only very weakly dependent on voltage and a higher voltage phase (H) where the breakdown time became much more dependent on voltage. George Herbert's old data had been obtained with 1-cos waveforms and this had obscured the details of the two breakdown stages and provided an intermediate power for voltage dependency. Given in Fig. 7d-1 are two approximate relations which give differential expressions which can be integrated, with the additional assumption that the two breakdown modes swap over when the two velocities become equal, to give the curves shown in Fig. 7d-1. The reader is warned that because a good fit has been reasonably obtained using the differential relations derivable from the integral ones, this does not prove that either is necessarily applicable to any other waveform than square ones.

Also the data obtained with very much smaller gaps using the small system do not agree exactly with the quoted fits. However, for the most important low voltage relation the error is not more than 15% in t for spacings changing from 0.1 cm to 6 cm and this difference may in part be due to the effect of the damped ring on the front on the large system waveform.

The disaster for the original plan is that for low voltages the voltage power dependency is 0.25, not 0.6. This indicates that the streamer transit times are much shorter than were initially calculated. In particular, as the needle is shielded there will be little improvement in breakdown field over that of a point-plane gap, even if the stick-out is very small.

However, all was not lost (or so it seemed) since also shown in Fig. 7d-1 there is a threshold voltage and a region just above this where there was a definite phase of transition to the low voltage relationship.

A word of explanation about the effects at threshold is necessary. Breakdown would take place at or just below this voltage, but it was very delayed (typically to 50 μsec or more), the final closure occurring a long way down the decaying waveform. If the voltage was dropped a little more, no complete breakdowns occurred, but light was emitted at the tip, showing that some streamer formation was occurring ,but the intrinsic breakdown processes that were taking place were freezing in space before reaching right across the gap. In view of the low voltage photographic data, it is considered likely that the very late closures involved a long bubble expansion phase with a final fast streamer closure when the bubble had closed over most of the gap.

LIQUID BREAKDOWN

NEGATIVE POINT DATA, SQUARE WAVEFORM

This data was obtained early on and should be treated with a little reserve. Again, a threshold voltage was observed and above this a distance-time relation found not very different from that of the high voltage positive breakdown. Again, closure can be obtained a little below the "threshold" voltage, but only for very long times compared with those shown in Figure 7d-2. Photo 1 shows a small gap negative point operated just below threshold and indicates a very big bubble growth phase.

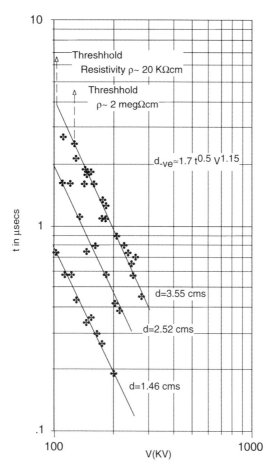

Fig. 7d-2 Breakdown times in water with negative point square waveform.

DEPENDENCE OF "THRESHOLD" VOLTAGE

For square waveforms the threshold voltage had a weak dependency on the water resistivity, the threshold voltage decreasing slowly as the resistivity was reduced. For positive points the threshold voltage is:

$$V_{T\ 100\ kV} = 0.2\ d^{0.4} \quad 0.1 < d < 6\ \text{cm for 2 megohm water*}$$

and

$$V_{T\ 100\ kV} = 0.15\ d^{0.4} \quad \text{for 20 kilohm water.}$$

For the negative points, based on much less data:

$$V_{T\ 100\ kV} \simeq 0.9\ d^{0.3} \quad \text{for 2 megohm water and for } d = 0.2\ \text{cm}$$

and $1.4 < d < 3.6$ cm.

Again, a weak dependency on water resistivity was observed of roughly the same magnitude as is given for the positive point.

All the results discussed from this point on are for positive points.

POINT-PLANE GAP, 1-cos WAVEFORM

Figure 7d-3 shows the data obtained on the large system with a 1-cos waveform, with $\sqrt{LC} = 0.44$ μsec. In this the breakdown voltage is plotted against time (measured from the start of the waveform) in microseconds. Also shown as the solid curve is the breakdown calculated from the integral relations given in Fig. 7d-1. The agreement is quite good, especially for the larger spacings and higher voltages. Apart from the highest voltages, essentially only the low voltage relationship applies. However, for voltages about 400 kV the theory suggests the streamer will start in the low voltage phase, switch over to the high voltage one and then switch back to the low again. Thus the point-plane breakdown will become more dependent on voltage and shorter closure times will result for a given spacing than are indicated by extrapolating the gently sloping portions of the curves in Fig. 7d-3.

---

* Editor's Note: In these emperical equations, note that voltage is in units of 100 kV.

LIQUID BREAKDOWN

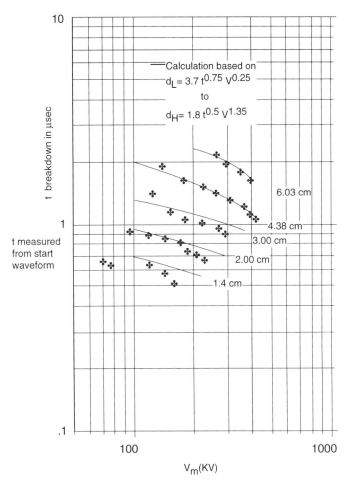

Fig. 7d-3 Breakdown times in water with position point 1-cos waveform ($\sqrt{LC} = 0.44$ μsec).

A rather smaller body of data was obtained for $\sqrt{LC}$ = .24 μsec and $\sqrt{LC}$ = .77. These are given in Figs. 7d-4 and 7d-5. The calculated curves are shown for 2, 4 and 6 cm spacings. For the 0.24 μsec data the fit is not good and in this graph the experimental points for 2, 3, 4, 5, and 6 cm are joined by dotted curves. For $\sqrt{LC}$ .77 the agreement is reasonable.

At this stage in the experiments with the large system some suppressed point plane gap data was obtained. The scatter of this data was significant, in part due to the fact that the improved coated sharp needle had not been developed. This data is referred to towards the end of this note. However, before this some of the aspects of breakdown near the threshold voltage will be briefly discussed.

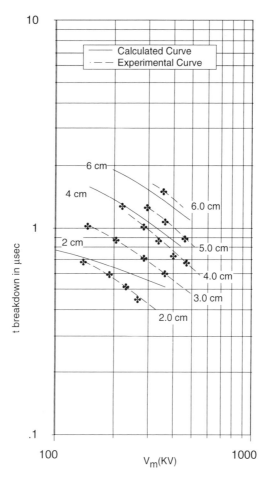

Fig. 7d-4 Breakdown times in water with positive point, 1-cos waveform (RC = 0.24 μsec).

BUBBLE PHASE

Referring to Fig. 7d-1, just above threshold voltage there is a region where the breakdown time is falling towards the low voltage streamer phase curve. It was felt that this was due to a bubble phase, as indeed it turned out to be. The small system was then built and 20X magnification optics developed. This showed that there was indeed a bubble phase which was observable for voltages over threshold and relatively short times (tens of ns) - see photo 2. If the square pulse was left on for a longer time a branched streamer, moving much faster, appeared from it. See photo 3.

LIQUID BREAKDOWN

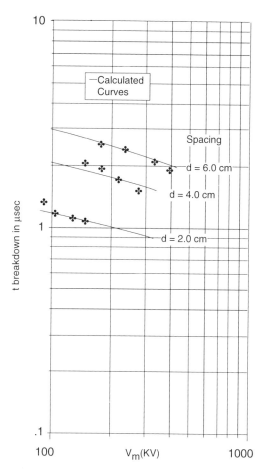

Fig. 7d-5 Breakdown times in water with positive point 1-cos waveform ($\sqrt{LC} = 0.77$ μsec).

The apparent velocity of expansion of the bubble was between 1 and 2×10^5 cm/sec, depending on the voltage applied. The photographs of the bubbles never showed specular light reflections but looked opaque. It is considered that the bubbles were very corrugated, rather like a shaving brush probably, and that the velocities of 1 to 2×10^5 were peak velocities of the gas/vapour interface. The actual water velocities (at right angles to the advance of the "shaving brush" hairs) could well have been only 1 to 2×10^4 cm/sec, which would agree with rough estimates of the stresses available in the electrostatic fields.

The situation is thus a sort of Rayleigh-Taylor instability one, with the gas/vapour interface moving much faster than the actual mass velocity of the water. At some stage in the development of the bushy bubble the field on the tip of the gas/vapour tapering channel reaches the very high value necessary to start the intrinsic water breakdown process. This process could be a u.v. one or, more plausibly, the protonic breakdown suggested by Yanshin et al. (1). The origin of the bubble would seem to be either cavitation behind the shock emitted from the metal/water interface as the voltage is applied, or, much less likely, local boiling.

In practice, the breakdown curve just above threshold had a considerable scatter and this was considered to be due to variability in the degree of crumpling of the bubble and hence the time at which the intrinsic breakdown threshold field was reached. Because of this, further efforts were made to sharpen the needle and indeed the scatter then reduced, with the breakdown times becoming shorter, as would be expected. It appeared that with sharp enough needles a significant bubble growth phase was necessary and represented a limiting phase which could be used in calculating the suppressed plane breakdown results. Some effort was also applied to finding the water conditions (resistivity and degree of air saturation) which led to the promptest breakdown, i.e. shortest bubble phase. The conditions for fast transfer from bubble growth to intrinsic breakdown were tentatively established as high water resistivity and either high water temperature or, more reasonably, highest degree of air saturation of the water. However, at this point in the investigation, during a rather large number of shots showing the bubble phase, two pictures showed streamer development without any significant bubble being apparent. There were also a couple of breakdown times significantly below the "bubble development" curve. Thus it became apparent that the relatively large bubbles were not essential and that intrinsic streamer breakdown could develop without much, if any bubble growth at all.

It was at this time that the shellac coating technique was developed and this regularly gave rise to well developed streamers at times when sharp but uncoated needles had only a bubble on their tips, see photo 4. Indeed, at voltages below half the threshold field significant streamer development could be observed. Even more interesting, streamers could be seen developing off the shank of the coated needles, where the fields would be way down on that at the tip. With uncoated needles bubbles always formed at the tip in the highest field: even bubbles intentionally introduced at the shank did not grow, crumple or give rise to streamers. Thus the potential delay in bubble formation bit the dust.

Purely as a speculation, the above observation may explain the results of pressurised water experiments. These showed that for small areas, pressures of 10 to 100 atmospheres increased the breakdown fields (particularly for long times). However, as the area is increased, the improvement disappears. The effect of pressurisation is to stabilise the bubble by suppressing its initiation

LIQUID BREAKDOWN

or by reducing its growth so that it cannot crumple enough to give rise to the extra field enhancement factor necessary to initiate intrinsic breakdown. However, as the area of the plate increases enough it will be possible to find a whisker sharp enough to provide the necessary fields on its own without any significant bubble phase. Another possibility is that a dielectric inclusion may have somewhat the same effect as the shellac trick. In any event it is to be expected that under these circumstances pressurisation will cease to have any significant effect for large areas, as is observed.

Anyway, since we were still intent on trying to establish some limit to the breakdown field for large areas, we stopped investigating the bubble phase, fascinating as it is from the physics point of view.

SUPPRESSED POINT PLANE BREAKDOWN, 1-cos WAVEFORM

While some results for small gap suppressed point breakdown were obtained with square waveforms, most of the data has been obtained with a 1-cos waveform and only with this waveform for cm gaps and above. These experiments are fairly tedious, as new coated needles have to be used for each shot. At the time of writing this note, only $\sqrt{LC} = .44$ μsec has been re-done with these points. This data agrees reasonably with the older data referred to earlier. The picture for different stick outs and different breakdown times is complex and has not yet been fully analysed, however the scatter of the data is satisfactorily small, indicating that the coated needle has regularised things satisfactorily.

The voltage necessary to give breakdown exactly at waveform peak has been extracted from the data for a relatively small range of gap spacings and is shown in Fig. 7d-6. The geometry is also shown in the figure.

As is to be expected, as the stick out (δ) is reduced, the on-peak breakdown voltage increases. A few trial calculations using the streamer relations and the threshold fields have shown reasonable agreement with the observed curves and in one case with the time dependency when the breakdown is earlier than peak volts. Over peak volts the picture is more complex. For the smaller spacings, as the voltage begins to fall the breakage over peak becomes very unlikely. For larger spacings there can be some breakdowns over peak.

In an attempt to provide similar curves for $\sqrt{LC} = .24$ and .77 μsec, the earlier and less consistent data obtained with blunter uncoated needles has been massaged and the results are

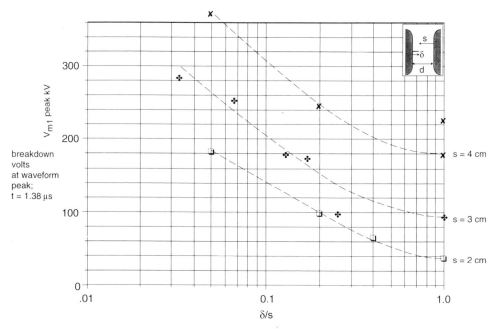

Fig. 7d-6 Suppressed point 1-cos waveform, $\sqrt{LC}$ = 0.44 μsec.

tentatively given in Figs. 7d-7 and 8. These should not be relied on too much and the original experiments will be repeated.

If they are correct, the curves strongly suggest that for a constant small stick out the limiting breakdown field is a much more powerful function of $\sqrt{LC}$ than the traditional $t^{-1/3}$. Also, if the curves are wildly extrapolated to small values of stick out, the observed breakdown fields for t ~ 1 μsec and areas of the order of 1000 cm² correspond to very sharp whiskers of about 10^{-2} long. This deduction is supported by three shots where the water capacitor mentioned earlier was intentionally broken. This had an area of 1000 cm² and gave a mean $Ft^{1/3}$ of 153 (t here is the old 63% effective time). The area of 1000 cm² was in the uniform field part of the plates, where the brass anode had been severely roughened by three different methods. This we felt ensured some stick outs of considerable sharpness with lengths up to, but no more than, a few thou. (mils). The plate separation was 2 cm and the breakdowns were near peak volts with a 1-cos waveform with $\sqrt{LC}$ = 0.44 μsec. It is planned to repeat the tests some time, thinly shellacking the plates so as to ensure the initiation of intrinsic breakdown very early in the waveform.

LIQUID BREAKDOWN

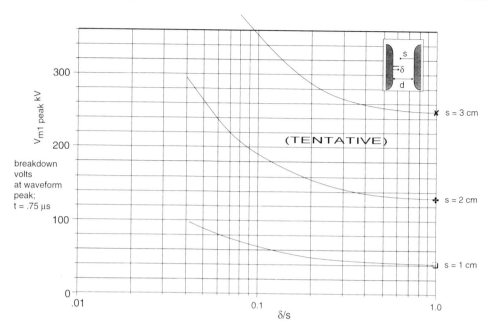

Fig. 7d-7 Suppressed point 1-cos waveform $\sqrt{LC}$ 0.24 μsec.

BRIEF DISCUSSION

It is felt plausible, but not certain, that we have developed a technique that ensures that electronic or protonic breakdown processes start at a very low voltage. Curves are being obtained which will show the effect of moving a sharp point back towards the plane of the electrode. It is still faintly hoped that the experimental curves can be matched by crude calculations based on the integral streamer closure time relations. Attempts to do this using the threshold voltage look very mildly encouraging. The reason for trying to do this is to give added credence to the claim that we have found the fastest breakdown mode possible (barring someone leaving a monkey wrench in the water capacitor). It appears as if it will be necessary to ensure that whiskers longer than 10^{-2} cm or so will have to be either removed or blunted to start with and then subsequently prevented from regrowing. The objective must therefore be to devise means of ensuring for areas of many thousands of square metres of plates that there is not one sharp protruberance longer than, say, 10^{-2} cm.

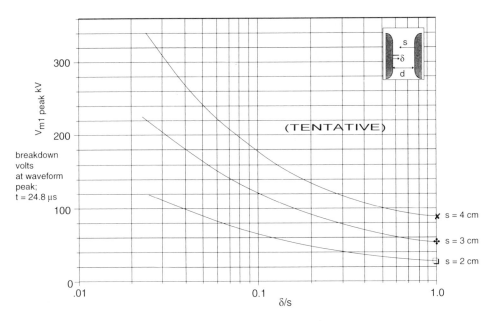

Fig. 7d-8 Suppressed point, 1-cos waveform $\sqrt{LC} = 0.77$ μsec.

Before moving on to this phase of the experiments we hope to establish a bit more about the threshold voltage and what growth-controlling mechanisms there are which cause the intrinsic streamer to "freeze" on its way out from the point. This is to aid in deducing the relevant threshold voltage for a point backed by a plate. At the moment we are using a length equal to half the physical stick-out of the whisker in our crude calculations.

CONCLUSIONS

While we are still some way from achieving our original objective of establishing large area limiting breakdown fields and their dependency on time and spacing, we feel that we have made progress. Various potential limiting processes such as streamer velocity, bubble phase, etc., have failed to provide reliable or useful methods of raising these fields substantially over that of a point-plane gap; a combination of "threshold" field and the intrinsic streamer transit phase appears to offer hope. As we see it at the moment, the consequence of this picture is that it must be ensured

---

Editor's Note: Originals of Figures 7d-6 through 7d-8 had δ/d instead of δ/s.

that the plates are free of very sharp projections of lengths greater than about 0.01 cm at all times. The function dependence of the limiting fields is much more strongly dependent on the time of the pulse than the traditional dependency of $t^{-1/3}$, suggesting that fast charging of the water for very large areas pays off more than was originally thought, or vice-versa.

A wealth of interesting physics has been either confirmed or discovered and it is frustrating in the extreme not to be able to investigate these. Finally the warning is given that all the tests we have done so far have been with voltages of 400 kV or less. The square waveform tests confirm that there is a transition from the low voltage relatively slow breakdown to a much faster higher voltage process which is fairly closely similar to the negative streamer. This implies that for high voltages the limiting fields may not be very good unless whisker suppression or blunting methods are very effective. Pressurising or degassing the water would be expected to have beneficial results, as is observed, but as the areas are increased these techniques are predicted to have less and less effect, as breakdown not involving a significant bubble phase becomes statistically more likely.

ACKNOWLEDGMENT

As ever, our warmest thanks go to Mrs Vikki Horne, who is not only having to cope with the usual last minute desperate rush but also an unfamiliar electric typewriter. Life is not fair, but all we can do is offer our deepest condolences.

REFERENCES

1. E V Yanshin, I T Ovchinnikov, Y N Vershinin, Sov. Ph. Tech. Phys. 13, 1974, 1303-1306.

216   CHAPTER 7

Needle shape and scale.

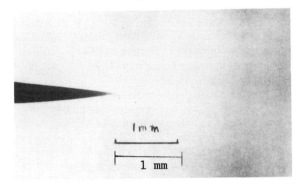

(point plane gap = 3 mm)

Photo 1   -ve point; 20 kV, 4 μs

+ ve pt at 11 kV:

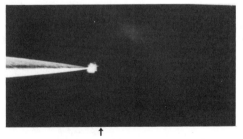

↑
Bubble Phase

Photo 2 ~ 70 ns

LIQUID BREAKDOWN

+ ve pt at 11 kV:

Photo 3 ~ 120 ns

Shellac-coated +ve point:

Photo 4 11 kV, ~ 20 ns   6 kV, ~ 35 ns

APPENDIX

Included at the last moment with this note are 12 further microphotographs by way of illustration of some points in the text. all were taken with the small 50 d0 kV system, using an auxiliary crowbar gap, and are all photos of nominal .0005 cm tip radius needles, 3 mm from a plane cathode.

Photos 1-4 show four stages in streamer growth across the gap, originating from the bubble phase (photo1) and terminating in the growth of current in a completed streamer (photo 4). Each shot in the series, as elsewhere, involved a separate application of the voltage pulse and a different setting of the auxiliary needle gap: they are not consecutive frames in the growth of a single streamer system, as would be produced by a framing camera.

Photos 5-8 shoe the remnants of streamers which have 'run out of steam' due to the application of a sub-threshold level of voltage (1-cos waveform). The growth of bubbles along the tracks can be seen progressing as the sample time is extended. The needles in these shots were shellac-coated.

Photos 9, 10, and 11 show the distinction in the very early growth between 'bubble' and 'intrinsic' mode; note in photo 11 that static air bubbles on the needle shank do not give rise to growths. (Uncoated needles).

LIQUID BREAKDOWN

PHOTOS

STREAMER GROWTH FORM THE BUBBLE PHASE

+ ve step voltage, 20 kV.
point-plane gap - 3mm
scale - 20:1 (2 cm = 1 mm)

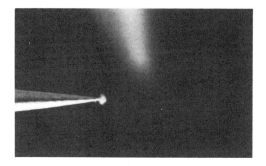

Photo 1   .018 μs

Photo 2   .035 μs

Photo 3 .046 μs

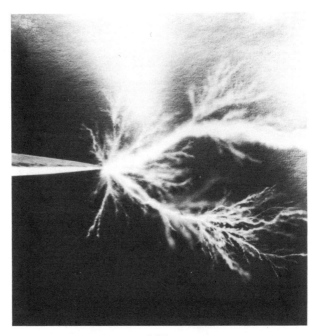

Fig. 4 .067 μs

LIQUID BREAKDOWN

STREAMER 'DEATH' BELOW V THRESHOLD (1-COS WAVEFORM)
point-plane gap, 3mm; Scale 20:1

Applied Voltage

Photo 5

1.06 μs

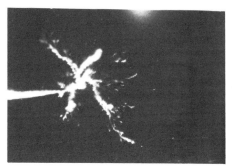

Photo 6

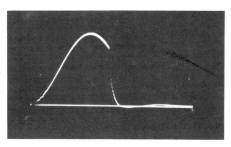

1.7 μs

Photo 7

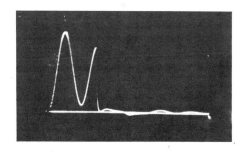

3.8 μs

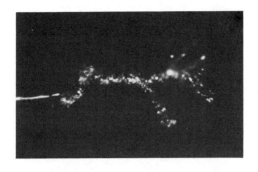
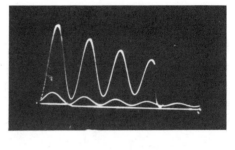

Photo 8

9.4 μs

Various Growths from 5 μ rad. point, 20 ns after + 11 kV step

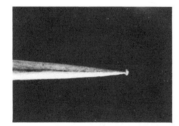
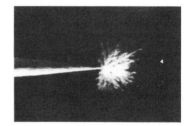

Photo 9 'Bubble' Phase; vel. ~ $2 \cdot 10^5$ cm/sec

Photo 10 Ocassional fast intrinsic' mode; vel ~ $3:10^6$ cm/sec (~10% of shots)

LIQUID BREAKDOWN 223

Photo 11 Fast mode in warm,
          (~ 50° C) gassy water
          (~ 60% of shots)

Section 7e

POINT PLANE BREAKDOWN OF OIL AT VOLTAGES ABOVE A COUPLE OF MEGAVOLTS*

J.C. Martin

I have excavated the graph in which I plotted the breakdown data of point plane tests in oil. On it is our old data for +ve to -ve points. Then there is some data I derived from a point plane dump gap in the Facility 1 large area oil breakdown tests done at P.I. during the Aurora tests. In addition there are some higher voltage points which come I think from PI but I have been unable to locate further data about them. I have a faint recollection that there were some prepulse oil gap data which would mean t_{eff} would be small and d ~ 10 cms. I do recollect that at least some of the data was for both polarities which led me to speculate that the breakdown mean velocity became polarity independent for a high enough voltage.

The fit of the $\bar{u}\, d^{1/4} = 80\, V^{1.6}$ is not the most excellent in the world but it should be borne in mind while looking at the PI data for gaps between 35-65 cm and 50-60 cm that the gaps get longer in those bands as the voltage is increased and hence the slopes of these curves are larger than the ones shown on the figure.

Sorry about the rather unsatisfactory nature of the data, maybe there is some better stuff around or maybe new machines can be made to yield some reasonably clean data. Incidentally, I would be very grateful for any water data in the 100-1000 ns range for voltages over 2 MV for reasonably unshielded points of adequate sharpness. By this I mean they should not be too sharp if there is any prepulse around. If there is no prepulse, then the sharper the better.

---

* Hand Written Note
   No Date

226   CHAPTER 7

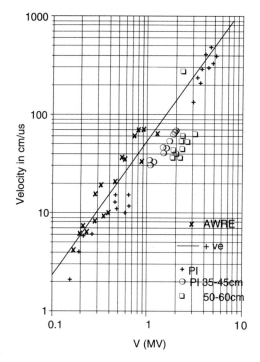

Fig. 7e-1   Streamer velocity in oil.

Chapter 8

Solid Breakdown

Section 8a

VOLUME EFFECT OF THE PULSE BREAKDOWN VOLTAGE OF PLASTICS*

J.C. Martin

"Intrinsic" Breakdown. This rather ill-defined quantity is practically the largest field that any sample of the material can be made to stand. Typically, the determination is made with D.C. and the applied voltage stresses a few thousandths of an inch thickness of the material. Thicker samples are not used because heating within a thicker sample leads to a runaway condition where the temperature rises: this leads to larger conduction currents causing more heating, etc. This difficulty can, of course, be avoided by using pulse voltages, as can corona-induced surface degradation, chemical reactions, and mechanical instabilities. A further effect that pulse voltage avoids is a sort of stress relieving which conduction currents at near breakdown fields can cause. In this effect, a localised defect conducts more readily than its surrounding material and this leads to a field lower than average in its vicinity. Incidentally, if the D.C. voltage is removed rapidly and applied the other way round, a reduction in breakdown voltage is observed, because the stored charge now adds to the field. This effect can raise the D.C. breakdown by up to 20% compared with the pulse values, while the "reversed field" breakdown after D.C. conditioning can fall to well under one half of the pulse value. The time scale of this effect is a few milliseconds.

---

* 16 November 1965
AFWL Dielectric Strength Notes, Note 3
SSWA/JCM/6511/A

Work at AWRE has shown that the mean breakdown voltage of a given sample does not depend on the rate of application of voltage when this is reduced from the microsecond range to that of 10 nanoseconds. If some transit time effect were happening (such as is observed in liquids), reducing the time scale by this large factor would produce a large increase in the measured breakdown field. In fact, any observed correction was less than 10%, implying that in the microsecond time scale no transit time effects are occurring. Thus pulse charge methods of determining the "intrinsic" breakdown voltage are to be desired because of the number of interfering effects it avoids. In addition, if pulse voltages are employed, the difficulty of applying the voltage to the sample (which is another major source of error in the normal D.C. methods) can be largely removed by using dilute copper sulphate solution. By selecting the conductivity, the thin film of solution under the flat electrodes ensures that at least 99% of the voltage is across the insulator, while at the same time conduction grading at the edges of the plate electrodes can be used to prevent any field enhancement there. When proper precautions are taken, all the breakdowns can be made to occur at random over the area stressed, with no tendency to cluster at the edges of the electrodes.

Using the above methods, the mean breakdown field for various areas of four plastics were measured. Because pulse voltages were used, it was possible to test large volumes and up to 10 litres of polythene could be stressed in any one test. The results obtained are shown in Figure 8a-1. All the four materials gave values a lot lower than those quoted in the literature for "intrinsic" breakdown and in addition showed a considerable area (or more reasonably volume) effect.

Consideration of this data led to the idea that if the mean breakdown fields, measured for a reasonable number of samples, have a standard deviation which is not a result of purely measuremental errors, then in the region of the measured point there must be a change of mean breakdown field with the volume of the test substance. This can be shown by considering an ideal experiment where the mean breakdown field of a large number of samples is known as well as the intrinsic standard deviation. The samples are then stacked or laid out in sets of ten and the question is asked what the mean breakdown voltages of the sets of ten units are. Obviously in each group the unit with the lowest strength will break as the voltage is applied and the new mean field will correspond to the voltage which broke only 6.5% of the samples when they were tested one at a time. This means that the mean breakdown field must decrease as the volume of the sample increases, providing there is any non-instrumental scatter of measured values of the samples.

SOLID BREAKDOWN

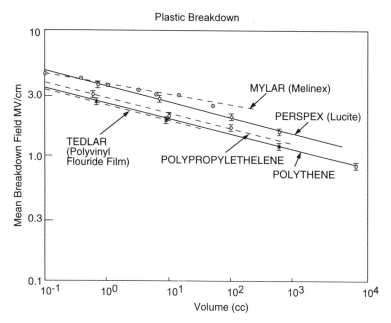

Fig. 8a-1 Plastic Breakdown*.

It seems a reasonable assumption to make that the probability distribution curve of the breakdown voltage of a set of samples is independent of the number of units in the subsets, at least over a restricted range of number of units. If $\bar{V}_1$ is the mean breakdown voltage of a set of samples and $p(V/\bar{V}_1)$ is the probability that a given sample will have broken when a voltage V is applied, then 1 - p is the probability that it will not have broken. Then, if the samples are taken N at a time, the probability that the group won't have broken is $(1 - p)^N$ and the following relation is required where V_N is the mean breakdown voltage of the subgroup of N samples per set:

$$\{1 - p\ (V/\bar{V}_1)\}^N = 1 - p\ (V/V_N).$$

It is not true that there is only one function p which satisfies this relation, but <u>a</u> function satisfying it can be easily generated by two methods at least, which each lead to the same answer. One is by taking a Gaussian distribution and considering successive sets of, say, 4 samples at a time. After three or four iterations a self-replicating probability curve is obtained. The other method is to take an approximate curve and use the top half of this to

---

* Editor's Note: Original has "Mean Breakdown *Voltages*" on vertical axis.

generate the bottom portion of the curve for a larger volume sample and the lower half to give that for a smaller volume. Figure 8a-2 shows an integral and differential probability curve which satisfies the requirement that it will self-replicate, which has been generated in this way. For an increase or decrease of a factor of four, the shift of the mean value is 1.18 times the standard deviation of the distribution. Both AWRE and other workers obtain an intrinsic standard deviation of about 11%, when allowance is made for the small number of samples tested, where this applies. Using the observed slope of the breakdown curve of 1.15 for an increase of volume of a factor of 4 gives a calculated standard deviation of 13 1/2, showing a reasonable agreement. If a reasonable number of samples are tested, two pieces of data can be obtained; the mean breakdown field and the slope of the curve of breakdown field against the volume, in the vicinity of the volume of the sample.

Figure 8a-3 shows breakdown data applying to a large range of volumes. It would have been desirable to use only pulse voltage values but these are limited in number and two sets of D.C. data are included, as well as some pulse values showing a volume effect obtained by Cooper et al. The two points obtained by J. Mason should be treated with reserve, as the yield strength of the polythene is close to the electric stress produced by the observed fields and hence the radius may be in doubt. In the case of the stabs and needles the positive breakdown values are quoted, since there is a polarity effect. The interesting feature of Figure 8a-3

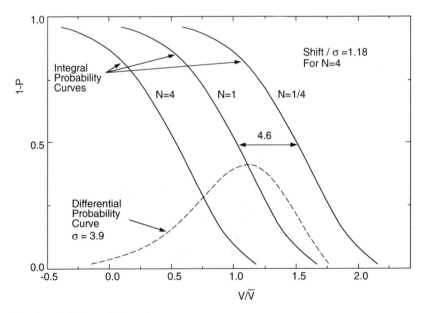

Fig. 8a-2 Self replicating curve.

SOLID BREAKDOWN

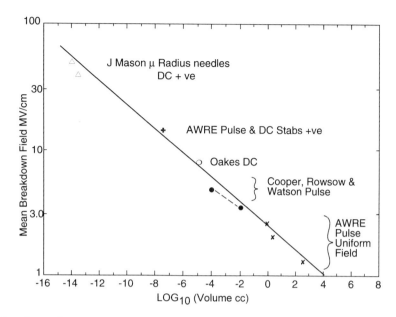

Fig. 8a-3 Polythene Breakdown.*

is that the slope is apparently constant over a huge range. Indeed, if the curve is taken back to atomic volumes, a reasonable value of 2 eV is obtained for the gradient across a molecule required to break it.

One point that has yet to be mentioned is the evidence that the effect is a volume one and not an area dependence. Two sets of data, one for mylar film 1 to 10 thou. and another for polythene from 1/16" to 3/8", show the expected dependency on thickness and hence volume. The suggested dependency is also made reasonable by consideration of the inherent defects likely to be in the plastic which are either crystals or chain molecules, particularly orientated with regard to the field. For liquids one would expect (and the experimental evidence suggests) an area effect since the discharges are seen to originate on the electrodes.

Additional evidence of the volume dependence of the mean breakdown voltage is provided by some experiments with a thin polythene film over a plane electrode with about 1 cm of water between it and a second electrode. A negative voltage pulse is applied to this electrode, when about half the voltage appears across the 5 thou. polythene film because of the large ratio of dielectric

---

* Editor's Note: Original has "Mean Breakdown "<u>Voltage</u>" on the vertical axis.

constants. However, when the polythene film first breaks at a voltage of 40 kV across it, the voltage pulse still rises because the slow moving streamer in the water takes several tenths of a microsecond to cross the gap. Thus it is possible to double the applied voltage before the increase of capacity loading and final closure of the water breakdown channels takes the voltage off the system. Using the breakdown against volume curve it is seen that doubling the field will reduce the volume for each break by 1000. In fact, when twice the breakdown field was applied to an area of about 200 cm^2, over 600 breaks were counted and it is to be expected that self-shielding of some of the breaks will reduce the number below that given by the curve. The same effect has been used in making multichannel solid dielectric switches, where very rapidly rising pulses with rates of increase of 10^{14} volts per second overdrive spark channels, and enable the pulse to continue to rise even though breakdown processes are already under way at the lowest voltage breakdown site. The fact that the pulse over-stresses the dielectric enables a number of breakdown channels to be produced (in some systems up to 200) by a single very fast pulse.

The relation between mean breakdown voltage and volume can be used in a number of ways.

One application is to explain why scratches and other surface defects do not cause a sheet of plastic to have a low breakdown strength under single pulse conditions. There is, of course, a field enhancement at the bottom of a fairly bad scratch of, say, 5 thou. depth and radius 1/2 thou. The field at the tip of the scratch will be about 3 times that in the main body of the material but the ratio of the volumes stressed can easily be 10^8 to 1 and this implies a ratio of breakdown fields of 6 to 1. Thus the breakdown will not, in general, originate at the scratch.

In a case where the stressed volume goes up while the field decreases, say where a ball bearing is pushed into a plastic, as in one of the "intrinsic" breakdown tests, the effective volume stressed can be calculated and the radial distribution of breaks away from the point of maximum field also obtained.

If an additional relation is assumed, the maximum life can be obtained of any volume of polythene used under pulse charging conditions. Table 8a-I shows the data on three generators of varying volume of polythene and gives the voltage at which a life of about 40 firings per broken line was obtained.

This value is to be compared with a single shot breakdown voltage of 2.4 MV/cm, thus a life of about 40 shots is obtained for an under voltage of 0.56. For a number of systems (including condensers) the ratio of the mean breakdown voltage to the pulse

SOLID BREAKDOWN

Table 8a-I
Life Test of Polythene

Generator	kV Charging for 40 Firing per Break	Corresponding Field MV/cm	Volume cc	Field for 1 cc	Mean MV/cm
Dagwood	130 kV	0.42	3×10^4	1.22	
Polaris	200 kV	0.43	1.3×10^5	1.40	1.32
SMOG	160 kV	0.34	6×10^5	1.34	

voltage under consideration raised to the 7th power gives the life roughly and the above relation applies approximately. Needless to say, the life may be less than this if conditions are unfavourable, such as D.C. charging with pulse reversal.

It should be mentioned that we do not have extensive data on the point at which the volume effect begins to decrease. In the absence of any measurement of the extent of the defect at any volume it is not possible to estimate this point. For instance, if the volume were in the form of a very large area of thin foil, then at some critical volume the defect would reach through the foil and from then on the breakdown would become independent of volume. This latter statement only applies to the assumed structural intrinsic mode; it would still be only too easy to have hairs, dust, or indeed a hole right the way through and these mechanical defects would still give a volume effect.

There is also one case where a possibly significantly higher breakdown field has been observed than would have been expected from the curve given in Figure 8a-3. This was with polythene cable, where, even allowing for the fact that the effective stressed volume is considerably smaller than the volume of polythene in the cable, a pulse breakdown strength some 20% to 30% higher than expected was obtained. This may be due to the mode of manufacture of the cable, including rate of coiling and compressional effects of the outer layers on that next to the inner core, where the field strength is greatest and from which the breakdowns originate. However, the experiments have very poor statistics and the effect may well not be a real one.

Section 8b

PULSE LIFE OF MYLAR*

J.C. Martin

INTRODUCTION

This work was performed by A. MacAulay and the author and stemmed out of an experimental setup where a series of sheets of mylar which had been Evostuck together broke at a very low mean field after a short number of oscillatory discharges. This led to some simple experiments aimed at roughly determining the life voltage relation for single sheets and multiple sheets, stuck and unstuck.

It had been assumed for a long time that, in common with condenser data, the life could be expressed as a power law of the ratio of the single pulse breakdown voltage to the working voltage. For the ICSE condenser the power was known to be about 8 and this had been used liberally in various other situations to provide a guess as to the likely life. In impregnated paper condensers (such as the ICSE condenser) the life is a function of the peak to peak voltage swing. Simple theories exist to account for this where breakdown occurs in small gas-filled voids which progressively elongate over many pulses. However, for lives of 10 to 100 reversals these theories are not very applicable. In the case of polythene, charge annealing of the defects occurs and if a sample is stressed close to its DC breakdown (which is about 20% greater than the pulse value) and then pulsed through zero volts to the opposite

---
* Printed as SSWA/JCM/661/106
  AFWL Dielectric Strength Notes
  Note 11
  4 Nov 1966

voltage, the polythene will break at a voltage of only 20% of the unidirection pulse voltage strength. Thus in polythene (and possibly some other plastics) there is good reason to consider that it is still the peak to peak voltage that is relevant for life determinations. Mylar fortunately shows very little, if any, tendency to follow suit and the DC and pulse strengths seem quite similar, also a sheet DC charged near to breakdown can be rung through several reversals before it breaks, the reduced breakdown voltage in the case of a reasonably undamped ringing circuit being attributable solely to the fact that the mylar has had a number of test pulses at one go and therefore would be expected to have a slightly reduced breakdown strength.

A small digression might be in order to give a tentative explanation of a somewhat puzzling phenomenon which has been observed on a number of occasions where a mylar sheet is broken at a voltage (DC or pulse) above the mean value for the volume stressed. It is then observed that several breakdown sites exist. For pulsed voltage it can easily be shown that unless the test pulse has a rise time of a few nanoseconds, there is a negligible chance of two sites' breakdown independently in the sheet; yet the phenomenon can be displayed with microsecond rise time pulses. Also such an explanation cannot account for DC or very slowly rising pulses at all.

If the sheet is highly stressed and by chance the voltage on it goes to a value above the mean when the primary break occurs, the capacity of the test plates ring via the inductance of the solid dielectric breakdown channel. In a case encountered in the work referred to in this note, the capacity was about 5×10^{-8} F and the inductance can be estimated as 5×10^{-10} H. Thus the $\sqrt{LC}$ is 5 ns and as the gradient in the mylar before closure of the channel in the solid was more than 3 MV/cm the resistive phase is about 2 ns. Thus the damping of the oscillations is not great and the circuit, if it were composed of lumped components, would ring to about 90% and carry on ringing with very little damping. Because the spark channel is within the capacitor plates, not all the volume of plastic will experience the 90% reversal, but the voltage drop across the inductance occurs largely with a few mms of the spark channel and the greater part of the volume of the plastic reverses the full amount. This ringing has been observed after sheets of mylar and polythene have been broken in pulse voltage tests as a high frequency, lightly damped oscillation after the break. If the sheet does not break on the subsequent many reversals of voltage the waveform remains simple. If, however, the plastic sheet breaks at a field above the mean for the volume stressed, subsequent breaks are quite likely and the waveform can be seen to get more complicated as they occur. Sheets have been seen to have at least six breaks that have carried reasonable currents, although it is usually easy to tell which is the original break because of the larger hole blown. By considering the number of cycles of reversal and the voltage on the sheet when it broke, it is possible to make a statistical estimate of the number of channels likely in any given situation. Equally, the fact that a mylar sheet that breaks at only a little below the mean field will have only one break in it shows that there is not likely to be any charge storage annealing, as is shown by polythene.

Returning to the main object of the experiments, the test circuit was designed to produce a lightly damped oscillation and in determining the life for single pulses it is desirable to estimate a multiplying factor (M.F.) to turn the number of oscillating waveforms into an equivalent number of single unidirectional pulses for estimating the life. For a waveform that reverses to about 92% per half cycle, an M.F. of 3 was taken. This can be approximately substantiated by raising the waveform to the 7 1/2th power, when the chain can be seen to be effectively about 3 reversals long. Thus the use of a reasonably high Q circuit (a) reverses the polarity many times, which is the worse case to be considered; (b) it increases the effective number of pulses by a small factor. The circuit chosen was an inductance of about 10^{-4}H, a constant capacity of 10^{-8} F and a series spark gap. The test mylar sheets were placed across the condenser, the waveform applied being largely controlled by the lumped constant circuit and except in the case of the largest area tested, it had a frequency of 160 kc.

The method of applying the voltage to the plastic is all important and over the past five years various techniques have been developed for both pulse and DC work. In this application it was required that the mylar start with a DC stress on it and then undergo a series of voltage reversals. Earlier work had shown that values equally or even slightly higher than the pulse ones could be obtained by leaving air in the thin space between the thin copper sheet electrodes and the sample. The air space is at most a few mils thick and can hold neither DC nor pulse volts to any significant extent. Tracking at the edge of the metal is of course the limitation to this simple technique and this is largely overcome by surrounding the sharp edges of the metal sheet by a strip of porous blotting paper lightly stuck to a backing sheet of plastic. Two such electrodes systems are sandwiched together with the sample between them under lead blocks, the localised forces of which are spread by a sheet of rigid plastic or wood with sponge rubber under it. In order to stop tracking it is essential that outside the blotting paper edge the plastic films are firmly pushed together, reducing any air films to very small dimensions. With this simple demountable system voltages up to about 60 kV can be held across a 5 mil sheet without tracking round edges located 3 inches away from the end of the electrodes. When a breakdown occurs the copper sheet electrodes (which are typically 10 mil thick) are dished in slightly and many breakdowns can take place between the electrodes before their scrofulous appearance eventually compels replacement.

DC BREAKDOWN VALUES

Data from these experiments and those of Ian Smith quoted in his "Note on DC and Pulse Breakdown of Thin Plastic Films"* are given in 8b-Table I and compared with the pulse values given in Ian's note: "Revision of Breakdown Data concerning Mylar".**

It is not considered that the difference between the pulse and DC values are really significant since the combined experimental errors are several percent and pulse values scatter about the line by about 10% as well. Thus it is taken that the slope of the mean breakdown field against volume is the same for DC as for pulse and this is consistent with the standard deviation of about 8% which was observed.

Table 8b-I
Breakdown of Mylar

Thickness mils	Area cm^2	Vol. cc	B.D. Voltage kV	Field D.C. MV/cm	Pulse Value MV/cm
2.1	20	0.11	23.2	4.4	4.4
2.1	180	0.95	17.2	3.25	3.7
5.5+	220	3	55	3.8	3.4
10.5	180	4.9	102	3.8	3.3
5	3,000	35	38	3.05	2.8
5.5+	3,000	38	44	3.1	2.8

*   Ref. SSWA/IDS/6610/105, AFWL Dielectric Strength Note 8
**   Ref. SSWA/IDS/6610/102, AFWL Dielectric Strength Note 5
\+   Two sheets, the rest being single sheets.

SOLID BREAKDOWN

LIFE MEASUREMENTS

These were carried out for three different areas of mylar, 52 cm^2, 220 cm^2, and 3000 cm^2. In the case of the first area, six separate test areas were laid side by side and the same test oscillatory voltage discharge applied to each when the spark gap broke down. When a sample broke, the number of spark gap firings was noted and the sample of mylar blanked off with a 10 mil sheet and the tests continued until all the samples in one 36 inch wide sheet had broken. For the other two larger areas, tests were conducted only one at a time. For these two systems the damping was slight and a multiplying factor of 3 was used to turn the number of applications of the ringing waveform into an equivalent single pulse life. For the array of six small areas, the damping was heavier and an MF of 2 was used for these results. The data is shown on the left hand side of Figure 8b-1 and indicates that over the range investigated the mean single pulse life is the 7 1/2 power of the ratio of the mean breakdown voltage to the operating voltage. The single pulse breakdown value for the 52 cm^2 area is a value estimated from Table 8b-I, because tracking made it difficult to get an experimental value.

After a large number of pulses, faint signs of erosion round the edges of the electrodes could be seen on the mylar test sheets. This is caused by the corona sparks flowing under the blotting paper surface during the oscillatory discharge. However, there was no tendency for the mylar to break on or outside the edge of the electrode. As there was no sign of erosion at all under the electrodes in the area where the vast majority of the breakdowns occurred, this is taken as evidence that chemical erosion and mechanical shock from corona was not causing degradation of the mylar. The discharges took place at the rate of about one every two seconds and in view of the very high volume resistance of mylar, heating is not considered to have affected the results; certainly at the end of several hundred discharges the mylar did not feel even faintly warm to the touch, although the system should have reached thermal equilibrium after 100 shots or so.

The next investigation was the effect of the method of sticking mylar sheets together on the life. Double sided Evo tape was tried but this reduced the D.C. breakdown strength from 4.4 MV/cm to under 1.7 MV/cm, the latter value ignoring the thickness of the double sided sticky tape. Evostick was then tried and after a period of some uncertainty, it was established that providing the Evostick film was not fluid, the thickness has little effect on the life at a given applied field. Again, the field is calculated as if the Evostick layer was not there; certainly for the D.C. breakdown it is very unlikely that the Evostick can have anything like the resistance of the mylar and hence the volts will be across the mylar. The D.C. breakdown was shown to be essentially the same as that of the unstuck sheets.

The life of the stuck sheets is given on the right hand side of Figure 8b-1 for different thicknesses of Evostick for the smallest area, and a couple of determinations for the largest area. There is a slight indication that for values of the field only 20-30% lower than the single shot breakdown field, the life of the Evostuck sheets is reduced on occasions to a single shot, but this is an area of no great practical interest, so this suspicion was not investigated further. It was, however, shown that the presence of liquid or near liquid puddles of Evostick produced by sealing the sheets together too soon can lead to great reduction in life. The original occurrence which gave rise to the investigations had a breakdown at just such a liquid lens. In order to get a good life with transmission lines that have to be stuck together, it is very desirable to form a thin, uniform layer of Evostick on both surfaces, wait until it is just dry and then stick them together. This process sounds delightfully simple but in my opinion requires considerable care and practice. It is considered that if the Evostick is fluid enough (guessed to have a viscosity less than one million or thereabouts) a streamer can form in it and punch through the mylar sheets.

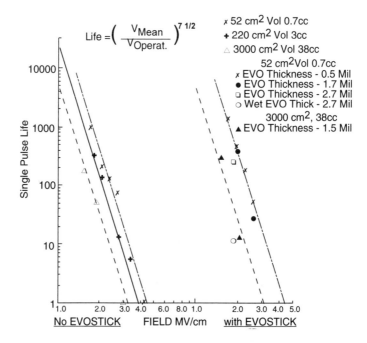

Fig. 8b-1 Life against pulse volts.

SOLID BREAKDOWN

Distribution of Lives for a Constant Applied DC Voltage

Experiments to obtain the distribution of lives at a constant pulse voltage, or applied DC volts as in these experiments, are tedious and quite tricky to carry out: consequently the results should be treated with care. Ian Smith determined the distribution for a pulse voltage chosen so that the mean life was a little under three shots. The results tended to suggest a simple radioactive decay type probability. An alternative possible form for the life distribution with a maximum differential probability situated near the mean life can be calculated from the distribution derived in "Volume Effect of the Pulse Breakdown Voltage of Plastics" (ref. SSWA/JCM/6511/A) by the author. This curve, for a standard deviation of 8% is shown in Figure 8b-2. The life relation found above has been used to obtain this curve and the life is given in terms of the mean life and the area under the curve has been normalised to unity. The life varies based on this theory simply because each sample has a breakdown voltage for a single pulse which has a distribution of values and this reflects directly into a life distribution.

The experimental values were obtained for two systems, one using 220 cm² of two sheets totaling a mean thickness of 5.5 mils and the other for 52 cm² of a single sheet of 5 mils thickness. Thirty-three samples were used in the first set and 36 in the second; all the observed values were included in the analysis. Care had to be taken that the spark gap did not vary its mean breakdown voltage because a 2% shift of this leads to a 16% change

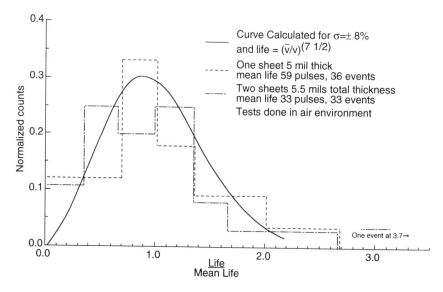

Fig. 8b-2 Life Distribution.

in the mean life. The same ratio applies to any variation of thickness of the mylar sheets. In fact, both sets of sheets varied by up to ± 3% and after the breakdown had happened, the thickness of the sheet near the breakdown site was measured and the sets of results nominalized to a constant thickness of sheet. This had the effect of sharpening the distributions slightly but not by more than 20%. It also brought the means of sets of results into line within scatter to be expected from the standard deviation of the results in each subset. The agreement between the two sets of results and the calculated curve is gratifyingly good and I consider that the experimental results differ from the exponential type probability by many times the standard deviations of the statistical errors in the results, when the combined results are taken.

Summarizing, given the standard deviation per shot of 8%, the mean breakdown voltage and distribution of such voltages can be calculated for DC or pulse breakdown for a given volume. With the addition of the life power law, the life can be calculated and the distribution of observed lives in a series of shots for any volume of mylar obtained. It is my personal belief that, given the standard deviation of single shot voltage breakdown values and the form of this distribution, it should be possible to obtain the power law deductively. If the probability of destruction had been the radio decay type, a simple argument can be used to obtain the power in the life relation, this value being a power of 12, which would of course have been incorrect. It is already interesting that, given the standard deviation for a set of measurements and the life law power, the life and distribution lives for a large range of volumes and test pulse amplitudes can be predicted with reasonable accuracy. It would be even better if the second piece of information was unnecessary.

It is probably worth repeating that this and all the other notes in these series are written to enable estimates to be made of the breakdown voltage life, etc. of various solid, liquid and gaseous insulators under various fairly ideal but experimentally realisable conditions. No great accuracy is claimed for the data and the estimates obtained by the methods outlined should be used with caution and checked experimentally as soon as possible. It is also always possible to do a lot worse than the values given by the simplified treatments if the experimental setup contains stress enhancements or serious tracking or multiple breakdown paths. It should also be mentioned that there is some experimental data on mylar condensers which seems to give values quite significantly higher than those obtained by applying the methods given in this note. A possible explanation might be that multiple very thin films are employed and that the effect occurs because either multiple foils have an advantage or because the individual foils are too thin to contain a defect greater than the foil thickness. In the latter case, the scatter of the breakdown of individual samples should be smaller than 8% and hence the volume effect on the breakdown voltage would be less.

With regard to the effect of multiple foils, Ian Smith has shown that for polythene the breakdown field of stacks of foils is greater than that of a single foil of equivalent thickness.

Some caution should be observed in interpreting these results, since there are sample to sample variations in polythene strength, but the effect seems to be too big to be accounted for by any likely variation of the properties of the plastic. Another way in which very different answers may occur is when the breakdown process is not an "intrinsic" volume process but something such as tracking at an edge. In this case a quite definite life with a very much smaller standard deviation than those reported here may occur. Such cases can usually be diagnosed because of the pattern in space of breakdowns which occur.

Section 8c

PULSE BREAKDOWN OF LARGE VOLUMES OF MYLAR IN THIN SHEETS*

J.C. Martin

This work follows that reported by Ian Smith in the notes: "Revision of Breakdown Data concerning Mylar" (SSWA/IDS/6610/102)** and "D.C. and Pulse Breakdown of Thin Plastic Films" (SSWA/IDS/-6610/105).*** These notes were concerned with fairly small volumes of mylar but also suggested that stacks of thin foils might have a significantly greater breakdown strength than thicker single films. This possibility was mentioned in the original note on the "Volume Effect of the Pulse Breakdown Voltage of Plastics", where it was suggested that for very large areas of thin foil, the intrinsic breakdown field might become independent of the volume since the defects could not be greater in length than the thickness of the foil. Owing to the encouragement (both moral and material) of Dr. Alan Kolb and his group at the Naval Research Group it was decided to investigate the breakdown field of large volumes of mylar to see if there was any sign of this happening. N.R.L. very kindly provided two large rolls of mylar, one of ten thou. and the other of 1/4 thou. thickness, with which the tests were done.

The material was laid out in a large shallow perspex tank, about 10 x 3 1/2 x 1/2 foot in size. The material was sandwiched between two or more 5 thou. copper sheets whose edges had been bent over. These electrodes stressed an area of 56 x 282 cms per layer

---

*First Printed as SSWA/JCM/673/27
AFWL Dielectric Strength Notes
Note 14
30 Mar 1967
** Dielectric Strength Note 5
*** Dielectric Strength Note 8

and it was checked that the breakdowns (except where noted for the 1/4 thou. tests) occurred at random over the electrodes. Flashover was prevented by using dilute copper sulphate solution with a resistivity of about 5 kilohm cm to give a grading length which was about 1 cm for the 10 thou. tests. Handling of the 1/4 thou. presented very considerable difficulties and after much confusion and helpful advice from everyone within range, it was found possible to guide the mylar out from the roll with perspex rods and then let surface tension pull it down onto the wetted surface of the layer previously deposited. To wet this the copper sulphate solution had to have detergent added to it and a fairly critical quantity did this without foaming too badly. After every 6 or 8 layers, the mass was debubbled by fairly gentle prodding and smoothing with the hands. Initially this process took a day to achieve 50 layers but practice speeded it up a bit. However, the very small quantity of data on the 1/4 thous. reflects on the tediousness of handling this material. After the mylar was wetted and copper electrodes were added and debubbled, polythene faced sheets of perspex weighted with lead were used to force the water out. Even in the case of 50 layers of mylar the fraction of the final thickness which was water was less then 25%. This, taken in conjunction with the fact that the water is rather conducting, means that better than 99% of the pulse voltage was applied across the plastic.

The voltage calibration of the system was carried out by two different methods which were within 1% of each other and the calibrations were repeated near the end of the measurements with results again within 1% of the previous determination. I consider the relative voltage measurements to be better than 2%, while the absolute values should be within 3% of the true ones.

As a check against values obtained a couple of years ago, twelve litres of polythene in the form of single sheets of 61 thou. average thickness was tested in the new set up. The results are shown in Table 8c-I. The mean breakdown field of 1.02 MV/cm obtained was slightly above the value given by the old curve of 0.85 MV/cm. The standard deviation deduced from four shots was 14% per shot. The difference between the two values is on the borderline of the statistics but it may reflect a change in quality of

Table 8c-I
Results for Polythene

No. of Shots	Thickness (cms)	Volume (ccs)	Mean BD (kV)	Field (MV/cm)	Life or S.D.
4	0.156	12000	159	1.02	± 14% per shot

SOLID BREAKDOWN

material. It is certainly now true that the polythene is much more uniform in thickness than it used to be.

The thickness of the ten thou. was remarkably constant and after a series of measurements which gave a value of 10.15 thou. with a standard deviation of .05 thou., most of which was estimated to be measurement error, the thickness measurement was stopped. It failed to register when halfway through the roll the mylar was spliced but later it was found that the inner mylar was 9.7 thou., again with a very constant thickness. The results have been corrected for the different thicknesses used in each test but the somewhat unusual variation between and within sets of tests was purely accidental and follows no pattern.

Table 8c-II gives the results for the 10 thou. material and the results are discussed below in the order given in the table.

The first set were two sheets in a single layer, the volume stressed being 800 cc.

The second set comprised four shots, each consisting of 5 layers of two sheets at a time. As these tests used the slightly thinner material mainly, the field quoted is a little larger than given in an earlier note. The ratio of the breakdown values for the two volumes (including those given in test 5) is $0.92 \pm .03$, whereas a standard deviation of 4% per shot would have been expected to give a ratio of 0.94. Using the standard deviations of all the tests a better value of 5% per shot might be a better choice and the agreement is then even better.

As is usual if the breakdown occurs at the mean voltage or lower, only one break is usually seen in the sheet, but when the breakdown voltage is rather above average, several breaks occur. These don't happen on the rising wave form but occur during the ringing after the first breakdown. The most spectacular example of this was during test 4, where a pair of sheets went to 201 kV before breaking and this resulted in a total of 13 breaks which carried very roughly equal currents. This effect gives a measure of how far over the mean any of a small number of shots are, since an excess of multi-breakdown shots shows that, for this set of measurements, the mean will be above the real average.

Test 3 was a very rough shot at the life of the material at a reduced voltage. The actual lives observed were 7, 2, 17, 1, averaging a mean life of abut 9 after allowance is made for the fact that the applied waveform was a slightly damped oscillatory one and hence each test voltage application corresponds roughly to 1 1/3 pulses. Using a life power of 16 rather than 8, gives better agreement for both this test and that described in test 6. While I still do not have a direct relation between the standard deviation

Table 8c-II
Results for 10 thou. Mylar

Test No.	No. of Shots	Thickness (cms)	Volume (ccs)	Soaked or Dry	Men BD (kV)	Field (MV cm)	Life or S.D.
1	6	.0514	800	D	161	3.14 ±.055	± 4% per shot
2	4	.0492	4000	D	145	2.95 ±.06	± 4% per shot
3	4	.0492	800	D	135 (test volts)	2.75	9
4	7	.0514	800	S	181	3.52 ±.09	± 7% per shot
5	3	.0492	800	D	163	3.34 ±.135	± 7% per shot
6	3	.0514	800	S	154 (test volts)	3.00	11 corrected

For Life data expectation on 16th power law is 11 for both tests 3 and 6.
Difference between tests 1 and 5 is 0.2 MV/cm ± 0.15 MV/cm.
Weighted mean for dry 800 ccs is 3.21 ± .06 MV/cm.
Increase on presoaking is 10% ± 3%.
Volume effect: Expected ratio 0.94 (based on 4% per shot)
                Observed ratio 0.92 ± 0.03
Errors are derived from the spread of data and do not include absolute errors.

of the breakdown fields and the life power it seems reasonable that the smaller the deviation, the higher the power in the life relation.

These results suggested that the volume effect is rapidly flattening out for the 10 thou. material and that the original curves were pessimistic for large volumes. It also suggested that the life might be significantly greater at any working voltage than had originally been predicted. All the tests had been conducted with material that was pulled off the roll into the copper sulphate solution and there had been no time for water to diffuse into the mylar. It was decided to repeat some of the tests with material which had been soaked in the dilute solution. The data for vapour transmission through mylar suggests that only a day or two is needed to saturate mylar to a value given in the literature of rather less than 1% of water. However, it has been suggested that rather longer times than this were needed and also that values of the dielectric constant of 5, after soaking, have been quoted. As a result of this we decided to soak most of the remaining material in 5 kilohm cm solution at an elevated temperature of about 45°C for some 12 days. Care was taken that the temperature couldn't go sailing up, in case embrittlement of the material occurred. In order to get the solution into the mylar, two 1" wide bands of 5 thou. porous paper were interleaved near either edge of the roll. When the material had been rewound with these spaces in it, it was put vertically into a column of solution, a previous test having shown that nearly all the bubbles would leave the setup in a few hours. Using some data on the temperature dependency of diffusion of gas through a plastic film, it seems possible that this elevated temperature soaking corresponds to 100 to 200 days immersion in room temperature solution.

On unwinding the material, a fairly strong, vaguely alcoholic, smell was obvious. In addition to this, the soaked material is very much more transparent than the dry material. The thickness of the material did not seem to increase by as much as 1% but this measurement was not an exact one. The dielectric constant was measured at 1.6 kc and compared with some dry material. The soaked material had a value only some 4% higher than the other. Another test with 2 thou. thick material failed to show any significant increase in dielectric constant. We are not very confident about these measurements but, taken in conjunction with the reported take up of water value given in the literature, it does raise the question as to how the dielectric constant can rise as high as 5.

Anyway, on the first shot of the soaked material we got a breakdown value of 85 kV where we had expected 160 kV - panic! However, the break had occurred very near the edge of the copper where the paper spacer had been. While rewinding the material we had taken care that grit did not get into the mylar rolls but we had been less careful about the paper and we suspect a bit of grit got wound in via the spacing paper and had punctured one of the two sheets of mylar. Anyway, in the subsequent ten firings of soaked material values about those expected were obtained. A test was also run with two mylar sheets in which breaks had been intentionally introduced. As expected, the breakdown originated from those in the sheet next to the positive electrode and the breakdown voltage recorded was 89 kV. Thus it is concluded that this shot had a hole in the sheet nearest the positive plate and has been excluded from the analysis. This is the only result to have been so excluded; all the others have been used in the analysis.

Test 4 gave a value significantly above those obtained with dry material. To check this the small remnant of remaining dry material was retested and a value slightly above that obtained in test 1 resulted. In these three tests, the two with the largest breakdown voltages showed considerable multiple breakdowns and this suggests that these were both well above average and hence the mean is likely to be above the true value. However, the difference between test 1 and test 5 results in 0.2 MV/cm and has an error of ± 0.15 MV/cm and thus is not too big to be chance. The weighted mean of the two tests has been used in the analysis - this is 3.21 MV/cm.

The difference between the dry and soaked material is 10%, with an error of ± 3%. Thus it is probable that soaking has slightly increased the strength of the material. It does not seem that this is unreasonable since presumably the water might have a small effect, if it is preferentially absorbed near to the defects, by reducing the field locally. We wouldn't go bail on the increased strength of soaked material but there might be another reason for pre-soaking, since it was found quite a bit easier to get bubbles out of the layers when the material had been soaked. The mylar also slides about more easily and also shows less tendency to float up to the surface when it has been pre-soaked. All in all, it is easier to handle, but neither of these reasons for pre-soaking is a strong one.

The final test on the 10 thou. material was a very rough life test. This was initially undertaken at the same voltage as was used in test 4, but the increased strength plus the fact that the thickness was 4 1/2% less in the earlier tests meant that no breakdowns were observed in 25, 25, and 16 shots on three samples of soaked material at a field of 2.65 MV/cm. The same material (after being left out overnight damp, but not under water) was retested at a mean field of 3.00 MV/cm. The life obtained then was 6, 1 and 2 shots. Using the 16th Power Law, the earlier testing is estimated to have added 6, 6, and 4 shots. Using a multiplying factor of 1 1/3 to allow for pulse ringing, the life at breakdown is estimated to be about 16, 8 and 8, with an average life of about 11. The expected life on a 16th power law is 11, an agreement which is obviously fortuitous. However, even without the second lot of high voltage life tests, 2,400 ccs had a life greater than 20 at 2.65 MV/cm.

The conclusions of these tests were that accelerated soaking has, if anything, made an improvement on breakdown strength and that the life relation again looks very roughly like a 16th power.

The 1/4 thou. mylar caused edge difficulty because the sheets adjacent to the electrodes showed a strong tendency to billow up and wrap themselves at least in part around the edge of the

SOLID BREAKDOWN

electrode. This reduced the copper sulphate solution to a very thin film and also caused larger voltages to appear across the raised bits of the thin mylar. Either of these effects ruins the grading of the solution and caused breakdowns to occur at or just outside the edge of the electrodes. A solution to this was found in small volume tests by masking the edges of the copper electrodes with two extra sheets of thicker mylar which only reached a couple of cms inside the electrode edges. On the small volume tests the breakdown occurred other than near the edges and also gave breakdown voltages with a satisfactorily small standard deviation. These are the values given in Test No. 1 of Table 8c-III. However, when the same technique was tried on a much larger volume, the breakdown occurred at the edges of the extra mylar that extended a couple of cms inside the electrodes. Several breaks occurred in each of the two tests, but all were round the edge of the masking mylar and hence the values obtained of 5.1 and 4.7 MV/cm are only lower limits. Unfortunately time and patience were both at an end and it was decided that there was no chance to develop a yet further improved edge field control, especially as it appeared that this would have to be done with the large area samples, each of which would take a great deal of effort, and so the experiments were discontinued. However, it is considered that the tests with 1/4 thou. had established that very thin foils of mylar could stand fields considerably above those that the ten thou. could support and that there was some indication that, as with ten thou., the field became largely independent of volume over some critical value.

It should be pointed out that in the 1/4 thou. tests the time taken to assemble the mylar was probably longer than the estimated time for the material to absorb 1% water. In the case of all the previous tests with thicker material this was probably not so and while the ten thou. tests had not proved that soaking raised the breakdown strength, they had made such an effect very likely. It is probably rather an academic point since any attempt in quantity to use very thin mylar under water will mean that the material is fully soaked but it might well be that the 1/4 thou. data is not

Table 8c-III
Results for 1/4 thou. Mylar

Test No.	No. of Shots	Thickness (cms)	Volume (ccs)	Mean BD (kV)	Field (MVcm)	Life or S.D.
1	3	.0127	1.3	79	6.2	~ 1% per shot
2	2	.0317	250	≥ 154	≥ 4.9	

strictly comparable with the rest of the data now to hand. This effect has not been allowed for in plotting Fig. 8c-1 because the magnitude as well as the effect itself is largely unknown.

The section that follows is highly speculative and is included more in the hope that someone will be stung to prove it wrong rather than as a claim to being factual. Certainly the experimental evidence which is relevant is very sketchy.

Figure 8c-1 replots the data listed in the table in "Revision of Breakdown Data concerning Mylar" and the present note. In this graph the data has been clumped into four groups of mean thicknesses .0006, .005, .025, and .10 cms. The point given in the table for 50 ccs at 3 thou. has been omitted for no very good reason except that the mode of testing used in this case was to roll a cylindrical specimen containing several test volumes. This differs drastically from the other tests and the value obtained may have been affected by the mode of testing. If the bald assumption is made that the large thickness breakdown-volume relation lies parallel to that of polythene, the curves shown for mylar can be drawn without doing to much violence to the data. An interesting question is whether polythene might show the same effect. Nearly all the breakdown data on polythene has been obtained with samples many times thicker than that used in the mylar tests because of the danger of mechanical damage to thin foils of this plastic. However, Ian Smith obtained some data for a stack of 4 one thou. sheets which gave a value well above the curve and the value he obtained was a breakdown field of 2.75 MV/cm for 4.4 cc. On Fig. 8c-1 the polythene curve for thick polythene is given as well as this one thou. point. If the mylar curves are used to extrapolate from this one point for polythene, it suggests that for 60 thou. samples the curve should flatten for volumes above 10 litres.

The fact that there is a critical thickness below which the volume dependency largely disappears does not prove that the initial defect size is the same as this thickness, unfortunately, since it may merely mean that the length associated with some portion of the breakdown process is equal to this thickness. For instance, the initiating defect might still be some hundreds of atoms long but if a multiplying process (similar to the Townsend avalanche length) has to intervene before a full streamer can be formed, the breakdown might still cease to be volume dependent for a thickness less than or equal to this pseudo avalanche length. If this is the case, then the curves suggested in Fig. 8c-1 might well depend on the presence and nature of the liquid layer. If the thin sheets of plastic were to be wrung together very well so as to exclude effectively all the liquid or air, the avalanche process might well carry on as if the plastic were a monolithic block and the volume dependency comes back with a vengeance.

SOLID BREAKDOWN

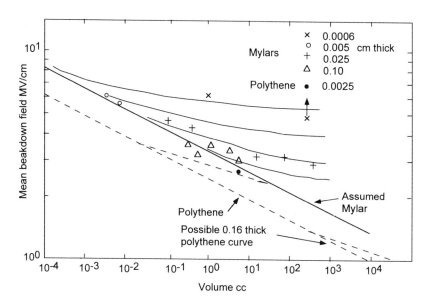

Fig. 8c-1 Breakdown fields for Mylar and Polythene.

From a purely practical point of view, however, mylar in the form of thin sheets has the greatest field strength, energy transmission and energy storage for reasonable quantities of electric energy. It would be very desirable for its properties to be investigated further than we have been able to so far.

This work was done by G. Herbert, D. Wilcox and J.C. Martin and we would like to repeat our thanks to N.R.L. for their encouragement and support in it.

Chapter 9

FAST PULSE VACUUM FLASHOVER*

J.C. Martin

The recent visit of Dave Hammer from NRL gave me occasion to analyse roughly the available small sample flashover data and I thought the results might be of minor interest to others. This re-analysis has been directed towards the region of interest to tube designers and concentrated on perspex (i.e. lucite) at about 45°, where it withstands the maximum gradient. The results from different experimentalists are in reasonable agreement, after manipulation, and in agreement with the performance of the best tube designs.

A few general comments are worth making before a brief resume of the available data. The first point concerns the value of t_{eff}. As is shown below, a $t^{1/6}$ power appears to apply reasonably well and as such the relevant fudge to convert an odd shaped waveform into an approximately rectangular pulse is the full width at 89% of full height. As with all fudges, this is only approximately correct but it is in the right ball park. Unfortunately it is not always clear from the reports how the quoted pulse times are in fact measured and some judgment has had to be used in estimating t_{eff}. However, as this is raised to the 1/6th power, errors are not strongly reflected in the final results.

The second point concerns the long pulse (of the order of 1 μsec) data, where the spread of results is considerably greater.

SSWA/JCM/713/157
* AFWL High Voltage Notes
  Note 2
  16 March 1971

This is, I think, because there is a greater dependency on the surface finish, etc. of the cones. In addition there appears to be a weak dependence of the breakdown field on the length of the sample. There may well be a residual effect for short pulses (of the order of tens of ns) apart from any area effect, but from Ian's results it would appear to be small. An additional piece of evidence is that for diaphragm tubes (ie ones with straight sheets of perspex orthogonal to the axis) the breakdown strength is very close to that estimated from the cone data, despite the fact that there are no intervening grading rings across the long surface. Fortunately, the short pulse data is the most interesting from the tube designer's point of view and here the data appears to be only weakly dependent on surface finish, fit and shape near the triple points, oiling, etc., when these are reasonably done.

I.D. Smith (1). The samples had various lengths but in general were 1" long and had areas of the order of 40 cm^2. Ian used as a criterion of breakdown the voltage at which the sample broke at the end of the pulse, as it passed through zero volts. This had the advantage of being very well defined but is perhaps a little less useful than taking the voltage at which the surface flashed over at peak volts. The easiest voltage to measure is the one on a linear ramp but I think this leans rather heavily on the t_{eff} fudge. Because of Ian's convention for the breakdown field, I have multiplied his results by 1.1 to get peak breakdown, probably a slightly conservative factor. The scatter was observed to be of the order of ± 10% per shot and the effect of oiling with MS F704 silicone oil small. The breakdown fields were 200 kV/cm for a t_{eff} of about 30 ns and 115 kV/cm for about 300 ns.

Watson and Shannon(2) tested glass and lexan. Unfortunately the lexan curve is incomplete and I have had to extrapolate it to give a very approximate value of 200 kV/cm for 70 ns at 45°. For long pulses, lexan behaves like perspex (see Ref. (4)) but its comparison with perspex can be questioned; however I feel it is probably rather similar. The area of the samples was of the order of 10 cm^2.

Glock and Linke (3) looked at perspex in some detail and in addition imposed a magnetic field on the sample with no significant effect. Their samples were about 25 cm^2 in surface area and they obtained 290, 200 and 114 kV/cm for times of 3, 30 and 300 ns approximately. At the shortest time they found the optimum breakdown gradient was obtained at an angle of about 30°, but I have used their value for 45° to be consistent. They also quote a scatter of ± 10% in single shot data.

O. Milton (4) has performed an extensive series of tests with a 5 µsec pulse with different materials and also quotes results for a short pulse for perspex. His samples were probably rather better

FAST PULSE VACUUM FLASHOVER

finished and seated than others and it appears he found a rather larger conditioning effect than others found for the short pulses. The question of what effective time to use is particularly difficult in this report. The pulse is quoted as 5 μsec for the long pulse tests, but the t_{eff} measured from a record given in the report is of order 0.6 μsec. If this is a typical record then the samples were broken on the faster rising front of the waveforms and the t_{eff}'s I have assumed apply. The area of the samples was of the order of 14 cm² and the mean breakdown fields were 304 kV/cm and 196 kV/cm for effective times of the order of 20 and 600 ns.

In analysing the above date I have assumed an $A^{1/10}$ effect (based on the ± 10% scatter of single shot results) and I have also used a $t^{1/6}$ power. This appears to be a reasonable first approximation to what is likely to be a fairly complicated situation and may only apply to the range 10 ns < t_{eff} < 200 ns. However, various real life tubes have shown a time dependency of this sort out to a hundred ns or so and in one test a diaphragm tube broke at a few microseconds on the slow Marx charging waveform at a time which fitted this power closely. However, for exactness the breakdown relation obtained really only applies to short pulses and should not be extrapolated to times over a couple of hundred ns, except in desperation.

Table 9-I summarises the data and includes the lengths of typical samples. The luxury of a third significant figure is obviously not justified in either this or the next table and I apologise. Table 9-II gives the values of $Ft^{1/6}A^{1/10}$ using the values given in Table 9-I.

Considering the internal data in each group, the $t^{1/6}$ power seems to be a reasonable average effect, subject to the remarks made above. The data of O. Milton is significantly higher but, as was mentioned above, his samples appear to be better made and mounted. For the short pulses, an average value of 173 is within a standard deviation of ± 15% in the individual data. The average for the longer pulses is much more uncertain and it is speculated that the fit of the cone to its electrodes is beginning to have a significant effect. For DC this factor is found to be very important.

However, for short pulses, the factor can be taken to be 175 and it is possible that where great care is taken with the finish and fit at the triple points, this number might increase by about 10%. Another 10% may be around for different plastics which are obtainable in large enough blocks to make tubes out of, but of course the material should be tested in sensible sizes. I.D. Smith at Physics International has found significant increases in certain samples of lucite made up as tube rings, but the cause of this increase is obscure.

Table 9-I
Vacuum Flashover Data at 45°
(All Values after Conditioning)

	Area (cm²)	Length (cm)	Breakdown Field (kV/cm)	t_{eff} (μsec)
I.D. Smith	40	2.5	220 128	0.03 0.3
Watson (Lexan)	10	0.95	~ 200	0.07
Glock	25	1.25	290 200 114	0.003 0.03 0.30
Milton	14	1.27	304 196	0.02 0.6

Table 9-II
Values of $Ft^{1/6} A^{1/10}$ (kV/cm, μsec, cm²)

	Short Pulse t ~ 0.03 μsec	Long Pulse t ~ 0.5 μse
Smith	178	154
Watson	154	
Glock	152 153	129
Milton	206	236
Average *	173	~ 170

* Giving equal weight to each experimentalist

Using a constant typical tube diameter of about 30 cm, the relation can be written as $Fd^{1/10} t^{1/6} = 100$ where the spacing rings are a small proportion of the thickness (~ 10%) and the tube is very well graded overall. This is the relation that we have used at AWRE for some time past. It applies, of course, to a clean tube. For real working conditions this would have to be reduced somewhat. During the P.I. Aurora programme, tests showed that short very well graded tubes achieved this level with overall lengths of 25 cm and using four sections. They also tested a well graded tube of about 83 cm length of 11 sections which went to 6.9 MV for a 100 ns pulse, giving a value of 90 for the constant. It should be repeated that this relation applies for tubes with an inside diameter of the order of 20 to 40 cm, and also the tube has to be well graded overall in order to reach this level.

Concluding, it may be objected that even the short pulse data in Table 9-II is not very consistent, but in view of the different conditions such as surface finish, thinness of the cone's edges, seating against the electrodes, voltage calibration and wave shape monitoring errors, etc, I think the agreement is rather good. This is confirmed by the fact that both axial and diaphragm tubes, when well designed and clean, closely approach the gradients predicted by the very small scale test results which have been obtained in considerably different geometries.

REFERENCES

1. I.D. Smith. Proceedings of the 1st International Symposium on Insulation of High Voltages in Vacuum. p. 261, Oct. 64.

2. A. Watson and I. Shannon. Proceedings of the 2nd International Symposium on Insulation of High Voltages in Vacuum. p. 245, Sept. 66.

3. W. Glock and S. Lenke. Pulsed High Voltage Flashover of Vacuum Dielectric Interfaces. Cornell Lab. of Plasma Studies, August 1969.

4. O. Milton. Pulsed Flashover of Insulators in Vacuum. SC-DC-70-5459.

Chapter 10

SWITCHING

Section 10a

SOLID, LIQUID AND GASEOUS SWITCHES[*]

J.C. Martin

INTRODUCTION

Some handy dandy formulas are given so as to enable the performance of various types of switches to be estimated. Occasional crude models of what is going on are outlined from time to time, although these are probably only notable for their lack of physical insight. Various possible layouts of switches are described and the performance of some of these are indicated although it is to be understood that these are typical best results. Reality is frequently/nearly always/inevitably worse than these (as usual). Various reasons why this may be so are hinted at in the talk. Finally a few novel, cheap and plausibly useless switches of unusual design are described.

BACKGROUND

When the small pulse power group at Aldermaston started work some twenty years ago, the world was our oyster and we could not go wrong (so long as we did not know anything about the current high voltage engineering practice). For instance our first effort was a 0.1 ohm 50 kV DC charged strip transmission Blumlein generator, the output of which was due to be stacked to provide a much higher voltage output. After operating the solid dielectric switch (a thumb tack) for the first and last time, it was observed that the line had broken down in some 20 places round its edge. All we then

---

[*] AWRE MOD (PE)
No Date

did was mentally remove the label "high voltage pulse generator" and substitute one reading "multichannel switch experiment".

In another experiment in about 1960 we had constructed a pulse charged water dielectric twenty ohm Blumlein generator named Triton. The output of this fed a polythene line (named Snail) of incredibly twisted topology which separated into four lines and stacked the 1 million volt output pulse of the generator to give a bit less than 4 million volts into a matched load. The switch of the water Blumlein was like a wire shaving brush and the rise time of this was calculated to be some 10 ns on the basis of its inductance. Our voltage monitoring showed it to be more like 40 ns. Doubting our voltage divider (it was early days) we used the Kerr effect in water as an independant way of measuring the pulse rise time. This confirmed that the pulse rose in four times longer than it should have done, thus was the resistive phase first found as far as we were concerned.

In such a fashion we progressed from disaster to disaster building some tens of different sorts of generators in a few years. Some of them we built twice, refusing to believe they were as bad as they really were when we first looked at them. All in all it was enormous fun.

We also generated fair quantities of breakdown data of use in designing high voltage short pulse generators. In addition we originated the two main forms of vacuum envelopes or diodes now in use, the multistage and diaphragm tubes. We also started looking at plasma cathodes and the control and transport of large current relativistic electron beams. About 1/3 of our effort was devoted to suppressing pick up and producing believable records of the voltage and current waveforms. We produced currents and voltages rising at the rate of about 5×10^{14} amps/volts per second which (a) were world records for a while and (b) proved that the ohm was a God given unit. We also introduced simplex and black tape to the embryonic pulse power community which was probably one of our major achievements.

As time passed it became clearer as a result of work on both sides of the Atlantic what was do-able and roughly how to go about doing it and some of the fun went out of pulse power generator design. However we are very grateful to the management of AWRE for backing us during those first few years, not asking too many questions, and not demanding that we lived up to our hopeful promises.

Whatever we achieved, we achieved as a group that was never more than seven strong. A very equitable arrangement was arrived at, the others did the work and I talked about it. I would like to record the names of the main members of the group over the years:

SWITCHING

Phil Champney (defected)
Dave Forster
Mike Goodman
George Herbert
Mike Hutchinson
Ted Russell
Ian Smith (defected)
Tommy Storr (retired and sadly missed).

This is partly so that then any credit coming can go to them for their sterling good work, but mainly so that I can blame them for any errors or mistakes.

Sliding gracefully towards the topic of this talk, for several years switching was a major concern of the group. Our interest involved DC charged spark gaps which were used in the feed to pulse transformers or in Marx generators which provided some or much of the voltage gain in the system. In addition we worked on pulse charged gaps which generated the high speed output pulses. For a fair period of time the main throttle in the electrical power flow was the output pulse charged switch and its limited performance. However the development of the necessary conditions for multichannel switch operation in principle removed this constraint. The fundamental limitation to the power flow then transferred to the vacuum envelope for most practical applications involving short pulses.

Many people have contributed to switch development, what I shall be talking about is mostly the work done at AWRE, and I apologise in advance for not giving a broader based view of the topic. As the pulse power group at AWRE has been effectively out of pulse power for the past four years doing other work the talk will also be more of an historical review than a topical talk about current progress and development. In particular I shall not be touching on important current interests such as high repetition rate gas gaps etc. Even so I hope the material covered will be of mild historical use and even occasionally of help.

ESTIMATION OF GAP BREAKDOWN VOLTAGE

Nanosecond Pulse Techniques (1), while very long in the tooth, provides means of estimating ahead of time the expected breakdown voltage of many types of gaps. These can be divided into three crude geometries, uniform field, mildly diverging and point or edge plane geometries. Most gaps fall into the second class. In addition they can be "DC" or pulse charged and the main dielectric can be gaseous liquid or solid. I know of no current gap that used liquid and is DC charged although I could imagine uses for such a gap.

DC Charged Gaps

N.P.T. outlines methods of calculating the ideal breakdown voltage for uniform air gaps to an accuracy of 1 or 2%. By considering the increase of field at one or both of the electrodes the standard relationships can be extended to real life mildly diverging field conditions again with ~ 2% accuracy. This treatment shows that for air the breakdown voltage is not exactly proportional to the pressure (or more accurately the density) of the air. When the pressure is raised above about 10 atmospheres or so the breakdown voltage can increase much more slowly than the pressure. It also becomes dependant on the nature of the electrodes, aluminium being poor and brass and steel better. However for gaps not taking high currents (or better coulombs) little surface damage occurs and experimental results should be close to the calculated ones. In addition the scatter of the self breakdown of a gap should be a few percent.

For very carefully polished and conditioned gaps (ones where many low energy discharges are used) scatters of a few tenths of a percent are possible. As the charge flowing through the gap is increased the scatter in breakdown voltage can become appreciable. However gaps passing many coulombs have been developed which still have an acceptable scatter.

The reason for calculating the expected breakdown curve is to compare this with the actual gap performance. If there is serious disagreement then some feature of the gap is wrong and it is unlikely that it will give usable operation. In addition it is nearly always desirable to use gaps at high mean fields where fast rise time output pulses are required. This is because the rise time is approximately inversely proportional to the field in the gap.

When testing a gap it is very desirable not only to measure the break down voltage as a function of say spacing and/or pressure but also to derive a crude measure of the standard deviation of the breakdown voltage. When the pressure of an air gap is raised above say 10 atmospheres, not only will the mean breakdown voltage increase much more slowly than the pressure, but the standard deviation will increase dramatically. In general it is not worthwhile to operate a gap in this region, it is better to open up the gap, recontour the electrodes or change the nature of the gas and/or electrode material.

Another use of the standard deviation measurement is to estimate the breakdown voltage of a set of such gaps, such as might be used on a large Marx generator. With a standard deviation of say ± 3% a set of 40 gaps will have a mean breakdown voltage of at best 90% of that of a single gap. Indeed the situation is frequently rather poorer than this since the gap breakdown probability curve can often have a two valued distribution curve. Figure 10a-1 shows such a probability curve.

The infrequent very low voltage breakdowns we call dropouts. For a case where these happen 2% of the time, a set of 40 gaps will more often prefire at low voltage than not. Drop outs can be caused by several phenomena. Among these are the growth of long whiskers well away from the main arc region. These eventually reach a great enough length and sharpness to cause low voltage breakdown. Another possibility is dust and debris in the body of the gap from previous firings. This particularly seems to happen with pressurised SF_6 gas and small compact gaps. Dust filters in the gas feed lines can help in this case. Another potential cause is incipient breakdowns along the insulating wall of the spark gap body. These not only irradiate the gap but at the same time alter the field distribution in between the gaps because of photo emitted electrons from the negative electrode and because of changing the field generally. This can cause the discharge to occur between the main electrodes before the wall discharge can complete along its much greater path. The cure of dropouts is a messy and sometimes time consuming business and I am afraid no general rules can be given as to how to go about it. It can to a large extent be avoided by copying a well tested gap or buying a commercial gap if a suitable one exists. Copying can still involve minor risks as exemplified by a case where versions of a well proved gap suddenly started to behave erratically. This was eventually traced to the use of brass with a percent or so of lead in it which had been accidentally supplied by mistake. A final and fairly heroic way of coping with dropouts is to monitor each of the many gaps in a set and when one prefires during the charging, the rest are triggered. This aborts that particular shot but can save a large system from damaging itself.

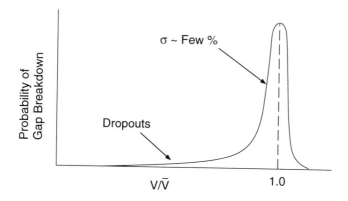

Fig. 10a-1 Probability Distribution of Gas Gap Breakdown.

Pulse Charged Gaps

N.P.T. outlines methods of estimating the breakdown field to be expected for gases, liquids and solids in uniform and near uniform conditions. All the three media can show polarity effects sometimes reaching a factor of 2, with the negative electrode usually being able to stand the greater field. A well designed gap can arrange to take advantage of this effect. For instance a self closing water gap pulse charged in about 300 ns had its electrodes designed so that over a range of gap spacings the field on the negative electrode was double that on the positive. This enabled it to work at a mean field of around 600 kV/cm with 1 MV on it producing a pulse with a rather good rise time. The question as to what the polarity effect is in high pressure SF_6 is an important but unfortunately not well resolved one.

In general pulse charged gas gaps will have a larger useful pressure range than DC charged ones, but even with short pulses the pressure dependance will turn over and the scatter become large. Ref. (2) was written after N.P.T. and may be of use for SF_6 gaps.

Pulse charge also enables point or preferably edge/plane gaps to operate at useful fields. This type of gap will operate at considerably lower mean fields than "uniform" gaps. However they can have a standard deviation of only a few tenths of a percent when fairly rapidly charged. This means that such gaps can be much more easily made to operate multichannel and hence have a useful performance with the additional advantage of being easy to construct in wide arrays. In addition point/edge plane gaps working in oil or water have found considerable application in the faster stages of larger generators such as those currently built at Sandia Laboratory Albuquerque and elsewhere.

Another great advantage of large arrays of edge or point plane gaps is their cheapness. Modern systems can contain thousands of such gaps or spark channels, something that would be prohibitively expensive if separate individual gas gaps were used.

Some work on edge plane gap operation in a range of gases is reported in Ref. (3) written after the N.P.T. note.

ESTIMATION OF THE PULSE RISE TIME

For design work, values good to 20 per cent are more than adequate and in order to be concise, various aspects which can occasionally be relevant will be omitted.

The rise time of the pulse, which is the same as the fall of voltage across the gap, of course, is largely controlled by two

SWITCHING

terms: the inductive term and the resistive phase term. The inductance of the gap is larger than that of the spark channel itself but in a well designed gap it is usually the major term. This inductance changes with time as the conducting channel expands and on occasions this can be important. However, in this note only a constant inductance will be considered. The second component of the rise time is the resistive phase. This is caused by the energy absorbed by the plasma channel as it heats and expands. In the case of channels working at reasonably high initial voltage gradients and driven by circuits of tolerably low impedance (i.e. less than 300 ohms or so), the plasma channel lowers its impedance mainly by expanding to become larger in cross section. The energy used up in doing this is obtained by the channel exhibiting a resistive impedance which falls with time. This expansion can be measured optically and correlated with the energy deposited in the plasma itself. A fairly wide range of experimental results have been collected and summarised in a semi-empirical relation which is easy to use.

The two terms in the rise time expressions both refer to e-folding times. For the case of a constant inductance L joined to a generator of impedance, Z the voltage falls exponentially with a time constant $\tau_L = L/Z$.

The accurate calculation of the inductance of the channel is difficult and would require accurate knowledge of the channel radius. Fortunately it is not necessary to do this, because an adequate approximation is to use that of a wire of radius a fed by a disc of radius b in which case the inductance is $L = (2\ell) \ln b/a$ nanohenries. In this expression $b \gg a$ in practice and the value of L is only weakly dependent on a. The rate of expansion of the channel is of the order of 3×10^5 cms per sec. for air at one atmosphere and maybe half an order of magnitude greater for liquid and solid gaps, depending on the field along the channel when it forms. In the above expression ℓ is the channel length and is in centimetres. For rise times of a few ns, the ln term is of the order of 7.

Incidentally, when care is taken in selecting approximations to the various sections of an array of conductors, simple inductance relations can be used to obtain the total inductance to an accuracy of 20 per cent or so.

The relation for the resistive phase is given in two forms, for convenience. For gases, it is

$$\tau_r = \frac{88}{Z^{1/3} E^{4/3}} \left(\rho/\rho_0\right)^{1/2} \text{ ns}$$

where E is the field along the channel at its closure in units of 10 kV/cm, ρ/ρ_0 is the ratio of the density of the gas to air at NTP, Z as before is the impedance of the generator driving the channel in ohms.

For solids with unit density

$$\tau_r = \frac{5}{Z^{1/3} E^{4/3}} \text{ ns},$$

where E is now in MV/cm.

To obtain the effective rise time of the pulse τ_{tot} the two times are added, i.e.

$$\tau_{tot} = \tau_L + \tau_r.$$

This time is the e-folding time, where the waveform is exponential and where the pulse rise is not, this parameter approximates to the time obtained when the maximum slope of the voltage waveform is extended to cut the 0 and 100 per cent values of the voltage, i.e.

$$\tau_{tot} \simeq V/(dV/dt)_{max}$$

Where the waveform is exponential, the e-folding time is multiplied by 2.2 in order to obtain the 10 to 90 per cent time. However, where a number of equal time constants are operated, one after the other, this factor falls rapidly and eventually tends to a little less than unity. Thus there is no unique relation between the e-folding time and the 10-90 per cent rise time and I would suggest that for many applications, the one used here is of more use than the standard parameter. In particular, it gives the value of di/dt max reasonably well and is the main parameter of use in calculating the energy deposited in the spark channel, $E \sim V^2/4Z \times \tau_r$ very approximately.*

After the main voltage fall has occurred, small voltages persist across the plasma channel as it cools, moves, bends, etc., as well as the more usual drops at the electrodes, but most of the energy is deposited in a time of the order of the resistive phase. In most gaps both terms are important, but for circuits with impedances of tens of ohms and higher, working at modest gradients, the resistive phase may be several times as large as the inductive one.

---

* Editor's Note: Original copy has $E \sim \frac{V^2}{4Z} \times \tau_R$

SWITCHING

Various very approximate treatments exist to cover branching, non-uniform conditions down the channel, etc., but these are beyond the scope of this talk (see Ref. (6)).

Table 10a-I gives an example of a calculation for a plasma channel of length 5 cms with a starting voltage across it of 100 kV, in air at one atmosphere pressure. The calculations are shown for three different values of Z.

Table 10a-I
Pulse Risetimes for Gas Gaps*

Z	τ_L	τ_r	τ_{tot}
100	.7	7.5	8
10	7	16	23
1	70	35	105

Table 10a-II shows a representative approximate calculation for a solid switch with 200 kV across a 3 mm solid dielectric, such as polythene

Table 10a-II
Pulse Risetimes for Solid Dielectric Switch*

Z	τ_L	τ_r	τ_{tot}
100	0.04	1.8	1.8
10	0.4	4.0	4.4
1	4	8.6	12.6

All the times are in ns.

As can be seen, calculations based on the inductive term alone can be badly wrong. The combined effect rise time is also used below in some brief comments on multichannel switching.

---

* Recomputed by Editors from original manuscript.

As Ian Smith has pointed out the empirical relation is not dimensionally correct. This can be rectified by adding $\ell^{1/3}$ (spark gap length) to the top of the relation. However in many cases the spark channel is not formed uniformly. For instance, in a liquid spark gap as the streamer progresses across the gap a significant energy from the local stray capacity can open up the initial portion of the channel albeit at a low velocity, consequently when final bridging between the electrodes occurs, some fraction of the length only has to be heated and the major energy loss of pushing the liquid apart with a shock does not occur. Such a picture would suggest that in some cases the full length of the channel is not the applicable parameter and this might largely remove the $\ell^{1/3}$ factor. Practically the empirical relation has been checked over a very wide range of parameters and not found badly wanting.

One study which is of importance for water gaps is reported in Ref. (4).

Again the major advantage of having a crude estimate of the expected pulse rise time (apart from the design phase) is that it can alert one to unexpected trouble. For instance if a rise time is much larger than expected maybe the spark gap channel is not going straight between the electrodes. Perhaps there is additional inductance in the system which has not been taken into account. Maybe the monitoring is up the creek. Anyway it would pay looking into various possibilities.

MULTICHANNEL OPERATION

Crudely speaking gases and liquids pulse charged cannot work at much more than .5 MV/cm mean fields. Thus single channel output switches have very rough inductances of at best 40 nanohenries per MV. Hence 1 ohm 1 MV switches have e-folding rise times of at best 40 ns. Solid dielectric switches can do better than this by a factor of 5 or so, but need replacing each time of course. Hence low impedance million volt generators (and high voltage 20 ohm or so units) need multichannel or parallel switches to provide short pulses.

Historically we first achieved multichannel operation with solid dielectric gaps. This was because it was easy to generate hundred kV pulses rising in a few ns, by using a solid master gap. Such a pulse fed into slave gaps would produce many tens of channels.

After considerably more effort we managed to make a pulse charged multichannel liquid switch perform (using water as the dielectric).

It was only after a lot more work and the following theory that we were able to make multichannel gas switching reliable.

Single channel gaps, transit time isolated, have been operated in parallel for several years, but these do not come within the rather strict definition of multichannel operation that I use. For me, multichannel operation of a gap occurs when all the electrodes are continuous sheets of conductor. Thus, while transit time isolation may play a small part, the main effect is that before the voltage across the first channel can fall very much, a number of other channels have closed. As such, it is the inductive and resistive phases which are important in allowing a very brief interval in which multichannel operation occurs. This time ΔT is given by

$$\Delta T = 0.1\ \tau_{tot} + 0.8\ \tau_{trans}.$$

τ_{tot} is as defined in the previous section and τ_{trans} is the distance between channels divided by the local velocity of light. Both of these terms are functions of n, the number of channels carrying currents comparable to the first one to form. ΔT is then placed equal to $2\ \delta$ where δ is the standard deviation of the time of closure of the gap on a rising trigger pulse. This is related to the standard deviation of voltage breakdown of the gap by the relation

$$\delta(t) = \sigma(V)\ V(dV/dt)^{-1}$$

where dV/dt is evaluated at the point on the rising trigger waveform at which the gap fires.

For ordinary gaps (both gaseous and liquid) a good jitter is 2 per cent or so, but for edge plane gaps charged quickly the jitter can be down to a few tenths of a per cent. Typically ΔT is a fraction of a nanosecond - ordinary gaps require trigger pulses rising in 10 ns or so. However, for edge plane gaps, the trigger pulse can rise in 100 or more ns and still give multichannel operation. In one experiment 140 channels were obtained from a continuous edge plane gap. The expression has also been checked for liquid gap operation and more approximately for solid gaps, and gives answers in agreement to some 20 per cent for ΔT. The way it is used is to define a rate of rise of trigger pulse and as such, this sort of accuracy is ample. Fast trigger pulse generators are required to obtain multichannel operation and these must have impedances of 100 ohms or less. Of course the pulse length produced by the trigger generator need not be very great and the generator can be physically quite small. Mr. T. James of Culham has done much work on high speed systems and gaps in general and Ref. (5) gives an example of a high current low inductance multichannel gap he has developed.

The above treatment has been applied to other systems, both by ourselves and by others and reasonable agreement found. Indeed the crude theory can be used to estimate the number of channels formed and this too usually agrees with experiment. For further details and examples the multichannel note should be consulted - Ref. (6).

The preceding section shows that the inductive rise time approximately reduces as the number of channels (this assumes the inductance of the feed to the spark channel can be kept small). However the resistive phase only decreases by the inverse 1/3rd power of the number of channels.

Consequently in any set up using multichannel switching the resistive phase will eventually dominate and then large increases in numbers of channels will be needed to reduce the rise time further.

There may, of course, be advantages (such as erosion, uniformity of the produced electromagnetic wavefront) in increasing the number of channels beyond this point, but improved rise time is not usually one of them.

There is a further pitfall for the unwary, and this is if in order to achieve multichannel operation the mean field in the gap has been reduced. This is usually necessary to do because the gap will need to be fired considerably below its self break voltage for reliable operation. Under the condition where the rise time is already dominated by the resistive phase some 10 channels may be necessary just to do as well as a well designed single channel gap.

All in all multichannel switching is best left as a last resort even though the resulting pictures are very pretty when operation in this mode is finally achieved.

This last comment is a general one, it always pays to use the simplest technology you can just scrape by with, I have seen a number of systems where imaginary or occasional requirements have driven the designer to more complex technology than need be. The results have nearly always been expensive, time consuming and a final operation usually no better than the simple approach would have given. Crudely put, don't make life more difficult than you have to for the poor bastard, you will pay for it if you do.

SOME COMMENTS ON THE BREAKDOWN PHYSICS

When the voltage on a uniform field gas gap is raised beyond the threshold voltage, breakdown starts with the emission of an electron, nearly always from the negative electrode, unless special arrangements have been made to produce ionisation in the body of the gas. This primary electron multiplies in a Townsend avalanche process. After some 15 to 20 generations the self field of the produced charge begins to dominate over the applied field and the avalanche process speeds up. The measurements of Felsenthal and Proud (Ref. (7)) basically cover the first part of this process. However as the later processes only usually take a small time, their results give a lower but often realistic limit to the breakdown time. After the full avalanche process is complete a weakly ionised column bridges the electrodes narrowing down where it approaches the site of the initiating electron. The resistance falls fairly fast but the current is still rather low. The heating of the weakly ionised column is distinctly non uniform, being mainly concentrated at the narrow base where both the field and current density are highest. It is in this region that the gas warms fastest and hence expands. A weak shock moves outwards and a rarefaction moves inwards and the pressure (density) drops by one or more orders of magnitude. While the energy being deposited in main the part of the column would take a very long time to heat this to plasma temperatures, it can heat the constricted waist low density region quickly. A brightly glowing streamer then moves up very rapidly until a relatively thin channel of plasma links the electrodes. At this point what has been referred to earlier as the resistive phase begins.

The main aspects of the above model are best seen in the work on streamer chambers, but are also clear in some sorts of surface guided sparks. The somewhat unusual phase is the hydrodynamic one which precedes the streamer transit one. On liquid breakdown a hydrodynamic phase also occurs where the dense liquid is pushed aside by a bubble formed at the top of a whisker or sharp point. Only when this bubble phase has elongated to about 10^{-3} cm in length can the fundamental intrinsic breakdown process start in the liquid. Returning to gas breakdown, when the thin channel of plasma has bridged between the electrodes, the resistivity of this cannot fall much without very high temperatures being reached. For gases other than hydrogen and helium the resistivity is roughly inversely proportional to the square root of the plasma temperature. Hence the way the resistance of the channel decreases further is by the cross section area increasing. This it does by an outwards moving shockwave, the temperature behind this being a few eV. For very high field gradients down the plasma channel this is possibly a radiation driven shockwave. When the resistance of the channel equals the effective one of the generator, energy is being delivered to the spark channel at its maximum rate. The shockwave continues to move radially out until the resistance of the channel is much lower than that of the generator and basically the switch has then closed. Thus the measurements of Ref. (7) are a lower limit to the breakdown time, but usually not grossly so. However under some circumstances the hydrodynamic and streamer phase may become the dominant term in the breakdown time. This happens when the spark gap is rapidly pulse charged. For this case the breakdown time falls in agreement with the results of Ref. (7) but when it reaches typically a few tens of nanoseconds the breakdown time stops decreasing as a powerful function of the applied field and only falls relatively slowly as the later processes become dominant.

Returning now to the original emission of the initiating electron, the breakdown of pulse charge gaps can sometimes be made much more regular by uv irradiating the electrodes. This ensures that there is no statistical wait for an initiating electron. However such irradiation can be over done. The gas discharge laser community go to great lengths to provide a large uniform flux of uv over the volume of their devices so that a uniform glow discharge establishes itself, and sparking is delayed for as long as possible. The converse can happen to pulse charged uv irradiated gaps when the applied pulse voltage will fall to about the DC breakdown voltage for some tens of nanoseconds and then collapse completely as a spark channel finally bridges across the gap. Such poor performance usually happens with an irradiated gap working at one or two atmospheres pressure. The cure is either to pressurise the gap to several atmospheres (when the glow phase is very short) or to make the uv irradiation non uniform over the electrode and not too intense. For highly pressurised SF_6 gaps uv irradiation is usually unnecessary as the microscopically rough surface emits copious electrons. For sharp edge plane gaps (and the trigger electrode of field distortion gaps) some form of field emission provides plenty of electrons provided the field local to the edge exceeds 300 kV/cm or thereabouts. This is one reason adequately sharp edge or point plane gaps have such a low scatter of closure voltages when pulse charges.

Laser triggered gaps usually operate by basically forming a sharp plasma column directly on one of the electrodes and the streamer process starts out from this. At much higher powers laser triggering may be able to bridge all or most of the gap with a plasma column in which case most of the resistive phase may vanish. However to do this in pressurised SF_6 requires very large laser powers indeed, since the laser must deposit many tens of joules of energy directly into the gas for each cm of spark gap length.

The above model of gas breakdown has rather more phases than has been described and also is a personal one which should not be taken too seriously. However it can sometimes be useful in explaining observations or devising unusual gaps. For instance one of the disadvantages of the pulse charged edge plane gap is that it works at a considerably lower field than the uniform field gap. Consequently we devised a gap which has some of the features of both gaps. This was known as the I.B. (intentionally buggered) gap. In this a near uniform field gap had a small thin cylinder

SWITCHING

let into one or both electrodes. For operation in air at atmospheric pressure this stuck out of the electrode about 5 mm for gaps with several cms separation. For SF_6 the stick out was 1-2 mm. This indeed gave pulse breakdown voltages close to the uniform field ones, but with much reduced scatter. The difficulty was that the edge had to be very sharp to provide a field of about 300 kV/cm on its surface and fairly quickly became eroded with use. A series of thin sharp cylinders let into the electrode could have extended its working life but was a bit too much trouble.

GENERAL COMMENTS ON SPARK GAPS

Even restricting the field to switches using gases, liquids and solids there are a very wide range of possible types. In the next three sections I shall concentrate on aspects of gaps with which we have had experience. However a few general personal comments may be of interest, but too much notice should not be taken of them.

Firstly the higher the voltage, the easier it is to operate the gap, the difficulty is usually in preventing it operating prematurely such as by tracking. An additional pit fall is trying to operate a gas switch beyond the region where the breakdown voltage is roughly proportional to pressure. As has been mentioned earlier in this region, the average breakdown voltage falls well below the expected curve and the scatter increases dramatically. If the gap is operated at say twice the pressure at which the breakdown curve turns over, only an extra 10% or 20% increase in safe working voltage may result. However the experimenter is trying to trigger the gap at about half its proper "intrinsic" breakdown field, and hence fast low jitter operation is unlikely.

Low voltage gaps (say working at 10 kV and below) are very stable in operation usually, but relatively difficult to trigger quickly and consistently. Usually, but not always, uv irradiation is required since in general the difficulty is to initiate the breakdown process.

Turning now to some types of gaps, these can be formed from a series of gaps in series (such as the rope or chain switch). Indeed most three element gaps are two gaps in series. However there exist types with 10 or 20 units. In these the gaps after the first are usually two element ones although this is not necessarily so. In general such gaps have application for high voltage pulse charged use and will not be considered here further, as I have had no real experience with them.

Being rather ruthless, common triggered gas gaps can be listed as

Laser Triggered
Cascade Gaps
Trigatrons
Field Distortion Gaps.

Some years ago we were fairly strongly against laser triggered gaps unless you were working in laser laboratory. However lasers have become much more available than was the case then. I would still maintain that for 90% of spark gap usage, a properly designed standard gap can do the job in a tenth of the time, cost and effort. However, there is one class of application where in my opinion this may not be any longer true. This is pulse charged high voltage ($\simeq$ 1 MV) switching where very low jitter is required. One difficulty is the optical plumbing necessary to keep the laser focussed on an electrode of the gap (big machines jump around a lot). However plumbing few hundred kV trigger signals is not an easy problem, although to my mind, more soluble. Also in the laser gap the discharges are generally located along one axis, and erosion (such as in the trigatron) may be a problem. I personally still regard this area of application as being an open one, but this area apart I would urge people to stay with standard gaps.

Cascade gaps were developed by the plasma physics groups where large banks of tens of kV capacitors had to be reliably triggered. For most work they need a small irradiation gap in the triggered electrode, have a higher inductance than need be and are also fairly bulky and expensive.

Trigatrons have a small electrode in or near one of the main electrodes. A discharge is struck between these two and then causes breakdown in the main gap. In most versions the discharge goes via the trigger pin and erosion is a serious problem. There are versions however where this does not happen, these tend to look like very asymmetric versions of the field distortion gap and can indeed be considered as such. For most work I consider the trigatron has little to offer over the field distortion gap and has a number of potential disadvantages. If the trigger electrode is small (such as a square cut rod or pin) the location of this for reliable prompt operation is critical and also usually its sharpness. As such erosion is a serious worry, as it can alter the position of this electrode and blunt it. For high voltage pulse charged operation, the trigatron has the advantage that the triggering electrode is mounted on one of the electrodes and can be fed from behind this. However there are versions of the field distortion gas (the V/n gap, see below) that also has this advantage, if real advantage it be. As with the field distortion gap the length is a minimum and hence its inductance, but all in all the

SWITCHING

perceptive reader will have noticed the author's faint preference for the latter gap.

For most applications I prefer the field distortion gap which, because of the small radius on the trigger electrode, does not need any extra irradiation, as field emission from the rough trigger provides the necessary initiating electrons. In addition, the length of the gap, which is closely related to the inductance, is a minimum in this gap and it can be cheaply and quickly made. It can also lead to the spark channels being well distributed along or around the main electrodes, so erosion is a smaller problem where this is a factor of importance. Figure 10a-2 shows a cross section of a field distortion gas gap (which also goes by the name mid plane gap). The gap can use either ball bearings as spherical electrodes or rods for a cylindrical electrode gap. The trigger electrode need not be in the middle of the gap and, indeed, in one mode of operation it should be offset. The field distortion or mid plane gap can be easily modified for use with liquids or solids and indeed for solids this form of construction is so simple as to be positively moronic.

TWO ELEMENT GAPS

Figure 10a-2 without the triggered electrode represents two element gaps, either in axial cross section using balls or in strip line configurations using cylinders. In either event it is well worthwhile filing down the top of the balls or the rods so as to

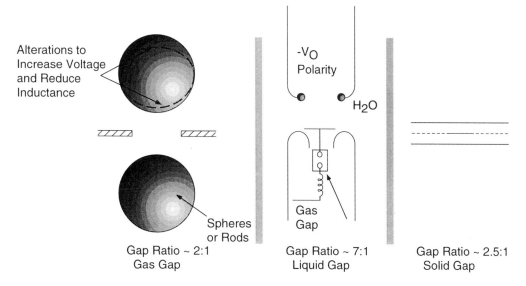

Fig. 10a-2 Field Distortion Gaps.

make the gap closer to a uniform field gap. Also the back of the sphere or cylinder can be cut off to decrease the gap length still further. However care must be exercised that the tracking length along the wall (if the gap is pressurised) does not become too small.

Pressurisation of the gap is relatively simple and enables the operating voltage of the gap to be changed from a distance.

Two element rail or cylinder gaps can have low inductances. The single channel breakdown is arranged to happen in the central region of the rails and the rest of the electrodes act as pressurised gas insulated parallel-wire feeds to the spark channel. For few hundred kV operation inductances of less than 20 nH are obtainable with 30 cm wide gaps. Also a 5 cm OD body with 10 cm long fins stuck on it to prevent external tracking operated at up to 800 kV with a pulse charge time of a few hundred ns.

One very simple and highly effective solid gap is a piece of polythene or other plastic with few mil copper electrodes stuck on it. A blunt thumb tack driven into the outside of the metal with a hammer deforms this and the plastic, thinning this down until intrinsic breakdown occurs at a field of about 8 MV/cm. The current then flows into the spark channel through a biconic tapered transmission line from the undeformed disc feed. Altogether it is simple and has about the best performance that I know.

Apart from laser triggering the two element gas gaps can be made triggerable with a neat trick. A saturable inductance (ferrite is a good material) is placed round the feed to one electrode. For short pulses applied on the gap side of this, the inductance enables the gap to be over-volted and triggered. After gap closure the inductance drops greatly when the magnetic core material saturates.

THREE ELEMENT GAPS

Gas Gaps

Maxwell Laboratories, Ion Physics of Burlington, Mass. and Sandia Corporation have all extended the use of pulse charged trigatrons to 3 MV or so, and operated several switches in very close proximity in what is essentially a multichannel mode of operation. Mr. I.D. Smith of Physics International has originated a version of the field distortion gap known as the V/n gap. This gap is pulse charged and uses pressurised SF_6 and the field distortion trigger electrode is very close to one electrode, typically at one tenth or one twentieth of the total gap spacing. Thus, a 2 MV gap needs only 100 kV or so of trigger volts to fire it.

SWITCHING

Descending from the rarified levels of very high voltage pulse technology to the medium voltage level, the three element field distortion gap operates over a wide range DC charged. Versions have been used over the range 200 kV down to 10 kV with no difficulty. Working at 80% of the self break voltage the gaps can be triggered in ten nanoseconds or less with a jitter of fractions of a nanosecond. The breakdown time of such gaps can be calculated using point/edge plane breakdown relation given in Ref (1). For the most rapid breakdown the off set or simultaneous mode of breakdown is best where the triggering electrode is spaced at roughly 2 : 1 in the gap and the two parts of the gap are broken at approximately the same time. A fast rising trigger pulse of amplitude roughly equal to the self break of the gap is needed, but as the capacity it is driving is low, this does not present serious problems.

Recently, versions of a triggered rail gap with a self break voltage in air of 15 kV have been triggered down to 6 kV and it is hoped to extend the working range of these gaps downwards significantly further in the near future. Work in the States has shown that similar gaps pressurised can have jitters of $\pm$ 0.2 ns and it is considered that by using pressurised hydrogen, this could be reduced further, if it was ever necessary.

On the practical side pulse charged three electrode gaps need careful balancing, if capacitors are used to hold the trigger electrode at its proper potential so that the gap does not operate early because the potential division of this has been disturbed. This difficulty can be obviated by putting a small gap in between the trigger and the trigger pulse generator. Of course, this has to hold off the pulse charging waveform of the trigger electrode but this is not too difficult to arrange. When the trigger pulse arrives it breaks this isolating gap and operates the gap. The isolating gap can be used to pulse sharpen the trigger pulse as well if this is desirable. Some form of auxiliary surface discharge uv irradiator may be necessary to break the isolating gap at reasonable voltages and so avoid having to provide an unnecessarily large trigger pulse. The spacing of the isolating gap can also be halved by back biasing it with a half voltage DC supply of the appropriate polarity.

Three element field distortion gaps can also be used as clamp gaps, i.e. ones that fire at nearly zero voltage but this application is a wide area where much work has been done in the plasma physics research establishments and the reports of these should be consulted.

As an example of the triggering range and fast firing possible with field distortion gaps, Figs 10a-3 and 10a-4 show the self breakdown voltage of a gap with spherical 1" diameter ball bearing

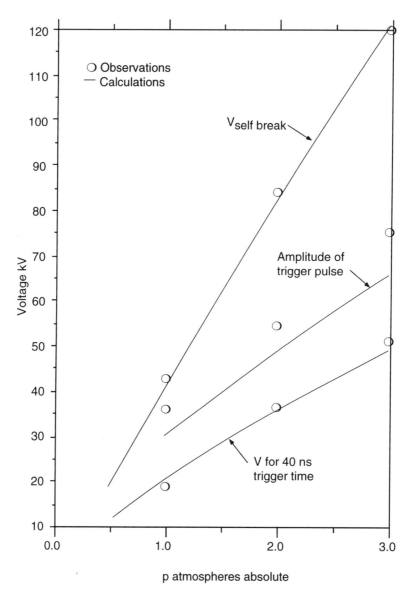

Fig. 10a-3 Trigger voltage vs pressure for SF_6 gap with 0.5 cm gap spacing and optimum trigger spacing.

electrodes operated with a 0.5 cm gap for a range of SF_6 pressures. The figure gives the breakdown voltage and also the voltage at which the gap closes after a delay of 40 ns. The curve labelled amplitude of trigger pulse is the value at which triggering starts in the gap. In a real system a trigger pulse more than this voltage would be needed, especially as it is desirable to fire on a fairly rapidly rising portion of the trigger waveform.

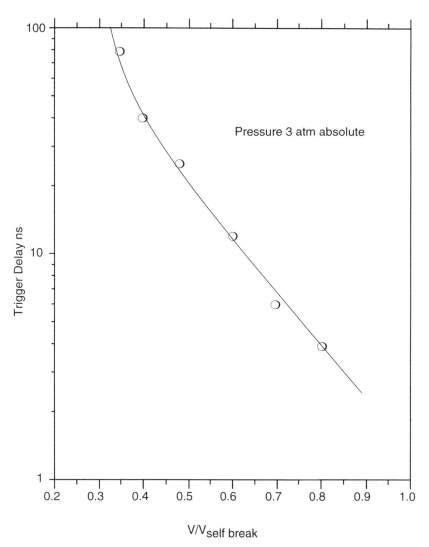

Fig. 10a-4 Trigger delay vs voltage for SF_6 gap with 0.5 cm gap spacing and optimum trigger spacing.

A mention should be made that with atmospheric SF_6 in a gap with fairly small spacing, breakdown voltages above that deduced from the values in the literature can be observed. These occur after fairly frequent low energy shots and are believed to be due to the formation of insulating layers (possibly sulphur) on the surface of the electrodes. These layers, if they exist, can hold up to ten kV or so in extreme cases. Filling with air and sparking removes most of the effect until the layer reforms. Also shown in Figure 10a-3 are the calculated curves as well as the observed data and the agreement is quite reasonable. The calculations are made using the edge plane breakdown relations. Figure 10a-4 shows the triggering delay for the gap at 3 atmospheres, the calculated values are not shown in this case but have the general form of the observed curve.

The optimum trigger spacing ratio in the case was 0.3 : 0.7 and essentially independant of pressure over the pressure range investigated. Jitter measurements were not made for this gap (they are quite tricky in the sub ns range) but would be expected to be ~ ± 5% of triggering time.

Liquid Gap

Figure 10a-2 shows the general form that has been used for pulse charged liquid gaps. Water gaps have been operated up to 5 MV and have shown modest multichannel operation at these levels. They have also been operated with oil by Mr I.D. Smith of Physics International, San Leandro, at levels up to 10 MV, where again they have performed well. Mr Smith has contributed greatly to many aspects of pulsed high voltage technology and it is a pleasure to record this.

In the version shown, the trigger pulse is simply derived from the energy stored within the gap, but better multichannel performance would be obtained by using an external pulse to operate the gap. In the case of water, the gap ratio is more like 7:1 but significant expertise is still required to operate liquid gaps at the multimegavolt level.

For simple, over-volted, untriggered gaps, both oil and water perform well and fast rising pulses can be simply produced. With care, jitters in the voltage breakdown can be 3 or 4 per cent and closure gradients of 400 kV/cm in oil and 300 kV/cm in water obtained for microsecond pulse charge times. These can be doubled for faster charge times and with good design.

Liquid gaps represent an attractive approach for switching at the multimegavolt level and the only other type gap probably worth looking at is pressurised SF_6 or similar gas mixtures.

Solid Gaps

Two versions of intrinsic breakdown solid gaps were developed at AWRE some 16 years ago. The first type uses an array of 50 needles to stab holes in, say, a 60 thou. polythene sheet. By varying the depth of stab, the operating range could be changed from 40 to 150 kV with the stabs positive. As there is a strong polarity effect on polythene with the stabs negative, the range was

SWITCHING

from 120 to 250 kV or more. This switch could be stacked for higher voltages and a pair of them made a simple triggerable switch. The standard deviation of the breakdown voltage could be as low as 2 per cent and the switches had very similar characteristics when used from DC down to 10 ns charging times. In one quite old system, 40 such switches were regularly fired simultaneously at 200 kV, producing a 4 MA current rising in 8 ns or so, i.e. a di/dt of 5×10^{14} amps per second. Such switches are still in use in various systems and have been operated up to 1 1/2 MV pulse charged in a system with one master switch and two slave gaps of the next type.

A second and even lower inductance solid dielectric switch was developed along the lines shown in Figure 10a-2. Two mylar sheets of different thicknesses enclose a trigger foil whose edges are sealed with a thin film of transformer oil or silicon grease. Copper or aluminium main electrodes are added to complete the switch. The inductance of a single channel switch of this kind is very low and the resistive phase is small as well. For DC charged high current banks a multichannel version of this switch is easily made and by injecting a pulse rising to, say, 40 kV in a few ns along the line formed by the trigger foil, hundreds of current carrying channels can be made to occur. Currents of tens of megamps can be switched by a couple of such switches and their inductance is extremely small. Pulse charged versions of this gap have been operated up to 1.5 MV.

Solid dielectric gaps have to be replaced each time, of course, but in high performance banks with a relatively low rate of usage, they can be a very cheap substitute for tens or hundreds of more orthodox gaps.

FOUR ELEMENT GAPS

The reader should be warned that low voltage spark gaps have always been a hang up of mine. The difficulty is that they have to use some trick to get the initiating electrons into the gap. However they are cheap and easy to build and can replace a mass of electronics.

An early three element version called the corona gap used a surface discharge across a thin film of plastic to provide the start electrons (Ref. (8), HUN2). This gap needed about a 2-3 kV trigger pulse and fired in 40 ns or more when working at 10 kV or thereabouts. The disadvantage of this gap is that the trigger capacity was around 40 pf and it could not carry heavy currents.

The later 4 element gap relies on uv irradiation from a small surface flashover sub gap, has a much quicker response and needs of the same order trigger pulse into only 3 pf. It can also carry substantial currents.

The gap is constructed with 4 electrodes and three gaps. For lowest trigger operation the trigger pulse is applied to the smallest gap which is the middle one. After firing this the voltage is transferred to the next largest spacing and then the final biggest gap is broken by the applied DC volts.

In the most sensitive mode (and there are many versions of it) a 0.55 kV pulse into 3 pf will fire a 9 kV gap (breakdown 10 kV) in 17 ns with about 1 ns standard deviation (see Ref (9)).

A wide operating range version works from 4 to 9 kV with a self break voltage of 10 kV. In this case about a 5 kV trigger pulse is needed and the overall delay at 8 kV is about 8 ns with subnanosecond jitter.

Tommy Storr devised a further pair of variants (Ref. (10)) which included a Marx coupled avalanche transistor circuit firing with a 20 volt input. One of these (an 8 kV version) had an overall delay of 25-32 ns and a sub nanosecond jitter. The other was a 15 kV gap and this was used to provide a 300 kV Marx trigger pulse via an air cored pulse transformer. The delay of the 15 kV gap from 20 volt trigger pulse in was 37 ns and the standard deviation about 1 ns. This gap had to carry substantial current and had a low inductance.

I honestly don't know whether these have any real use nowadays, but the basic gap takes half a day to build and they are rather fun to play with.

They also exhibited multichannel operation over quite wide ranges of the parameters and hence might be very low inductance indeed. Operated under pressurised hydrogen I believe they would be very fast and have very low jitters.

REFERENCES

The group has written a fair number of notes. These were never complete enough to justify publication it seemed to us and we could not spare the time to make them better. A fairly complete set of references can be found in the HUN lecture notes (Ref. (8)) which is also an update to some extent of the old Nanosecond Pulse Techniques (Ref (1)). Karl Baum has done a magnificent job of reproducing many of our notes in a more legible form and his references are given first in the list below, the original reference numbers being given second.

REFERENCES:

1. "Nanosecond Pulse Techniques Circuit and Electromagnetic System Design, Note 4, J.C. Martin, April 1970, original SSWA/JCM/704/49. (Chapter 4)
2. "High Speed Breakdown of Pressurised Sulphur Hexaflouride and Air In Nearly Uniform Gaps, J.C. Martin, SSWA/JCM/732/380.
3. "Results from Two Pressurised Edge Plane Gaps", J.C. Martin, and I. Grimshaw, SSWA/JCM/729/319. (Section 10 d)
4. "The Resistive Phase of a High Voltage Water Spark", J. Pace Van Devender, J Appl. Phys. 49 (5), May 1978.
5. "A High Current 60 kV Multiple Arc Spark Gap Switch of 1.7 nH Inductance", T.E. James, CLM-P212, Culham Labs, 1969.
6. "Multichannel Gaps", J.C. Martin, AFWL Switching Note No. 10, SSWA/ JCM/703/27. (Section 10c)
7. "Nanosecond Pulse Breakdown in Gases", D. Felsenthal and J.M. Proud, Tech. Report No. RADC-TR-65-142, Rome Air Development Centre, Griffins Air Force Base, June 1965.
8. HUN Lecture Notes, J.C. Martin, DS27, SN22, C&EMN16, RP60, C&EMN17, originally HUN 1-5 and an additional update, HUN 1A, 1980. (Section 5a)
9. "Four Element Low Voltage Irradiated Spark Gap", J.C. Martin, August 1977, AFWL Switching Note No. 26, originally SSWA/JCM/778/ 514.(Section 10e)
10. "Further Four Element Spark Gaps", T.H. Storr, February 1979, SSWA/THS/792/29.

Section 10b

DURATION OF THE RESISTIVE PHASE AND INDUCTANCE OF SPARK CHANNELS[*]

J.C. Martin

The following is a brief resume of the formulae used for the resistive phase of a hot spark channel and some other comments relating to spark channels formed under various conditions.

It should be stated that while a considerable quantity of work was done on the duration of the resistive phase in gas sparks, considerably less data has been obtained on spark channels in liquids and only a little on those in solids. However the underlying theoretical explanation and calculations give some confidence that the answers given by the relation are reasonably correct.

The basic assumption is that the circuit impedance feeding the spark channel is of reasonably low impedance (less than 1000 ohms, say) and that after the normal ionization build up has taken place the warm weakly-ionised channel expands. This leads to the generation of a lower pressure filament at the center of the warm (~ 5000°) gas channel and the main current starts to flow in this. The low density material can then be heated to temperatures above 20,000° and this leads to the formation of a plasma column. The above sequence has been photographed at AWRE and typically takes tens of nanoseconds to complete in gases. The plasma column at first sight can lower its resistance by heating itself up, but this causes the degree of ionisation of the air (or other gases) to increase and in fact, for a temperature range from 2×10^4 to over

---

[*] SSWA/JCM/1065/25
AFWL Switching Notes
Note 9

10^5 the resistance is only proportional to $T^{-1/2}$. Thus only a modest reduction of resistivity is possible with a considerable increase in temperature. With gas channels of dimensions of more than a few 10^{-2} cms at these temperatures the column is blackbody and this leads to considerable radiative transfer of energy which prevents the current column becoming unstable. For instance, for the current to collapse in a column of half the diameter, the temperature of this must increase by a factor of about 16. This leads to an enormous increase in radiative transfer which returns the plasma column to a uniform temperature. It is thus reasonable to take the current carying column to be of approximately uniform temperature and resistivity up to the edge of the plasma.

The spark channel can now only decrease its resistance easily by increasing its area, which it does by a cylindrical explosion, the velocity of which is governed by the energy density in the plasma driving the shock outwards. Measurements of the velocity of expansion of the plasma sheath in air by means of 10 ns Kerr cell photographs have given velocities in good agreement with these calculated from the energy density measured to be in the spark channel. In general the terms $L(\partial i/\partial t)$ and $i(\partial L/\partial t)$ are small corrections to the measurements of the voltage across the channel until its impedance has fallen to a low level. The formula for the resistive phase duration calculated from the above considerations agrees tolerably well with that found experimentally and given below in the next section. It might be surprising that the same formula should cover a wide range of different substances but this is explained by the fact that to the first order the equation of state of multiply ionised low Z atoms does not differ to any great extent, except for hydrogen and helium.

The resistive phase measurements have all been done using transmission lines to drive the spark channels. This experimental detail seems important to me, since it is possible to obtain V and i from the same measurement providing the inductive terms are kept small by good layout. When the spark channel is formed across the transmission line, the voltage falls more or less exponentially and it is only when the voltage has dropped to about 10% of its starting value that deviations from the exponential shape may become detectable.* By this time nearly all the prompt energy has been delivered to the channel. Also in a pulse forming application, this fall of voltage is one of the limits on the pulse front available from any system.

---

* Editor's Note: This is as shown in Fig. 10b-1.

SWITCHING

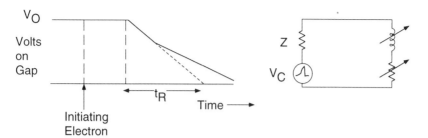

Fig. 10b-1 Definition of t_r and approximate circuit.

In the experiments with gas sparks transmission lines (or cables) of 100, 10 and 1 ohms were used and spark channel fields of 3×10^4 down to 5×10^3 volts/cm were used in air. Other pressures than atmospheric were employed and also gases of other densities. From these experiments (supported by the calculations) the following relation for the resistive phase was obtained:

$$\tau_r = \frac{88}{Z^{1/3} E^{4/3}} \left(\frac{\rho}{\rho_0}\right)^{1/2} \text{ ns}$$

where Z is the impedance driving the channel in ohms, E_1 the field in units of 10 kV/cm, ρ the density of the gas and ρ_0 the density of air at about N.T.P. The accuracy with which τ_r was measured was reasonable and the fit to the observations was to 10% or better in most of the measurements.

Measurements have since been made on spark channels in oil, water and polythene. These measurements are rather more difficult since the resistive phase in solids tends to be a few ns while that in gas channels can easily be made several hundred ns. The measurements in polythene gave a formula

$$\tau_r = \frac{5}{Z^{1/3} F_2^{4/3}} \text{ ns where } F_2 \text{ is now in MV/cm.}$$

This agrees surprisingly well with the gas observations where the value of ρ for polythene is inserted. The solid breakdown measurements could not clearly show the power of F_2 but this was assumed to be that given by the theoretical calculations.

The velocity of expansion of the plasma channel is not strongly dependent on the resistance of the circuit feeding it and for a rough estimation tends to be between 3×10^5 and 10^6 cm per sec. While an accurate value of the radus-time history can be obtained from the calculations during resistive phase, a crude value suffices to obtain the inductance of the channel. Thus an 8 ohm line feeding a spark channel with a means field gradient to start with of 1 MV/cm has a resistive phase of 2.5 ns. This gives a radius of roughly about 1.5×10^{-3} when the channel impedance has fallen to slightly less than the generator impedance. If the line feeding it is 10 cms radius and the spark channel is located in the line in a replaceable solid dielectric switch, the inductance is given by L = (2 ℓ) ln b/a where ℓ is the length of the spark channel, a = 1.5×1^{-3} and b = 10. This gives L $\simeq$ 18 ℓ nanohenry. For a switch working at 100 kV, and 1 MV/cm, ℓ is 10^{-1} cm and the inductance is 1.8 nH. This gives a τ_L, L/Z = 0.2 ns, very small compared with the resistive phase. The $i(\partial L/\partial t)$ term can be estimated for the same case by the fact that a further increase of channel diameter by (i.e. 2.718) will decrease L by 0.2 nH and that this will take a time of the order of 5 ns and hence $\partial L/\partial t \sim 0.04$ ohm which is very small compared with 8 ohms. Since the radius only occurs in the logarithmic term in the inductance calculations, errors in it have next to no effect on the value calculated. The energy deposed in the spark channel during the resistive phase is given by $(V^2/4Z)\tau_r$ where V is the voltage before the spark closes*. Measurements of the volume blown in materials by spark channels fed by short current pulses give values in reasonable agreement with this, considering the errors inherent in these measurements.

There are a number of circumstances where τ_r may appear longer or shorter than that calculated above. If the spark channel is formed along a curved field line (say at the edge of a metal conductor) the mean field may be significantly lower than that calculated from the voltage divided by the separation of the metal plates, and hence the resistive phase longer.

If the spark channel is formed by a long corona across a surface, then only the end of the corona is in a resistive phase because the time of advancement of the top of the streamer may permit the root of the channel to expand to a considerable cross section. I have a rough theory of corona tracking based on this point which is not wildly at variance with some measurements taken under carefully controlled conditions. Thus long channels whose conditions vary considerably along their length can give lower values of τ_r.

---

* Editor's Note: (Martin)
Discussion with Charlie concerning the switch channel energy loss resulted in some interesting observations. First, the channel energy loss in this equation should be $V^2/(2\ Z)* \ T_r$ given that the current and voltage are exponential. Second, more recent energy loss estimations are now available. Third, the main thrust of switching research at this time was to predict the rise time. These concepts, as shown in this and other notes, provided that capability.

A much more likely way that a low τ_r value may be obtained is when the current carrying breakdown channel branches. In compressed gases at 20 atmospheres and above and with liquids and solids it is frequently possible to show that the current is carried in a number of channels after a short single channel. Typically 5 or more heavy current carrying branches may reach the far electrode with many more channels visible which do not carry significant current. To show the effect of the τ_r, I will estimate the resistive phase and inductance for such a branching system where it is assumed for simplicity that one channel is 0.03 cm long and then 8 channels carry the current for a remaining distance of 0.07 cms. Again 100 kV starting voltage is assumed and a generator impedance of 8 ohms. For the single branch of the channel

$$\tau_1 = \frac{5}{2 \times (3.3f)^{4/3}} \text{ ns}$$

where $f \times 10^5$ is the voltage at the branching point.

The resistive phase of the 8 branches is given by

$$\tau_2 = \frac{5}{4 \times [1.4 (1-f)]^{4/3}} .$$

These two times have to be equal of course and this gives $f = 0.4$ and $\tau_1 = \tau_2$ 1.7 ns which is to be compared with the value previously calculated for the unbranched channel of 2.5 ns. This ratio seems to be fairly typical and for many applications a resistive phase of about 0.7 τ_r will be found to occur.

The inductance of the branched case can be estimated to be in the region of 1.4 nH compared with 1.8 nH previously obtained.

As will doubtless be apparent, no great claims for accuracy are made in any of these calculations; they are only to indicate the sort of conditions likely to be obtained in planning an experimental set up. The best that can be said for them is that we have not yet found any of the estimates we have made to be badly wrong.

Some difficulty may be found in determining Z in various circumstances. One of these is a condenser bank application. The first point to be made is that Z only appears as a low power and hence even crude calculations will suffice. In the case of a bank whose $\sqrt{LC}$ is comparable or lower than τ_r, Z may be taken equal to $\sqrt{L/C}$. For the usual case of banks which are slow, the impedance feeding the switch may frequently be estimated from the transmission line linking the switch with the capacitors, providing its electric length is of the order of or greater than τ_r.

A more complex case is provided by a transmission line switched by liquid gap with a rather low value of the mean field. In this case, r_L can be comparable to or greater than r_r calculated directly from the transmission line impedance. The impedance feeding the spark channel is obviously quite a lot higher than the impedance of the transmission line in this case. By estimating L (which is only slightly dependent on r_r) an effective impedance is obtained by putting

$$Z_{eff} \simeq Z + \omega L \simeq \frac{1.7\ L}{r_r} + Z.$$

This number is now used to calculate the reduced r and this is then combined with r_L. For the cases where $r_r \sim r_L$ and if these effects are assumed to be in series, the resulting $r \sim r_r + r_L$. This analysis is of course extremely crude but in a couple of instances it has been used to give answers in tolerable agreement with experiment. Mostly these difficulties only arise with solid or liquid switches working at poor field strengths and/or in very low impedance circuits.

I hope the above outline will be of use to you in estimating the rise time of your pulses and that you will not be too repelled by the violence done to the physics of the situation. I have considered a number of effects that might change the hydrodynamic explosion of the plasma column. Firstly pinching; in general it is difficult to get a high enough rate of rise of current to make the magnetic pressure exceed the particle pressure in the plasma column. The best chance seems to be in an overdriven gas gap when it may occur if the impedance driving the system is low but under these circumstances the inductance and rate of change of inductance can no longer be ignored and establishing the conditions for a pinch seem to me to be pretty difficult. Thermal conductivity expansion of the column; for very small channels, thermal conductivity may be a faster way of transferring energy outwards than by a hydrodynamic shock. But the self magnetic field reduces this effect and it only applies for very small channels well below anything considered in the above calculations. Radiative fronts; it is most possible that radiation transfer may outrun the hydrodynamic shock in those cases where very over-driven gaps are fed by low impedances (i.e. just the conditions considered above for a possible pinch effect). Thus even if pinch conditions were to be set up, energy expansion by a radiation front may prevent it occurring. In the case of sparks fed by very high impedances a fully fledged plasma column may never come into being and then of course the calculations are inapplicable. It is not easy to be sure at what level of impedance this will take place but I suspect it to be above 1000 ohms. It is of relevance that with liquid and solid systems the fields normally required to cause breakdown mean that enough energy is stored locally in the electrostatic field to provide a fairly low impedance to drive the channel, so that the question of fast spark channels fed by high impedance can only really occur in gases at lowish pressures.

After times of few τ_r, energy continues to be deposited in the channels and, more importantly, in the electrodes. Blow up of the electrodes in the case of a solid dielectric gap fed by short lines provides additional proof that the current densities are very high and hence the spark channels are initially very small in diameter. Energy is also deposited in motion of the channel if this is not coaxial in its return current, and even if it is, eventually magnetic wriggling occurs. However, nearly all these effects are quite small and for quite long times the major energy in the channel is that deposited during the resistive phase, when the impedance collapses from a few times Z to one over a few times this value. Mention should be made of underwater sparks which are used to provide shock working of metals. The banks that drive this are very slow and the resistive phase of the spark channel does not take much energy out of the system. The real energy seems to me to be damping in a period of 10 to 20 microseconds, when a stabilised arc is formed down the axis of a cylindrical gas bubble. Continued influx of cold atoms from the water interface and thermal conductivity outwards provide a cooling effect on the arc and this provides a long term, rather low, impedance, which eventually drains most of the energy out of the bank, provided its internal resistance is reasonably low.

Editors Note:

Attempts to validate the concept of the resistive phase by one of the editors (T.H. Martin) has resulted in further understanding of this phenomenon. Ref. Proceedings of IEEE International Pulsed Power Conference, 1993.

Section 10c

MULTICHANNEL GAPS*

J.C. MARTIN

> "It is too rash, too unadvised, too sudden,
> Too like the lightning, which doth cease to be
> Ere one can say it lightens."
> (Romeo and Juliet)

INTRODUCTION

In the general field of high speed pulse generators, the breakdown strengths of gases, liquids and solids in uniform and non-uniform fields can be estimated well enough for design purposes, using approximate relations and measurements obtained in the past few years. This suffices to do a first design for the high speed section and the feed from this to the load. If a single channel gap is used, approximate treatments exist to enable the rise time of the pulse to be obtained to a rough but adequate accuracy. However, if this rise time is inadequate, the position becomes more complicated. Single gaps, transit time separated, can be employed in some designs, but then the resulting combined pulse must have a rise time comparable with the jitter of the individual gaps. Such an approach is usually expensive and where low impedances and high voltages are required, it may be physically impossible to realise.

Several years ago, when the stabbed solid dielectric switch was developed, multichannel operation of a number of closely spaced spark channels was readily obtained. The corresponding development of liquid and gaseous multichannel gaps has only occurred in the last couple of years. The reason why multichannel operation of solid switches was obtained so easily is a combination of the

---

* SSWA/JCM/703/27
  AFWL Switching Notes
  NOTE 10
  5 MAR 1970

relatively long resistive phase of the spark channels and the fact that trigger pulses of several hundred kilovolts rising in a few nanoseconds were easily generated by using a solid master gap.

In contrast to an array of single gaps, where transit time isolation is used, multichannel gaps have continuous electrodes. Transit time isolation may play some part in enabling a number of spark channels to take comparable currents, but this is usually of minor importance and is only included in the analysis for completeness. The continuous electrodes of the gap may be serrated or have localised raised areas on them, but this is used to stabilise the number of channels formed at a number lower than that which would have been obtained with continuous electrodes.

There are several reasons why multichannel gaps may be useful, but in general they fall into two classes. Firstly a lower inductance is obtained and to a lesser extent a reduced resistive phase. Thus faster rise time pulses may be generated and in high voltage, low impedance systems, the rise time can be made to decrease an order of magnitude. Secondly electrode erosion may be very serious in a single channel gap and this is reduced drastically where the current is carried in a reasonable number of channels. A secondary effect of using large continuous electrodes is that each erosion as does take place, causes little change in the geometry of the gap and, in addition, magnetic forces and high current density blow up are considerably reduced.

In preliminary experiments I assumed that if the jitter in closure of the channels in the gap was of the order of the fall time of the voltage across the spark channels, multiple operation should result. A couple of failures over a period of a year convinced me that a time more like a tenth of this was required before the desired end could be achieved. Consideration of the operation of a couple of different solid dielectric gaps and of a surface air gap confirmed that in my early experiments to obtain multichannel gas gaps I was being wildly optimistic and, sadder but possibly wiser, I designed an overdriven mid plane gap which indeed managed to achieve a number of channels. Needless to say, this number was always less than my expectations but at least I was making progress. Over a couple of years the analysis of a number of cases where multichannel operation was intentionally or accidentally observed enabled a rough relation to be developed which worked crudely over a wide range of conditions.

The work reported here was brought on by an attack of conscience because while I was happy my approximate relation worked, I was reluctant to write it up without something more respectable in the way of experimental backing. Within the limits of the experiments reported, I consider that this has now been done. As usual, accuracies of a few percent are neither sought nor needed for design purposes and all the following should be read in the light of this requirement.

If the relationship is accepted, and if the jitter of a single gap or portion of a gap can be estimated or measured, the trigger pulse required to obtain any number of channels can then be calculated. In general, some threshold rate of rise of the trigger pulse is required and provided the trigger pulse circuit can exceed this, adequate performance will result. Thus it is hoped that the treatment suggested can be used to solve, at the design stage, the switch problems of advanced pulse generators.

In addition to the experimental work reported, a number of other points are covered, including some of the many ways of not getting multichannel operation. I live in hope that I have uncovered most of these, but continued experience convinces me that in the field of multichannel operation Dame Nature is unusually inventive in finding evasive tactics and frustrating the ardent experimenter. However, as in another field of endeavor, success is even sweeter when it is ultimately obtained.

THE APPROXIMATE RELATION

In a multichannel gas gap as the first closing channel starts to take substantial current, the voltage on the electrodes begins to collapse. Other channels closing after the first will have less volts across them and hence will finish up carrying significantly less current when the gap voltage has fallen to a low value. After this phase is over, the currents carried by the various channels will redistribute themselves over a very much longer time scale. The current eventually finishes up flowing in one channel, after a time typically three or more orders of magnitude longer than in the first phase. The analysis given here is restricted in general to the first phase, when the gap voltage falls to a low value, although some data on the long term current rearrangement is given towards the end of this note.

In the case of channels where the current is largely controlled by the inductance of the spark, a naive approximate theoretical treatment is given in the next section which shows that if a second channel closes after the voltage has fallen by 15 per cent, the ultimate current in the two channels is divided roughly in the ratio of two to one at the time of major interest. The resistive phase is even more voltage dependent and a channel going over to the plasma phase after such a 15 per cent voltage drop has taken place has an even greater ratio of currents carried. Unfortunately the approximate relation which gives the duration of the resistive phase does not give any information about the form of the early voltage time history. However, as the plasma channel has a strong negative resistance, it is reasonable that small drops of voltage across later channels should have a disproportionately large effect on the currents they finally carry.

The rate of rise of a voltage pulse from a single spark channel (and hence fall of voltage across the electrodes) has been treated in "Duration of the Resistive Phase and Inductance of Spark Channels", reference SSWA/JCM/1065/25.* In this, two phases are considered, the inductive (τ_L) and the resistive phase (τ_r). These are both e-folding times and to obtain the resulting τ_{tot} they are added. The treatment only applies to sparks where reasonable currents are carried and is restricted to those cases where the impedance feeding each channel is a few hundred ohms at most.

In any multichannel gap a whole spectrum of current carrying channels close and, in obtaining an effective number of channels, it is necessary to decide at what fraction of the current carried by the earliest channel a cut off is to be applied. Some of the difficulties about this are mentioned in the next section but even if these were solved in any given application, in order to decide the effective number of full current channels any particular gap is giving, the distribution function of the actual currents would have to be known. In order to simplify a very complex situation, two cut off levels of current are used below, one counting all channels with about 45 per cent or more of the channel with maximum current. The other level corresponds roughly to 35 per cent of this current. I personally prefer to use the first cut off level in calculations, but in order to increase the statistics in most of the experiments described the second lower level has been used, to give a larger number of channels to compare with the calculations. The two criteria are referred to as A, and B, respectively, in what follows. While it is realised that the two extreme cases of purely inductively and resistively controlled spark channels are different, the relatively crude analysis proposed does not take this into account and the time span during which effective additional channels can complete is obtained by taking a fraction of the total τ_{tot}. Thus the time ΔT during which useful channels may form is given by

$$\Delta T = f\ \tau_{tot}$$

For condition A, f = 0.1 and for condition B, f = 0.15. To this time must be added an allowance for transit time isolation. If the electrodes are of length ℓ then the velocity of light isolation is $\ell \div nc$ where n is the number of channels and c is the local velocity of light in the dielectric. The whole of this time (τ_{trans}) is not added to ΔT because the channels are not uniformly distributed and hence only 0.8 of it is added to make a rough

---

* AFWL Switching Note 9 (Footnote)

SWITCHING

allowance for this and other effects. Thus the expression for the useful interval during which additional channels may close is (for condition A)

$$\Delta T = 0.1\ \tau_{tot} + 0.8\ \tau_{trans} \tag{1}$$

Now τ_{tot} depends on the effective number of channels carrying current (n), the inductive term going as $1/n$ and the resistive phase as $1/n^{1/3}$. The transit time isolation term also goes as $1/n$, of course.

Any gap operated in a single channel mode has a scatter in its breakdown voltage which can be characterised by a standard deviation $\sigma(V)$ and this is the quantity experimentally measured, as a rule. For a trigger pulse rising at a reasonable rate, this can be turned into a real deviation in time, $\delta(t)$, by using the following relation

$$\delta(t) = \sigma(V)\ V\ (dV/dt)^{-1} = \sigma(t)\ T. \tag{1a}*$$

where dV/dt is evaluated at the point on the rising wave form at which the gap fires.

To an adequate degree of accuracy the value of ΔT calculated from Equation (1a) is equated to $2\sigma(t)T$.*

For example, an edge plane gap pulse charged linearly in about 100 ns can have a $\sigma(V)$ of the order of 0.3 per cent. Thus ΔT has a value of 0.6 ns. Given the physical characteristics of the gap and the electrical circuit feeding it, the number of channels which give values of the terms on the R.H.S. of the equation which satisfy Equation (1) can now be obtained.

To obtain multichannel operation, two obvious ways of improving any system are: (a) to reduce $\sigma(V)$ of the gap; and (b) to increase dV/dt; both of these reducing ΔT. In typical applications, initially the inductive term dominates the value of τ_{tot} and hence n increases inversely as ΔT. As the number of channels increases, the resistive phase term begins to dominate and n goes inversely as ΔT^3 eventually. In point or edge plane gaps in gases, increasing the rate of rise of the voltage also causes the $\sigma(V)$ to decrease and the two effects combine to improve the gap performance quickly. With edge plane gaps in good condition, fed from a reasonably low impedance transmission line, multichannel operation can set in with pulses rising in 200 or 300 ns and a large number of channels can result from pulses rising in several tens of nanoseconds. With ordinary roughish gaps in gases, or with good liquid gaps, $\sigma(V)$ is more ordinarily of the order of 2 per cent and pulse rise times of tens of nanoseconds are needed to start multichannel operation. Solid gaps, well made, have similar standard deviations to those of

---

*   Modied by editors

liquids and pulses rising in 10 ns can cause substantial numbers of breakdown sites. As was mentioned earlier, the first strict multi-channel operation achieved at AWRE was with solid gaps and pulse rise times of this order and less were certainly generated by the solid master gap.

Several examples of calculations using the proposed relation are given in the experimental and application sections below. It should be explained that in these examples, when evaluating Equation (1) for various values of n, the value of r_{trans} is evaluated for n = 1. This is obviously an absurdity and it is included partly because it is difficult to remember to leave it out and partly because it enables estimation of the operation where, say, 1 1/2 effective channels are produced. This of course is only a statistic average over a number of shots. The reader's indulgence is begged for the author's idiosynractic behaviour in this matter.

INDUCTIVE CALCULATIONS

The simple inductive fall time for a constant diameter conductor driven by a circuit of impedance Z is given by r_L = L/Z where L is the inductance of the conductor. The extra inductance from the transmission line to the conducting channel is ignored, of course, something that in practical cases is usually inaccurate. In this simplification that verges on the moronic, the effect of closing a second similar channel at an arbitrary time after the first one can be easily calculated. In Fig. 10c-1, curves A and B show the normalized voltage collapse for one and two channels respectively, where in the latter case both channels exist from t = 0. The units of time are normalized to L/Z. Curve C shows the voltage collapse when the second inductive path is added to the first at a normalized time of 0.2. The current division between the two channels at a late time is in the ratio of 1.0 to 0.68. From this simple treatment a serious difficulty is already apparent, when the question is asked how much has the rise time been improved by the later addition of a second conductor. If the maximum slope of the voltage curve is taken as a criterion, the case shown is only a 10 per cent or so improvement on a single channel. If the criterion is the 10 percent to 90 per cent pulse rise time, i.e. 90 per cent to 10 percent fall time, the curve C is much better and is quite close to the value given by two independent conductors in parallel from t = 0. It is not unreasonable that a single parametric measurement of a complex waveform can give rise to paradoxes and considerably subjective decisions have to be made as to which parameter, if either, is applicable to a particular case.

For instance, for EMP and radiative work, the dV/dt maximum is probably most applicable, whereas for high voltage X-ray production the shape of the curve late on may be much more important. As this

SWITCHING

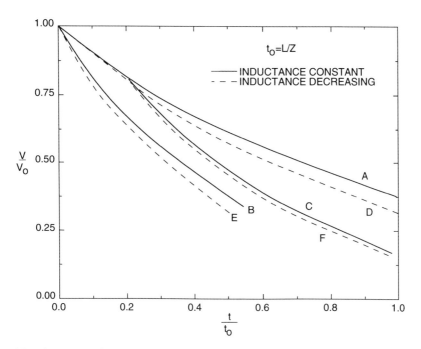

Fig. 10c-1 Normalized voltage collapse vs normalized time.

note attempts to give guidance to designers of pulse equipment, detailed and probably sterile debate is not relevant and I have no shame in declaring that in this instance I propose to take the time at which half the voltage fall has occurred as an approximate parameter. As such, the closure at $t/t_0 = 0.2$ is rather over half as effective as a second channel from $t/t_0 = 0$.

Also shown in Fig. 10c-1 are curves D, E and F. These refer to the case where the inductance L is allowed to fall to 75 per cent of its initial value during a time $t/t_0 = 1$. Something like this happens in real life as the plasma channel expands rapidly in its early phase and hence the assumption of a constant L is unrealistic. The value taken for the drop in inductance is fairly typical for spark channels of interest to this report. These curves show that the gap closing at $t = 0.2$ is about half as effective as it would have been if it had closed at $t = 0$. The current ratio in this case is 1.0 to 0.55. Now the rate of expansion of the channels is dependent on the energy flowing into it and hence the later closing channels will have a significantly higher inductance and smaller rate of change of inductance than the first channel. This effect again tends to reduce the effect of the later closing channels, but is difficult to calculate. However, some estimations suggest that closure at $t/t_0 = 0.2$ will be less than half effective, and that closure at about $t = 0.15$ would be needed to produce an effect on the fall time of a half of that produced by two independent channels and that the current division will then be in the ratio of 1 : 0.5 very approximately.

It will be remembered that earlier it was suggested that channels closing at 0.1 of τ_{tot} would be taken as the cut off point in calculating the effective numbers and this corresponds to a current of some 45 per cent of that in the first forming channel. This statement applies to cases where both inductive and resistive phases are significant and it is very probable that in the resistive phase, where the heating of the channel is strongly dependent on the voltage across it, this case will be more sensitive to voltage drop than the inductive one estimated here. If the reader wishes, he can separate τ_{tot} into two parts and multiply τ_L by 0.15 and τ_r by 0.1, but it is my advice that such accuracy is scarcely justified by the simplicity of the analysis in this section. It is also the case that in nearly all practical calculations the net effect of this elaboration on the value of ΔT is 10 per cent or less and this is well within any accuracy claimed for the relation.

COMMENTS ON THE RESISTIVE PHASE FORMULA

The relation for the duration of the resistive phase given in the note referred to in earlier has the form for gases of

$$\tau_r = \frac{88}{Z^{1/3} F^{4/3}} \left(\frac{\rho}{\rho_0}\right)^{1/2}$$

where the time is in nanoseconds and Z is the impedance driving the plasma channel in ohms, and F is the field along the channel in units of 10 kV per cm, and ρ is the density of the gas used, while ρ_0 is the density of air at NTP.

In a report from Maxwell Laboratories, MLR24, volume 2, Dr. Ray O'Rourke has theoretically derived a relationship closely resembling the above and this has the form

$$\tau_r' \propto \frac{\ell^{1/3}}{Z^{1/3} F^{4/3}} \left(\frac{\rho}{\rho_0}\right)^{1/3}.$$

A significant difference between the two relations is the power of the density of the medium in which the spark channel forms. In the original measurements on which the relation was based, the density effect of various gases from hydrogen to freon and sulphur hexafluoride at various pressures was investigated and the density variation was quite closely a power of 1/2. In addition, if the large jump is made from densities like 10^{-3} to densities like 1, the original relation rather accurately predicts the resistive phase duration for water, transformer oil and polythene, using the density to the 1/2 power. This touches on a rather important point as far as I am concerned, and that is that the larger the range over which a formula holds, the better it is, and that rather than try very accurate experiments over small ranges of variables, it is better to look less accurately at cases where big jumps in the variables are involved. Thus for density one materials, the 4/3 power of the field was checked rather accurately in an underwater experiment, where a resistive phase of nearly 2000 ns was measured for a series of parallel channels with a mean gradient of only 8 kV/cm down them, a result in agreement with the formula. Thus, while it is very pleasant to have a firm theoretical underpinning to the expression, and I would like to express my appreciation of the fine work of Ray O'Rourke, I feel that some additional effects such as the slow variation of some assumed constant in practice may give rise to the slightly different power. Thus I shall continue in this note to use the original relation.

Ian Smith of Physics International has also independently pointed out that the original expression is dimensionally undesirable, lacking an $\ell^{1/3}$ where ℓ is the length of the spark channel. He points out that if a gap is divided in two and it is assumed that at all stages the voltage at the mid point stays at half that of the top, then two half generators, each of impedance Z/2, can be joined across the channel and now a resistive phase $2^{1/3}$ longer is calculated. Ian Smith tentatively suggests an earlier from of the expression that was used at AWRE, of

$$\tau_r{}'' \propto \frac{\ell^{1/3}}{Z^{1/3}} \left(\frac{\rho}{\rho_0}\right)^{1/2}$$

which does not suffer from this possible defect. For the reasons given above I would prefer a slightly different from of Dr. O'Rourke's expression

$$\tau_r{}'' \propto \frac{\ell^{1/3}}{Z^{1/3} F^{4/3}} \left(\frac{\rho}{\rho_0}\right)^{1/2}$$

It is true that in the original experiments no large change of length was attempted. The longest gaps to which the calculation was applied was some 30 cm surface gaps which operated at voltages down to 15 kV, giving very long resistive phases once again. However, the accuracy was not high and while I am sure an $\ell^{1/3}$ term would have been found, it can be argued that surface gaps may be a bit odd. I have applied the original relation to data from some 3 metre gaps, driven by slightly ill defined impedance of an impulse Marx, and found quite reasonable agreement. Once again I think that new small effects may come in to reduce, or even remove, the suggested $\ell^{1/3}$. For instance, in breakdowns at 3 MV in a water coaxial system, the damage done to the metal walls was considerably different whether the streamers were moving inwards or outwards during the µsecond or so that the line was being pulse charged to breakdown. In this case photographs showed that uncompleted streamer bushes had main trunks that were plasma filled. This plasma was generated by energy from the dC/dt* term as the bushes grew outwards and by the time they were well under half way across the main gap, the trunk of the bush had expanded to several x 10^{-2} cm and while not hot enough to conduct, probably did not have to expand any more when final closure occurred. As such, the voltage drop down the channel was nowhere near uniform; indeed I estimated that well over three-quarters of the voltage was dropped across the last half of the channel. Thus the resistive phase would be lower than in a uniform channel and the energy deposited in forming the channel would be largely concentrated away from the electrode from which the streamer had originated, in agreement with the observations.

In the section on applications of the multichannel relation, a case where a large dC/dt term is believed to have loaded the generator prior to streamer closure is referred to. The photographs of this set up showed partly completed streamers where the glowing core was seen projecting part way from the very long knife edge. This is the only time I have seen this in gases but this again would show that uniform conditions along the channel may not exist in all cases and hence the initial assumption may well be a bit idealistic. Once again I will continue to use the original relationship and while being very happy to acknowledge its theoretical blemishes, I believe in its practical performance. Again I would like to thank Ian Smith for his interest and active efforts to improve the prediction of rise times from spark gaps.

It is probably worth pointing out that there are a number of known circumstances where rise times faster than that given by τ_{tot} have probably been observed. One of these is the non uniform channel conditions mentioned above, which can lead to reductions of rise time of up to a factor of 2. Another case is branching, where the rise time may be better by about the same factor. A particular case of this could exist in some laser triggered gaps, where very extensive branching can be seen for at least half the spark channel and this branching may of course exist all the way, but so closely spaced that the expansion of the plasma channel after the resistive phase is over joins them up into one photographically recordable channel. Again, with laser triggering it is possible that enough energy can be coupled into the gas to heat a wide plasma column which would then have an essentially zero resistive phase, since there is no need for an expansion phase to occur at all.

---

* Editor's Note: The dC/dt refers to change in capacitance with time for the streamer closing the gap.

SWITCHING

JITTER MEASUREMENTS

Initially it was hoped to obtain the data for this section from the literature but it rapidly became obvious that such data as existed was fragmentary and sometimes contradictory and in any case was restricted for the most part to pulses longer than a microsecond. A very useful summary of data is contained in an ERA report No. 5080 entitled "The Switching Surge Strength of Insulating Arrangements for Systems Operating at Voltages Above 100 kV", by A. Morris Thomas. An additional source of much good material is IEEE Transactions paper 63-1040, "Sparkover Characteristics of Large Gap Spaces and Long Insulation Strings", by T. Udo. This covers the longer pulse jitters reasonably well, but the picture for the jitter of a negative point plane for long times is rather confused, to put it mildly. Indeed the review article declined to produce a summary for this particular quantity because of the scatter of the quoted experimental results. In addition to the uncertainty in long pulse data, I had some evidence that the jitter fell as the pulse length was reduced and published data for this did not seem to exist. With manifest reluctance a grand research programme was drawn up. At this point I must apologise for the lack of new data in practically every respect. It had been intended to measure the scatter of both positive points and edges as well as the negative ones. However, as the whole experimental programme, including the building of various experimental set ups, had to be done in a period of six weeks without assistance (except where noted), the grand design shrank irresistably under the pressure of time. As such, no useful positive data was collected for air and this was the only gas investigated. For liquids the position was even more extreme, where liquid data was collected for only one effective pulse length and again only for negative pulses. The predilection for negative pulses is explained by the fact that negative edges will in general hold more voltage for a given gap length that positive ones and sometimes considerably more. Hence r_{tot} can be quite a lot down on the value for a positive point or edge. However, to set against this is the suggestion that under unfavourable conditions the jitter of a negative gap can be artificially increased by backfiring, a matter dealt with below. Still, this was the reason for the concentration on the negative data in the experiments to be described.

The measurement of a jitter of a fraction of a percent presents some difficulties. It is relatively easy to make sure the spacing is held constant to a tenth of a percent or so, but to ensure that the oscilloscope measurements do not have too much drift and inherent scatter is rather more difficult. In the case of the experimental data given below, the longer period points were obtained with a 'scope with a spot that provided an image diameter on the polaroid print of 0.2 mm width and side experiments indicated that the error in measurement due to the resistor chain and 'scope together was about ± 0.2 per cent. The measurements were performed with both points and short edges. In the case of the liquid measurements, some significant difficulty was experienced in not getting several channels; this was of course because multichannel operation was not wanted.

The test set ups were driven from a Marx generator of stacked capacity 1.2 nanofarad and inductance 7 microhenries. This was capable of going to about 500 kV in air and had a decay period of about 100 microseconds. The different pulse lengths were obtained by feeding the output of the Marx into an air insulated low inductance capacitor and then varying inductances. The value of the capacitor was 400 pF and additional inductances up to 40 microhenries were used to slow up the pulse rise time. Relatively small value resistors were included in the circuit to damp out some ringing and to limit current reversal when the point plane closed. The additional air insulated capacity was closely positioned across the point plane gap to which it was joined in a reasonably low inductive way. The height of this capacitor was 6" and for short pulses it could be rung up to over 400 kV.

Table 10c-I gives the results for negative pulses for different values of t_{eff}. This single parameter measurement of the effective pulse length is traditionally the full time width at 63 per cent of the peak of the voltage. There is no justification for its selection for gases and only a little for the liquid. However, being traditional, it enables various quantities to be compared easily with previous tabulations and relations.

Another point for inclusion is one at 0.03 μseconds where a jitter of 0.3 percent was measured. The field data does show a small and almost certainly real difference between point plane and edge plane results, with the edge plane requiring more volts to break it down. This effect was confirmed more fully with liquids,

Table 10c-I
Jitter Measurements for air-negative polarity

t_{eff} (μsec)	d (cm)	V (kV)	σ (V) (%)	F (kV/cm)		
11	4.15	108	± 1.7	24.0	)	
2	7.14	184	± 2.0	25.8	)	Point Plane
0 75	7.14	200	± 1.3	28.0	)	
0.23	7.14	208	± 0.6	29.2	)	
2	7.14	191	± 1.2	26.8	)	Edge Plane
0.4	7.14	205	± 0.5	28.8	)	

where it is somewhat bigger. This result at first sight is rather surprising, since an edge can apparently be considered as an array of points. If this were so, the lowest voltage point would break first and it would then be the case that the edge plane would hold a lower voltage and have a better standard deviation. However, both of these conclusions do not follow. Regarding the breakdown voltage effect first, what probably happens is that in a point, the energy from a diverging cone is fed back to heat up the channel near the point and this enables it to progress slightly faster than streamers heated by energy flowing into a wedge shaped sector. A similar phenomenon was observed some years ago in a series of tracking experiments. In these a square piece of copper foil was stuck onto thin mylar, which was in turn stuck onto an extensive ground plane. Where the surface had been properly discharged (something that was not easy to do) and when an adequately low inductance capacitor was used to pulse charge the top square electrode, streamers ran out from all round the edges to a quite regular, finite distance. This distance had a powerful dependency on the pulse voltage. The point of interest here was that much stronger streamers ran out from the corners of the square and obviously charged a much bigger area than the closely packed uniform array along the edge ones, indeed going a bit more than $\sqrt{2}$ their length. With regard to the jitter of an edge when viewed as a large number of points, this might be supposed to have a jitter several times smaller than that of a single point, but for this to be so the distribution of breakdown voltages has to be Gaussian. This is in general not so, although it requires fairly refined measurements to prove this. If, as is much more likely, the distribution is of the self-replication form discussed in the note on the "Volume Effect of the Pulse Breakdown Voltage of Plastics", then an array of points or edges will have the same jitter as a single point. Indeed, the measurements tend to show that the situation approximates much closer to this case than to that appertaining when the breakdown of a single point is truly Gaussian.

Figure 10c-2 shows the new data plotted; in addition some data from T. Udo for negative points and an average number for a positive point is shown. The errors shown for the data quoted in Table 9c-I are not firm statistical ones, which are in fact rather less than those shown, but attempt to include factors for consistency, etc. While there appears to be some small difference between points and edges, the curve shown probably is as good as the accuracy of the data warrants.

While performing these measurements, a number of incidental points came up and these will be dealt with separately below.

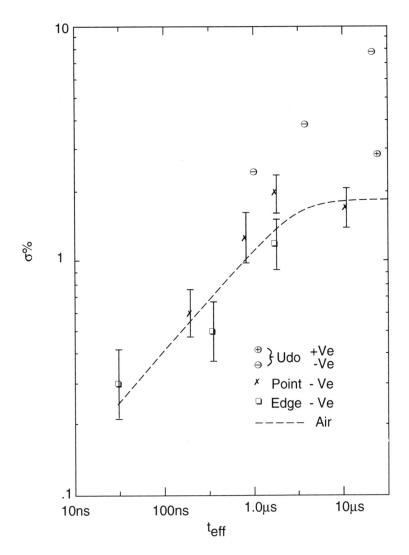

Fig. 10c-2 Jitter vs t_{eff}.

Extra Zigzag Length Estimation

As the effective pulse duration was reduced, it was observed that the path of the streamer got a lot straighter. Indeed, by roughly following the zigzag for pulses with times longer than a couple of microseconds, it could be estimated that some had path lengths at least 3 per cent longer than the shortest route. As the pulse decreased to significantly less than 1 μsecond, the channels became much straighter, until for pulses of tens of nanoseconds duration the channels were nearly all dead straight. Thus if the

SWITCHING

set up is geometrically stable and no backfiring is occurring, it seems possible to look at an open shutter photograph of the streamer channel and to estimate the jitter to a factor of 2 or so.

Backfiring

For pulses of duration in the region of 2 μseconds or longer it was found occasionally that a fraction of breakdowns would occur at voltages significantly less than the mean of the usual group. The spark channel on these occasions seemed usually to have a pronounced kink a cm or so off the plane and then the track went straight down. It was found that when the plane was roughly cleaned and dust, etc. removed, this second population vanished, and indeed when half the discharges were of this lower well-scattered variety, a couple of wipes would eliminate the effect completely. The explanation advanced is that a positive bush has grown out from the plane and met the negative one on its way down at a point fairly close to the plane's surface. The result of this is two populations, the regular zigzagged channel type and breakdowns depending on two bushes. When both populations are included in the analysis, the jitter can get as big as 8 per cent. It is suggested that in out-of-door experiments such as by T. Udo, dirt on the earth plane could well have caused a variable number of occasions on which backfiring took place, leading to large and variable jitters. The reason that the phenomenon does not seem to take place with negative pulses of less than 1 μsecond or positive pulses of any duration is interesting to speculate on and may have consequences of importance to triggered gaps in general. For negative point plane breakdowns for short pulses, an approximate relation has been derived in "Pressure Dependency of Pulse Breakdown of Gases" (SSWA/JCM/679/71).* There was an earlier paper on the subject that seems to have vanished from the face of the earth. This approximate relationship has a weak dependency on t (a 1/6th power, in fact) but for negative points this dependency disappears for times in the region of a microsecond. The positive point plane relation continues to fall out to times of several hundred microseconds, at which times it will only hold half the voltage that a negative point will. The point at which the negative breakdown mean field becomes time independent appears to depend on the length of the gap being about 7 μseconds for 6 metre gaps and about 0.1 μsecond for 7 cm gaps, suggesting a relationship of the form t_{crit} = .013 d where d is the point plane gap in cms and t_{crit} is the pulse duration at which the negative pulse becomes time independent. For times greater than this, the strength of the positive gap becomes significantly less. Thus as a positive bush grows out during a long pulse, any dirt or defects on the negative

---

* Dielectric Strength Note 15 (Footnote)

plane will be much less likely to cause a bush to grow back during the last stages of closure than in the case with reversed polarities. For pressurised hard gases the negative point is always stronger (usually by 50 per cent or more) than the positive one, so that if a rough "plane" surface exists, some form of backfiring may occur. In this case a point or edge plane gap pulse charged positive will have a lower breakdown voltage but may well have a somewhat smaller jitter. I gather from Ian Smith that this phenomenon has indeed been observed.

dC/dt Loading

This is of more importance to multichannel operation with long gaps and indeed is mainly dealt with in the very long gap described near the end of this note. It is also very much more applicable to liquid gaps and in particular to water gaps. The name by which the effect is described may be misleading since it has yet to be really proved that streamers exist on the time scales in question and while I feel that those from negative points are a reality, I am a little less certain about positive points in gases. However, assuming that streamers do move in the way the approximate relations already mentioned suggest, then a differential velocity for the streamer front can be found. For instance, for gases $dx/dt \propto V^6 x^{-4}$ where x is the distance from the tip and V the voltage on it. This shows that if it applies, the streamer moves out rapidly, slowing up as it goes and takes quite a long while to cross the last few percent of its travel. If the streamer is in air and it spreads out like it does in liquids and solids, then quite a large change in capacity occurs as the streamer is in the last phases of closure. For fast pulsed and/or high impedance driving circuits, the loading caused by the $(dC/dt)^{-1}$ term can be quite severe and cause the voltage on the point to fall before the streamers close and the real plasma channels start to form. This phenomenon has been observed by Harlan Aslin of Physics International and I am indebted to him for the observation. The phenomenon can also be seen in some oscillographic records of 6 million volt point plane gap work on a much longer time scale than the present observations. The dC/dt explanation is not the only one that is tenable since photoionisation of the gas ahead of the streamers could also account for it, but some very indirect evidence mentioned in the later sections tends to support the proposed explanation. In any case, which ever is the correct explanation, the rather slower fall in voltage before the main heavy current carrying channels close would be expected to increase the jitter somewhat and this has possibly been observed in the very long gap experiment described in a later section of this note. It is therefore important in jitter measurements to ensure that this does not happen appreciably by having a low impedance source coupled by the minimum inductance across the gap when fast breakdown measurements are attempted in gases and, in particular, in work with liquids.

Liquid Breakdown Jitter

The liquid employed in the multichannel edge plane experiments described in the next section was carbon tetrachloride and for the pulse duration used in these experiments the jitter was measured. As will be described in the following section, it is necessary not only to keep the spacing of the gap constant but also to have an adequately sharp edge, so that the rather large jitter in voltage at which the streamers start out will have a negligible effect on the final jitter at closure time. This effect also applies to gas edge plane gaps but here the radius of the edge can be ten mils. or more and is easily obtainable. However, with liquids, working at only 400 kV or so, the edge needs to have a radius of a couple of mils. or less. This matter is dealt with further in the section on the multichannel liquid gap. In the jitter experiments, as indeed with all the liquid point and edge plane experiments, the points and edges were frequently sharpened, to ensure that the streamers started out at a low voltage. Table 10c-II gives the breakdown voltages and jitters for an effective pulse time of 80 ns.

As can be seen, the jitter is approximately 1 percent and the edge plane holds off some 18 per cent more volts than the point plane. As was mentioned earlier, even though the edge was only some 3 cms long (excluding the roll up at its ends) multichannel breakdowns were frequently observed. The streamer channels were not straight but were mildly zigzagged in reasonable visual agreement with the jitter of 1 per cent. Figure 10c-3 shows the jitter curve for air again and also the carbon tetrachloride point. Also shown is a point which Mr. George Herbert of AWRE got in an experiment with water with both an edge plane gap and also with a version

TABLE 10c-II
Breakdown Time and Jitter in Carbon Tetrachloride
Gap 2.05 cms, t_{eff} = 80 ns

Experiment No.	Gap Type	Voltage (kV)	Jitter (%)
1	Point Plane	428	± 1.3
3		420	± 1.1
2	Edge Plane	504	± 1.4
4		492	± 1.1

Mean point plane breakdown voltage   424 kV
Mean edge plane breakdown voltage   498 kV
Ratio point to edge   1: 1.18

312   CHAPTER 10

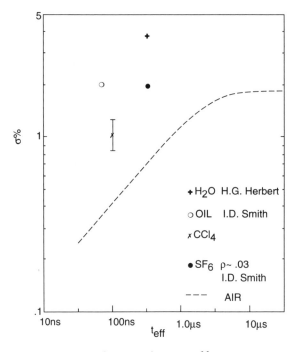

Fig. 10c-3 Jitter vs t_{eff} for various media.

of this called the I.B. Gap. This stands for intentionally messed up gap and consists of a sharp edge sticking out a small controlled distance from a cylinder. Basically it was invented a year or so ago to obtain higher breakdown fields than the edge plane gap would give, but with the low jitter associated with these. Typically rod gaps in water can have jitters of 10 per cent. The IB gap can also be used in gases where the fin should stick out about 0.5 cms, or hard gases, where smaller projections are necessary to get fields up near those obtainable for uniform field breakdown. Also included is a point deduced from some data kindly provided by Ian Smith on a triggered multichannel oil gap. This point was partly obtained by using the proposed relationship in reverse, but the jitters calculated were in rough agreement with oscillographic measurements made of the variation of time of gap operation. Also on the graph is the jitter measured for some pressurised SF_6 gaps working at a density about midway between that of air and water; these too have been provided by Ian Smith to whom my thanks are again due. It will be a brave man who will draw curves through this meagre and rather scattered data and I for one will only do so in private.

It should be emphasised that the jitters shown in Figures 10c-2 and 10c-3 are the minimum that have been measured under good

SWITCHING

conditions. A gap poorly constructed or working with an inadequate driving circuit can well have considerable worse jitter and hence fail to operate in a multichannel mode when it is expected to. Triggered gaps in at least one form are just a pair of edge plane gaps back to back and an example of this type of gap is mentioned at the end of this note. If such a mid plane is used, then inadequate sharpness of the trigger edge may give enhanced jitter and very rough main electrodes may lead to backfiring, especially if the trigger edge is being driven negative.

MULTICHANNEL EXPERIMENTS

In this set of experiments the Marx generator mentioned in the previous section was used to pulse charge a polythene insulated transmission line. Ideally this line would have had an electrical length rather longer than the longest τ_{tot} it was to be used with, but this was unfortunately impracticable, especially in the case when the line's impedance was low. As a compromise, the line was made physically 5 feet long, corresponding to a two-way electrical length of some 16 ns. The line was made from a total of 4 sets of sheets of insulation and these were separated by three boards of 1" plywood which were covered with a thin layer of spongy rubber and then sheathed in thin copper foil. The distance between the two outside output lines was about 9 cms but the total thickness of insulator was only some 0.6 cms for the case of the line with 2 1/2 ohms impedance. The width of the line was 18 inches and loading with lead bricks pushed the assembly together. The capacity measured after a few shots (when electrostatic forces helped) was within about 10 per cent of that calculated from its dimensions. This result was achieved largely because of the way the layer of spongy rubber took up small imperfections and assisted in spreading the localised loads of the lead bricks. In the 8 ohm version, which had about 2 cm of solid insulation in it, the line could be pulse charged to over 400 kV and at no time was any flashover experienced, despite the fact that the line was in air. For fairly fast charging it is considered that in a 10 cms height of line some 600 kV could have been held off in air.

The particular technique described here is rather useful in providing pulse charged high voltage condensers with very low output inductances. For instance if the line described is used to feed a closely coupled load some 10 cms high along its 150 cms length, the output inductance of it, viewed as a capacity, would be under 1 nanohenry for the case with a total of 0.6 cms of insulation. Unfortunately there are not too many loads that can be closely coupled into such a condenser cum line and in the case of the multichannel gaps described below, the main inductance was in the feed from the line to the edge plane gap. Considerable efforts were made to reduce this feed inductance to as small a value as possible. However, it still remained a significant additional impedance and to allow for this in calculating the number of channels to be expected, an effective impedance of the system was deduced, given by the relation

$$Z_{eff} = Z_{line} + 2L_{excess} / \tau_{tot} ,$$

where τ_{tot} was an average value for the set of experiments under consideration. In general, the extra term was less than 40 per cent of the final assumed effective impedance and usually significantly less. In the case where a reasonable number of channels was being formed, the resistive phase term was dominant and this only uses the impedance of the generator to the one-third power, so any error introduced in the calculations was small. However, for cases with a small number of channels, the use of a constant Z_{eff}, might have been a bit pessimistic, since τ_{tot} was considerably bigger under these conditions. However, it is in just this case that the finite length of line might have begun to have an effect and this would presumably have slightly lowered the number of channels observed against the number calculated. Thus the two approximations tend to work in opposite directions and while it would be the wildest fluke if they canceled, to the accuracy to which this work has any pretensions, the combined effects were probably tolerable.

Multichannel Experiments in Air

In the experiments, lines of two impedances were used, 2 1/2 ohms and 8 ohms, respectively, and also two lengths of edges were employed, namely 40 and 80 cms. In the case of the 80 cm length edge, the inductance of the feed out to the outer parts of the edge plane became important and as it was practically very difficult to reduce this to a small value, the gap was reduced by a few percent, so that a very roughly uniform distribution of channels was obtained. This correction was not made very accurately and the effective length of the edge was therefore probably rather smaller than its physical length. One important feature of the experiments was to decide the criterion on which to count the effective number of channels. Ideally this would have been based on measurements of the current in the individual channels and then counting all those with currents over 45 per cent or 35 per cent of the maximum recorded. This would have been prohibitively expensive in equipment and very time-consuming and a simpler technique was employed based on photographically recording the light output. In previous work it had been established that for channels driven by relatively short current pulses, nearly all the light was emitted by the plasma channel up until the time it had expanded and become self-transparent. At this stage it had also cooled somewhat and the combined effect of these two phenomena was that typically the full width at half height of the visible light output was well under one microsecond. This simple picture of the plasma channel suggests that if the energy deposited in the channel is doubled, the width of the cylindrical channel at which the luminosity falls sharply goes up by $\sqrt{2}$ and the duration of the light pulse scales up by the same factor. Thus the photographically integrated light output should be proportional to the energy deposited in the channel, since the luminosity at which the channel ceased to be black body is quite closely constant. This proportionality was checked by using a single spark channel and driving it by the lowest impedance version of the line with a number of different resistances in series with it. These resistances took the form of low inductive networks and had values of approximately 30, 60, and 120 ohms. From the previously mentioned note the prompt energy deposited in the channel scaled as $Z^{-4/3}$ and indeed the photo- graphically integrated light was in the ratio of 1 : 0.4 : 0.2 which was close enough. In order to avoid absolute densitometry, these test channels had 0.3 and 0.6 neutral density filters placed over 1.5 cms each of the 4.5 cms channel length. It had previously been established that the channels were uniformly bright along their lengths in all these and the following experiments and the use of neutral density filters placed close to the channels themselves simply calibrated the density of the photographic images on Polaroid type 55 negative/positive film. Film from the same batch was employed and was fully developed and fixed. Since in these subsidiary tests a range of channel brightnesses of 5:1 was covered, the additional filtered sections covered a total range of 20:1. In counting the effective number of channels in any shot, the brightness of the most luminous image was matched and then all those down to one-third of this were counted (criterion A) or down to one-quarter of this (criterion B). The effective aperture of the camera was changed so that the brightness of the most luminous channel was within the dynamic range of the print, since as the number of channels increased their brightness decreased, of course. As was expected from the simple theory of the plasma channel and its dynamics, the brightness of the liquid channels was the same as that of those in air under corresponding conditions, which, considering the very different conditions existing in the two cases, was reassuring that the simple theory of the channel was applicable. In deducing the approximate current being carried by a channel whose integrated luminosity was down by 3 and 4, respectively, use is made of the fact that the energy deposited in the channel is given by

$$E \propto V^{2/3} Z^{-4/3}$$

Thus where the voltage across the channel has fallen by 0.9 and 0.85 respectively, the following table (Table 10c-III) of very approximate data is obtained.

Table 10c-III
Luminosity Derived Data

Criterion	E	V	Z	i
A	0.33	0.9	2.1	0.43
B	0.25	0.85	2.6	0.33

The values given above are fractions of those applying to the heaviest current carrying channels. No great accuracy is claimed for these figures of course and it would have been considerably better to have measured the currents directly. However, in view of the fundamental uncertainty in how a given current ratio reflects into the effective number of channels and rise time, I do not consider them to be too bad.

One minor consequence of the above analysis is that unless the effective aperture of the camera is increased as the number of channels gets larger in a series of tests, an erroneous visual impression of the density of channels will result. This is because the impedance driving each channel increases as the number of channels increases and hence their image brightness falls. This means that for each doubling of their number the camera ought to be effectively opened one stop approximately. When this is done the mean brightness of the first channel should remain roughly constant.

One further point should be mentioned with regard to the photographic recording used and this is that the image should be faintly out of focus unless the actual channels are extremely fine. This is so that the image on the print is reasonably uniform across its width and that this width is more a function of camera lens than of the actual image. This requirement is not an exact one but dictates that it is desirable to put in or remove neutral density filters rather than simply closing or opening the iris of the camera by large amounts. This effect was allowed for to a considerable degree by the use of neutral density filters in front of the reference single channels, and the whole effect is made more complex by the non-uniformity of point image size across the image plane of the lens. Consequently a very detailed analysis of the effect would be necessary to obtain accurate answers by the use of the technique described above, but due to the limited density ranges used in the experiments and the approximate nature of these, this was deemed not necessary. In a series of shots, the same piece of film was used and the camera tilted upwards so that about seven images were placed one above the other, thus saving film and allowing each set to be internally more consistent.

Low Impedance Line Results

The impedance of the line was 2.5 ohms and Z_{eff} was 3 ohms. The length of the channels was 4.5 cms and the length of the edge was 40 cms. The following gives the details of the calculations for this case.*

L spark channels = 2 x 4.5 x 7 = 63 nH and

$\tau_L = 21/n$ ns.

For the resistive phase F = 30 kV/cm and

$\tau_r = 88/(1.4 \times 4.6 \times n^{1/3}) = 14/n^{1/3}$ ns

$\tau_{trans} = 40/(30 \times n) = 1.3/n$ ns.

Table 10c-IV
Times in ns for the Low Impedance Line

n	τ_L	τ_r	τ_{tot}	0.1 τ_{tot}	+ 0.8 τ_{trans}	ΔT
1	21	14	35	3.5	+ 1.1	4.6
2	10.5	11	23	2.3	+ 0.6	2.9
4	5.2	8.8	14	1.4	+ 0.3	1.7
8	2.5	7	9.5	.95	+ 0.15	1.1
16	1.3	5.6	7	.7	+ 0.07	.8

				0.15 τ_{tot}	+ 0.8 τ_{trans}	ΔT
				5.2	+ 1.1	6.3
				3.4	+ 0.6	4.0
				2.1	+ 0.3	2.4
				1.4	+ 0.15	1.5
				1.0	+ 0.07	1.1

For the two criteria used above the number of observed channels are given in Table 10c-V below. Each number generally represents the average of seven shots. Also given is the observed value of T from $T = V (dV/dt)^{-1}$ where dV/dt is evaluated at the point of channel closure on the charging waveform.

Editor's Note: A summary of the following calculations is given in Table 10c-IV.

Table 10c-V

Observed Number of Channels

Value of T (ns)	n criterion		n corrected for dead space		ΔT calc.	
	A	B	A	B	A	B
350	2.3	3.0	2.5	3.4	2.4	2.6
230	3.2	5.1	3.6	5.9	1.7	1.8
170	5.0	8.0	5.8	10.1	1.3	1.3

The correction for "dead space" is analogous to the correction for dead time in counting random events, say in radio-active decay. The distance between channels was plotted out and while there was a reasonable distribution of channels, the ones close together tended to be fainter. This meant that there was a dearth of small gaps and a correction is applied on the basis that the streamers do affect each other slightly when they are close together. The correction for a channel length of 4 1/2 cm is very approximately $N = P(1 - .025 P)^{-1}$ where P is the number of channels observed and N would be the number with no dead space. For the 80 cm long edge, the constant inside the bracket is halved. The correction is not large for all the experiments in this section, but in the last section the long edge gap has a fairly big correction which is certainly not well established. However, for the present experiments it is really included only for completeness and is very approximate. Table 10c-VI gives the values of ΔT obtained from the jitter experiments. The value of t_{eff} to be used with Figure 9c-2 is obtained from the pulse charging waveform, but is approximately equal to 0.4 T.

Table 10c-VI
ΔT from Jitter Measurements

Tns	σ (%)	ΔT = 2σT	ΔT from Table V
350	0.5	3.5	2.5
230	0.45	2.1	1.8
170	0.4	1.4	1.3

The agreement is quite reasonable, except for the first entry, and this is in the region previously discussed where the approximations used are not very good. However, to 30 per cent or so the agreement is satisfactory.

For all the subsequent measurements criterion B is used to improve the statistics. For the case of T = 170 ns a series of determinations of n were done at different times and these gave values of 8.0, 7.9, 8.4, 8.3 and 7.6 uncorrected for dead space effects. The reproducibility was, rather surprisingly, good. Considering the individual numbers of effective channels in each set of seven shots gives a standard deviation in the number per shot of ± 20 per cent for this case.

Summary of Results for Both Lines and Gaps

Table 10c-VII summarises the data for 40 and 80 cms gaps and for the two effective line impedances of 3 and 9 ohms. The observed effective number of channels has been corrected for dead space effects. These are compared with the number of channels predicted by the relation and jitter values given earlier.

Table 10c-VII
Number of Channels for 40 and 80 cm Gaps

Negative Edge - Criterion B
Gap 4.5 cms corrected dead space

T	N_{40}	N_{80}	N_{40}	N_{80}		
	Observed		Calculated			
350	3.4		2.3		)	
230	5.9		4.8		) Z_{eff} = 3 ohms
170	10.1	10.4	9.2	10.6	)	
220	1.4	2.0	2.3	3.1	)	
170	3.2	3.7	4.0	5.1	) Z_{eff} = 9 ohms
140	4.1	4.8	6.5	8.0	)	

Figure 10c-4 shows N_{40} data plotted out.

Comparing the number of channels is a more sensitive measure of agreement than comparing values of ΔT as the effective number is a higher power than one of ΔT. In particular the disagreement at

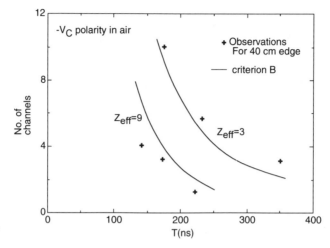

Fig. 10c-4 Channel Numbers vs time for 40 cm gap.

the bottom of the table reflects no more than a 40 per cent difference in ΔT. The disagreement is in part due to the fact that the effective length of the edge is less than the full 80 cms. In general the agreement is quite reasonable, certainly within the limits hoped for for design purposes.

Figs. 10c-5 and 10c-6 are typical open shutter records of a series of shots using the 40 cm and 80 cm long edges, respectively. Needless to say, the lowest impedance line and the smallest charging T given in Table 10c-VII were chosen in each case in order to give as many channels as possible but, this apart, the records are otherwise average ones. The faint set of fuzzy images down the left hand side of the prints are the spark gap columns in the Marx and should be ignored.

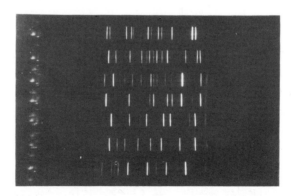

Fig 10c-5 Open shutter photographs of multichannel discharges (40 cm long edge).

SWITCHING 321

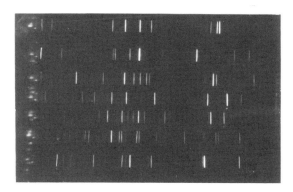

Fig. 10c-6 Open shutter photographs of multichannel discharges (80 cm long edge).

Multichannel Experiments in Liquid

The liquid chosen for these experiments was carbon tetrachloride. This was selected because in the compilation of mean streamer velocities given by George Herbert in "Velocity of Propagation of High Voltage Streamers in Various Liquids" (SSWA/HGH/6610/104)* This material has the highest velocity at voltages of about 400 kV. The high impedance line only was used in these tests in order to have as large a voltage available as possible. Even with these choices the gap was only 2.0 cms and extra care had to be taken to keep this rather small gap constant. In addition the higher impedance line was chosen to reduce the effect of the feed inductance which was slightly bigger in this arrangement, giving a Z_{eff} of about 10 ohms. The length of the edge was about 34 cms. As the experiments proceeded the carbon tetrachloride became quite dark with finely divided carbon, which was removed by filtering periodically or by the use of new liquid.

When the experimental set up was first used, multichannel operation was immediately obtained, but the channels came from both ends of the edge. These ends were gently curved up, rather like a double-ended ski in profile, as had been the edges used in air. There were incipient streamers along the main length of the edge, but few of these had closed in time. While it was not surprising that streamers could close first from square cut ends, because of the effects described earlier, it was felt that the "roll off" of the edge had been well done. Closer examination of the edge showed that the 9 mil. thick phosphor bronze out of which it was made had lipped sideways, leaving a very sharp edge where it had been cut with scissors to form the ends. Thus the closure at the ends was

---

* AFWL Dielectric Strength Note 10

being caused by the fact that the streamers were starting out earlier in the rising waveform than those from the blunter central region of the edge.

George Herbert's paper gives the mean streamer velocity from a negative point as

$$\overline{U} = 166 \, V^{1.71}$$

where the mean velocity is just the distance from the point to the plane in cms divided by the effective time of the pulse in microseconds. The amplitude of the pulse is in units of a million volts. This can be used to provide a differential velocity relation which on integrating with a linearly rising voltage pulse gives

$$d = 62 \, V^{1.71} \, t_2^{2.71} \, (1 - (t_1/t_2)^{2.71})$$

where t_2 is the time of closure and t_1 is the average time at which the streamers start out. This time is likely to be very variable and in order to ensure that 100 per cent jitter in this time has an acceptable effect on gap operation, the value of $(t_1/t_2)^{2.71}$ should be less than .005, i.e. $t_1/t_2 \leq 1/7$.

This does not at first sight seem to be a very strict requirement but closer examination proves that it requires a rather sharp edge to the electrode. For the longer of the two pulses used in the experiments described below, V ~ 0.3 MV and hence the streamers should have started by the time the voltage on the edge has risen to 40 kV. The effective area of the edge is about 0.1 cm² and the t_{eff} is of the order 100 ns. Unfortunately we don't have the area effect curve uniform breakdown in carbon tetrachloride but assuming that it is roughly the same as in transformer oil, this means that surface fields of the order of 1.5 MV/cm are required. For an edge to achieve this sort of field with only 40 kV on it requires a radius of the order of 2 mils. or less. In the experiments the edge was sharpened to better than this and then reliable operation occurred from the main length of the edge. In order to ensure that the voltage of streamer initiation was adequately low, the edge was periodically sharpened during all liquid breakdown tests.

The streamer velocity data given by George Herbert can be integrated to give the gap that will break with a linearly rising pulse and the calculated and observed values are given below in Table 10c-VIII.

The agreement shown in Table 10c-VIII is not as good as it seems, because George Herbert's data was obtained with points, while the experiments were conducted with edges. Thus it should have required voltages some 20 per cent more than were observed to

Table 10c-VIII
Gap Breakdown with Linearly Rising Pulse and Negative Polarity

Voltage (MV)	T (ns)	d Breakdown Calculated (cms)	
0.400	150	1.95	) Actual gap
0.286	280	2.04	) 1.95 cms.

get agreement. However, in view of the fact that the streamer velocity relation is only approximate and that a linearly rising waveform relies on the low voltage extrapolation of this relation, the agreement is acceptable. A second possibility is that the point in the liquid streamer work may not have been quite as sharp as the edge used in this work, although it too was regularly sharpened.

Liquid Gap Results

Two sets of experiments were conducted with different values of T. In the case of liquids, the breakdown field is much more heavily dependent on the value of T and this in its turn leads to a fairly strong dependency of τ_{tot} on T. Consequently fairly large changes in the rate of rise of the pulse have less effect on the number of channels, as they are partly compensated for by the change in field along the channel when it closes. This is shown by the calculations given below for the two cases used experimentally.

In obtaining the values for Table 10c-IX a couple of points have to be borne in mind. The equation for the resistive phase in materials of unit density has a constant of 5 in it. This has been increased to 6.5 to allow for the fact that the density of carbon tetrachloride is 1.60. In addition its local velocity of light is only two-thirds that of free space.

The jitter in closure of the gap was measured in the six separate sets of experiments and these averaged out at ± 1.0 per cent in reasonable agreement with the independent value determined in the experiments described in the jitter experiments section.

In the case of the liquid gap it is somewhat more difficult to determine the effective number of channels because branching occurs and it is not easy to decide how many effective channels a branched discharge corresponds to. Fig. 10c-7 shows an example of multi-channel liquid operation and shows the fairly extensive branching

Table 10c-IX
Breakdown Calculations for Multichannel Liquid Gaps
Times in ns
(Z_{eff} = 10 ohms)

n	τ_L	τ_r	τ_{tot}	0.15 τ_{tot}	+ 0.8 τ_{trans}	ΔT
T = 150 ns		$V_{breakdown}$	= 400 kV		F = 0.2 MV/cm	
1	3.0	27	30	4.5	+ 1.4	5.9
2	1.5	21.5	23	3.4	+ 0.7	4.1
4	0.8	17	18	2.7	+ 0.4	3.1
8	0.4	13.5	14	2.1	+ 0.2	2.3
T ≐ 290 ns		$V_{breakdown}$	= 286 kV		F = 0.14 MV/cm	
1	3.0	44	47	7.0	+ 1.4	8.4
2	1.5	35	36	5.4	+ 0.7	6.1
4	0.8	28	29	4.4	+ 0.4	4.8
8	0.4	22	22	3.3	+ 0.2	3.5

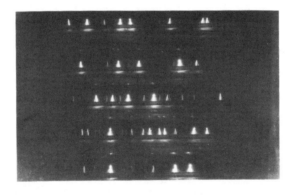

Fig. 10c-7 Multichannel liquid breakdowns.

that can occur. In practice a somewhat subjective increase to the number of channel roots has been made; this in general increased this number by some 30 per cent. The dead space correction has not been applied but if it were, the same as for a scaled-down air case, it would be less than 10 per cent.

Table 10c-X compares both the predicted and actual number of channels and also gives the values for ΔT. Once again, very good agreement seems to have been obtained.

Table 10c-X
Negative Edge - Criterion B
Theoretical and Experimental Values of n and Δ
Gap 2.0 cm
Uncorrected for any dead space

T (ns)	n observed	n calculated	ΔT from observation	ΔT $\sigma = 1.0\%$
290	2.8	2.4	5.4	5.8
150	6.0	4.3	2.6	3.0

One of the points which has yet to be raised is the question of the simultaneity of the channels. Where a system contains a considerable excess of energy which can ring on for a long time and where the single spark gap channel is a substantial fraction of the total inductance, then big oscillating voltages caused by the L di/dt term can cause channels to complete on later current reversals. The circumstances have to be rather special for this to occur but it has been observed to happen. In the present instance, the spark channels themselves form a load for most of the energy in the line and the inductance of the channels compared with that of the feed and line is very small. Thus the current that oscillates is well down and the fraction of the inductive voltage appearing across the gap is less than 10 per cent of the initial charging pulse. It also lasts for a shorter t_{eff} than that of the pulse. This fact was checked by the voltage monitor whose output had to be corrected because the feed inductance was included in its tap off loop. Making a reasonable allowance for the effect of this, the measurements suggested that less than 5 per cent of peak voltage appeared across the channels after firing in the experiments with both gases and liquids. In the case of the experiments in air, integration of the streamer shows that any subsequent distance of closure was very small because of the $t^{1/6}$ dependence of the relationship. This shows that the 5 per cent voltage pulse would have had to have gone on for thousands of nanoseconds to have caused any additional closures. Even if these had taken place and it did manage to close, there could be no significant voltage across the channel, and the current and hence light output would have been undetectable. However, it was felt necessary to check this point and the liquid tests were chosen because of their stronger dependence on t, which should lead to any effects being found in them ahead of the experiments in air. A set of experiments was done with a low inductance network of resistors whose value was 10 ohms in series with the gap, thus damping any oscillations in the line very quickly. Because the impedance of the generator was being doubled, the number of channels would of course be expected to decrease, but by nothing like as much, if most of the channels were due to late closure.

In the experiments with the damping resistor in series with the gap, the number of channels was 4.5 where previously it had been 6.0 without it. The calculated effect of increasing the generator resistance from 10 to 20 ohms would have been to drop the expected number to 3.7. Thus there was, as expected, no evidence at all of late closure of channels in the experiments most likely to show any such effect.

EXAMPLES OF MULTICHANNEL GAPS

As was mentioned earlier, while accurate experimental verification of an expression is of course useful, if this is limited to a relatively small range of parameters it may be less useful than a much rougher experimental agreement with many considerably different systems. As such, it is worth recording that, to a significantly lower accuracy than the work reported in this note, the suggested expression has been tested against three different multichannel gaps, three different liquid gaps, four different gaseous gaps and a surface air gap of vintage design. In general, the jitter is the factor which is least well determined and intelligent guesses have sometimes had to be made about this factor, but, with this reservation, reasonable agreement was obtained from these rough calculations. This in my opinion offers as much proof that the expression is valid as the data presented in the previous section of this note.

In this final section two examples of gaps will be briefly described, the first one because it bears on the effect of dC/dt loading and the second because it led to some work on the current stability and perseverance of heated plasma channels.

The Three Metre Edge Gap

This was a set up designed to demonstrate switching with about 200 channels. As ever, performance did not match up to expectations but in fact 140 channels were achieved with it. These channels closed within a span of 0.6 ns and the fall time of the voltage was about 5 ns. The measured time was about 8 ns but the location of the monitor and its response probably accounts for the difference. As the line had an effective impedance of 0.45 ohms and the switch closed at 105 kV, the short circuit current was about 230 kamps and the di/dt was about 5×10^{13} amps per second, achieved at a relatively modest voltage. The rise time of a single channel across the same impedance would have been about 110 ns: thus the multichannel switch outperforms the single channel gap by a factor of about 20.

In order to obtain many more channels than in the previous experiments, it was necessary to reduce the impedance of the line feeding the gap. At the same time the pulse rise time had to be decreased, if possible. The length of the edge had also to be increased considerably, in order to reduce the dead space correction to manageable proportions. The latter requirement led to a total width of edge of 3 metres and by folding it over on itself, so that there were two gaps back to back, the width of the lines could be held to 1.5 metres. The length of the line was made 84 cms so an insulator, 3 feet wide, could be used, leading to an electrical length of about 9 ns. The capacity of the first line was 8.5 nF, giving a transmission line impedance of about 0.5 Ω and a Z_{eff} of 0.8 Ω. In order to charge this capacity quickly, a new condenser set-up was built. This consisted of 8 condensers each 10 nF, 100 kV, in two sets of 4, and were plus and minus charged as usual. A rather low inductance 200 kV pressurised spark gap (which had either air or SF_6 in it, according to the working voltage) joined the hot terminals of the condensers together. When the pressure was released, it broke down. The output capacity of this 200 kV unit was 20 nF and its inductance, including the feeds to the line, was 110 nH. The bank was not charged over 180 kV because it was not necessary, but it was fired a number of times at this voltage. As the condensers were rather long (45 cms) and not very low inductance, they contributed over half the inductance of the unit.

With no output leads and with capacitors of good design and large capacity, the inductance of the unit could have been about 60 nH. Thus if these units were stacked, which was inconvenient but possible, an inductance of 300 nH per million volts would have resulted: this number is contemporary. Figure 10c-8 shows a schematic of the layout and also a cross-section of the line and the double sided gap. In the first set of experiments this was set at a gap of 3.7 cms and the breakdown voltage was 126 kV for fast charging. Table 10c-XI lists the total number of channels and compares the ΔT values deduced from these and the expected jitter.

The agreement is not too bad at the top of the table but gets progressively worse for faster and faster charging times. The reason was clear from the 'scope records, which showed that the charging waveform curled over and had fallen significantly before a sharp break occurred, presumably at the time the plasma channels began to conduct in earnest. The earlier phase is attributed to the dC/dt term. This loading meant that roughly 10 kAmps were flowing prior to channel closure. This was more or less in line with some earlier measurements, where rather smaller effects had been noticed corresponding to a few tens of amps per cm of edge, when a gap had been fast charged.

If the dC/dt term is the explanation, then it would be expected that the load impedance would be proportional to the charging

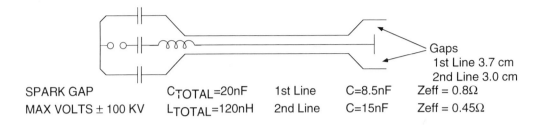

SPARK GAP	C_{TOTAL}=20nF	1st Line	C=8.5nF	Gaps 1st Line 3.7 cm 2nd Line 3.0 cm Z_{eff} = 0.8Ω
MAX VOLTS ± 100 KV	L_{TOTAL}=120nH	2nd Line	C=15nF	Z_{eff} = 0.45Ω

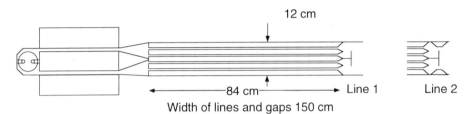

Width of lines and gaps 150 cm

Note: The insulation has been omitted for clarity

Fig. 10c-8 Experimental arrangement for wide edge (3m) gap test.

Table 10c-XI
Results from Wide Edge (3m) Gap Tests
Negative Polarity - Material Air
Criterion B
Z_{eff} = 0.8 ohms

T (ns)	n observed	n corrected dead space	σ assumed (%)	ΔT from observations	ΔT calculated
118	49	59	0.35	1.04	.83
86	60	72	0.3	.93	.52
79	61	74	0.3	.91	.48
73	65	79	0.3	.89	.44
68	82	102	0.3	.78	.41

time and inversely proportional to the length of the edge. In the extreme condition of fast charging given in Table 10c-XI, the voltage was falling to about 75 per cent before the streamers finally closed. It should be repeated that the dC/dt picture is being used in the section but an equally valid (and rather similar) set of explanations can be built up on photoionisation ahead of the

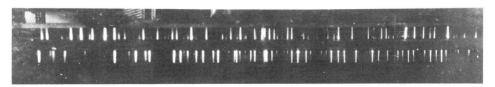

Fig. 10c-9 Multichannel breakdown for fast charging.

streamers. Figure 10c-9 shows a photographic record of the gap operating under these conditions. One point of interest is the partially formed channels reaching out from the centre edge towards the plane electrode. The fact that these are already fairly luminous shows that substantial currents are flowing before streamer closure. They also indicate, as was mentioned earlier, that the assumption of uniform conditions up the incipient channel is not always tenable. It should be stated that it was only under these conditions that these partial channels have been recorded photographically in the experiments recorded in this note.

At this stage a very short series of experiments was performed with positive polarity on the edge. These showed about half the number of channels that the negative polarity gave. Because of the very brief nature of the series of shots, this result should be treated with considerable reserve. As was expected for air the breakdown voltage was closely similar to that measured with negative pulse charging.

In order to counteract the dC/dt effects, it was decided to rebuild the line with a real impedance of 0.3 ohms and an effective one of 0.45 ohms. In addition a small width raised portion was built up on the plane electrodes opposite the knife edges. This closed down the gap spacing to 3 cms and hence the breakdown voltage decreased to 105 kV. The effect of these changes was beneficial and the slower fall on the waveforms was small and confined to the fastest charging conditions. It is difficult to be sure that both changes had an effect but I consider that the higher capacity of the line on its own could not have accounted for all the improvements.

Table 10c-XII lists the same parameters as Table 10c-XI but for the new line and edge plane gap, of course.

A small correction has been applied to the experiments with large values of T. This is because if a constant pulse were applied to a point or edge plane gap there will still be a finite jitter in closure time. Consequently when the streamers close to peak on the charging waveform, as T goes to infinity, ΔT remains finite.

Table 10c-XII
Results from Wide Edge (3m) Gap Tests
Z_{eff} = 0.45 ohms

T (ns)	n obs.	n corrected dead space	σ assumed (%)	ΔT raw cal.	ΔT from obs.	ΔT cal.
~220	54	64	0.4	1.8	1.2	1.6
~160	71	86	0.4	1.3	1.1	1.2
120	100	133	0.35	0.85	0.8	0.8
100	113	157	0.3	0.6	0.76	0.6
90	132	195	0.3	0.55	0.70	0.55
80	141	220	0.3	0.5	0.65	0.5

obs. - observed
cal. - calculated

The agreement shown by Table 10c-XII is much better than that in Table 10c-XI and since for very short values of T some residual dC/dt effect is still present, it probably helps to account for some of the difference here.

Figure 10c-10 shows the record corresponding to the entry in Table XII where something like 140 channels have been obtained. Using the rather ill-established dead space correction (after correcting the correction for the shorter gap and much greater edge length), this number corresponds to about 220 channels, if the edge length were much bigger.

One further quick experiment was performed to display the possibility of locating the channels reasonably uniformly along the length of the gap. This took the form of a 5 mil. mylar sheet placed over the plane electrodes in which 100 small holes had been punched. Figure 10c-11 shows a record taken with this arrangement. For various reasons the lower sheet of mylar sat down rather better

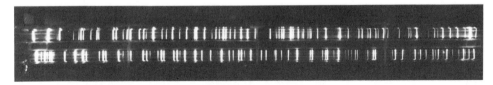

Fig. 10c-10 Multichannel breakdown corresponding to ~ 140 channel entry in Table 10c-XII.

SWITCHING

Fig. 10c-11 Multichannel breakdown for modified wide edge gap.

on the bottom electrode than did the top one and this accounts for the considerable difference between the two halves of the gap. It is my opinion that if more time had been available, the top gap could have been made to work as well as the bottom one. In a real gap, such a way would not have been employed, of course. Probably the best way would be to provide smoothly raised bumps on the plane electrode such as would be produced by rivet heads. This, with a judicious cutting back of the knife edge between required sites, should be effective in localising the channels. The cutting back of the edge would have to be smoothly done and could not be very pronounced.

I would like to thank Chris Richmond for his kind assistance in the experiments described in the above section.

Triggered High Current Rail Gap

This work was done rather over a year ago and was designed as a possible replacement for the solid dielectric multichannel gaps that had been in use for some years at AWRE. While it worked satisfactorily, the solid gaps have remained the preferred method of switching for the type of work that the division does. I would like to express my warmest appreciation to Rex Bealing and Pete Carpenter who were responsible for the experimental results briefly summarised in the section. The arrangement was almost perfect; they worked and I criticised.

A number of rail gaps were built, culminating in a four foot wide version which was tested in conjunction with one quarter of the megajoule bank. The gaps which were used differed only in length and consisted of two 3/4 inch brass rods bent back gently at the ends and separated by about 1.0 cms. A brass trigger electrode was set back from the central plane, in general with gaps of 0.8 and 0.4 cms to the rods from its edge. Three methods of producing the trigger pulse were employed, initially a Blumlein transmission line switched by a solid dielectric switch, then a low inductive mylar insulated capacity ringing into the capacity attached to the trigger edge. This was the method used in the large current tests and finally an improved double transmission feed system was developed to charge the capacity associated with the trigger in the shortest possible time. In general the gap was operated at 20 kV with a self break voltage of 30 kV. The trigger pulse used had a peak amplitude of about 40 kV and maximum rate of rise of about 8 ns. The trigger pulse was applied to the sharp trigger electrode in such a polarity as to break both portions of the gap simultaneously. Some care was taken to ensure that the pulse was applied to both portions of the gap, by having low impedance feeds to the main electrodes. In the final 4 foot wide version, some 40 and 70 channels were obtained in the two portions of the gaps when the feeds to it had an impedance of about 0.1 Ω. As such, the initial current established in the channels was about 200 kA and an important question was whether when the main current of some 4 MA was built up from the condensers several microseconds later, the current would still be well distributed. This question was important because it was known that the light pulse from channels driven by short bursts of current decayed in well under a microsecond. However, this merely meant that the hot channel had expanded into pressure equilibrium with the surrounding air and had become self-transparent in its own radiation. As such its emissivity dropped to a very low value even though the channel was still very hot. Various methods by which the channel might cool were investigated and it was concluded that thermal conductivity radially was probably the fastest mode and that this would take a time of the order of a few tens of microseconds. As such it was expected that the main bank current would establish itself satisfactorily in a number of channels. To check this a subsidiary experiment was performed with a 10 cm section of the gap fed by a 28 microfarad condenser and an inductance to give an 80 microsecond period. The peak current delivered by this intentionally slow system was about 50 kA. In this gap the trigger impedance was increased from a low value, when about 7 channels were observed to 14 ohms, when about 2 channels carried the main bank current. The trigger impedance at which the number of final channels was down by two was about 10 ohms. This meant that indeed even for very long bank periods the much larger bank current would still distribute itself reasonably between the briefly heated channels, even when these had been warmed to take only a few percent of the final current, and that the trigger impedance for the 4 foot gap should be a little under 1 ohm.

When the 4 foot gap was tested on the quarter clump it was no longer possible to photograph individual channels, because a diffuse glow filled some 20 per cent of the total rail length, but the relatively slight markings on the rail electrodes indicated that substantial areas carried the 4 MA, as expected. The inductance of the gap was about 1/2 nH, as calculated, and the DC hold off of the gap after carrying large current was within 10 per cent of its initial value. The rather flimsy nature of the feed to the trigger electrode suffered damage from the expanding hot gases, but, this apart, the gap survived well for a few shots, again indicating well distributed current carrying.

ENVOI

When I first became reasonably sure that my treatment of the problem was adequate for design work, I was tempted to write a short note and leave it at that. If any readers have survived as far as this point, they may well be in heartfelt agreement with my original intent. One person I am certain who must have wished I had followed my first inclinations is Mrs. Vikki Horne, who has nobly deciphered my impossible handwriting and disentangled my incredible syntax. If by any chance there are any sentences in English in this note at all, the whole credit belongs to Vikki. I would like to express my warmest gratitude to her for attending to her multitudinous duties and also keeping up with me in the four days it has taken to write this note. I must accept all responsibility for the doubtless numerous errors and omissions and can only plead that it has been prepared on a rather short time scale. If anyone wants enlightenment on any particularly opaque passage, I will be delighted to try and understand it myself.

- - - - - - - -

"The common cormorant or shag
Lays eggs inside a paper bag.
The reason you will see no doubt
It is to keep the lightning out.
But what these unobservant birds
Have never noticed is that herds
Of wandering bears may come with buns
And steal the bags to hold the crumbs."

(Anon).

Section 10d

HIGH SPEED BREAKDOWN OF PRESSURISED SULPHUR HEXAFLUORIDE AND AIR IN NEARLY UNIFORM GAPS*

J.C. Martin

INTRODUCTION

    This note records a short series of experiments on the pulse breakdown voltages of SF_6 and air in nearly uniform gaps. The initial motivation to start these experiments was the anomalous behaviour observed by Tommy Storr in a small pressurised SF_6 rail gap used as a switch in a 500 kV oil Blumlein generator. These results suggested that the breakdown voltage of the gap failed to increase with pressure above about 40 psig. This effect was fairly quickly traced to three different phenomena occurring in different versions of the gap. But before this had happened, a short series of tests was done with a rail gap in a 1 MV EMP generator (TOM) that was about to be delivered to the customer. These tests roughly confirmed a curve based on data obtained six years ago by Ian Smith, when he was at AWRE, for a sphere gap used in an oil Blumlein system called PLATO and showed that Tommy's gap was being influenced by other effects. Just as this position was reached and the original trouble resolved, data arrived from Ian Smith showing very large fields in a rapidly charged uniform field pressurised SF_6 gap, which had been obtained at PI. It was speculated that the effect could be due either to the rapid pulse charging of the gap or that there might be a gap length dependency of the field, the breakdown field increasing with gap length, the gap spacing being rather bigger in the PI tests than in any previous cases available.

---

* AFWL Switching Notes
   Note 21
   February 1973
   Originally published as
   SSWA/JCM/732/380

As a suitable test bed existed in the form of the EMP generator, a rather more extensive series of tests was mounted, involving air and SF_6 in two different rail gaps, one roughly a halfscale version of the other. In addition, Tommy Storr generated some completely independent data for SF_6 with his rail gap which was now operating very satisfactorily. As an exercise in scientific archaeology, Ian's original PLATO data was excavated from the record books and reduced to maximum field values. In addition, I recalled some data with a very small rail gap which was obtained in the prehistory era before written records were invented.

It seems to me that it is worth recording this information and on the way describing the performance of an output rail gap for EMP generators, which was mildly surprising from a different point of view, that of air flashover.

BRIEF DESCRIPTION OF GAPS

PLATO Gap

This was a cylindrical gap with an ID of 4-1/2 ". The spherical electrodes (phosphor bronze ball bearings) had their rear faces cut off and were situated in the cylindrical portion of the gap only. The spacing was 3.0 cm and the gap was located in between fairly extensive plates forming the transmission line, under oil. It had been operated up to 200 psig and on one momentus occasion it (or another small gap like it) had tracked, cracking the perspex pressure vessel. This would have done little damage in itself, but the 200 psi took out the face of the perspex oil container, dumping a cubic metre or two of oil on the floor.

Estimation of the field enhancement factor (FEF) presents a little difficulty, as it does in all the gaps described, but it is estimated at 1.3 because of the presence of the main line electrodes, giving a uniform equivalent gap of about 2.3 cm.

WEB GAP

This is the Tommy Storr gap mentioned earlier, the generator in which it is used having been christened (or at least named) for the purposes of this note. No prizes are offered for what it stands for, the generator being 500 kV, 50 ohms, 20 ns, and contained in a box 6-1/2 x 1-1/4 x 1 feet in size. The cylindrical electrodes are 5/8" OD and are mounted along the axis of a 2" OD perspex cylindrical body. The rods have been flattened on the back slightly and the gap is again mounted under oil between the transmission line electrodes. The electrode spacing is 1.3 cm and the FEF is estimated to be 1.24, giving an effective uniform gap of 1.04 cm or thereabouts.

SWITCHING

TOM GAPS

TOM is an EMP generator of the Marx peaking capacitor type and the long pulse charged output gaps are what were used in these experiments. JERRY, the low level pulser, had been delivered to the user some time ago. Prior to starting on the full generator, Messrs. George Herbert, Mike Hutchinson, and Denis Akers had tested some of the features of it on a half scale and for these tests they had built a "half scale" gap. This gap consisted of 5/8" cylindrical electrodes mounted inside a 2" OD perspex body. These electrodes had had approximately 1 mm removed from their inward-facing surfaces and then the cylinders contoured to give an effective radius of curvature of the order of 1.2 cm. The actual electrode separation was 1.25 cm and it is estimated the FEF was of the order of 1.20, giving a uniform field equivalent spacing of about 1.04 cm.

It was this gap which gave some interesting results as regards the flashover, so a further description of it will be given.

The one foot wide transmission line feeds which were attached to the gap were made from 3/4" wood contoured into "uniform field" profiles at the edges and wrapped in aluminium baking foil. Approximately 3 mm perspex sheet was wrapped around the lines and simplexed onto the 2" OD body of the gap. In addition, three fins of 3 mm perspex were bent and simplexed onto the body of the gap on each side of the gap. These fins stuck out 4" from the axis of the gap and were curved in the planes containing the gap axis. Figure 10d-1 shows a sketch of the gap. In the half-scale tests, the gap was taken up to about 450 kV in air, with the bottom plane of the transmission line 18" away, with no breakdowns across the gap or between the lines observed.

Subsequently the same gap was installed in TOM (which is a 2/3 MV user generator but has been tested up to a little over 1 MV) and the half-scale gap went up to a little under 800 kV in the tests described below. At this level a track occurred which flashed over the perspex wrapped feeds but punched through the base of the three fins where these were simplexed to the body. When the gap was examined, it was observed that there were some small bubbles in one of the joints and also that it was upside down. The small holes made by the track were drilled out, patches simplexed on and the gap reinstalled the right way up. Subsequently the same gap went to 800 kV without further trouble. Such a voltage on an air-insulated gap is progressive and was partly achieved because of the rapid charging in the peaking capacity circuit ($\sqrt{LC} \sim 28$ ns). The inductance of the gap, in a one foot line, is about 22 nH and in a 140 ohms transmission line would add 0.6 ns to the rise time (it was working at 100 psig SF_6). Of course the rise time of the output pulse is mainly controlled by the wave fronts rattling around

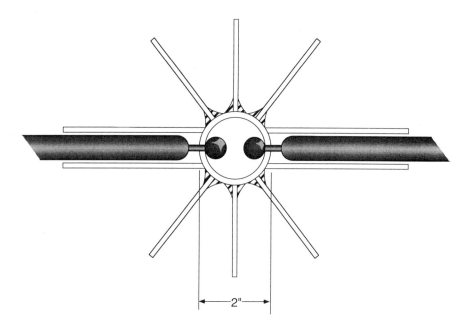

Fig. 10d-1 1/2 scale TOM gap, 800 kV in air.

across the depth and width of the line, but an 800 kV air-insulated gap with these characteristics which is simple to make is quite useful.

The full-scale TOM gap was similar in design but had a 4" OD perspex pressure vessel. The electrodes were about 50 cm over the uniform field part and were made out of 1-1/4" tube with solid end caps added at the end where the contouring was done. The physical gap was 3.05 cm but, just to make the problem of the FEF factor even more complex, brass was removed from both outwards and inwards facing surfaces of the rods. However, the field enhancement factor was estimated to be about 1.24, giving a uniform field equivalent spacing of about 2.42 cm. This factor would apply if the gap were remote from any earth plane and in practice the field is a bit crunched up against the electrode nearest the peaking capacitor. This factor has been crudely estimated and included with another factor (the prepulse, which slightly reduces the measured voltage) and this is briefly covered later.

Table 10d-I lists the relevant parameters and also gives a very crude estimate of the stressed area in cm^2, based on a standard deviation of breakdown of about ± 3 per cent. As can be appreciated from the above, the data which are used to obtain the maximum field on the electrode at breakdown are somewhat a matter of judgment, but it is felt that the equivalent uniform field gaps

Table 10d-I
Test Parameters

Gap	Electrode Shape	Diameter (in)	Operating Length (cm)	Material	Gap (cm)	Effective FEF	"Uniform" Gap Equivalent (cm)	Stressed Area $\sigma \sim 3\%$ (cm^2)
PLATO	Spheres	2	-	Phosphor Bronze	3.0	1.3	2.3	~ 1
WEB	Cylinders	5/8	10	Brass	1.3	1.24	1.04	~ 10
Half-Scale TOM	Cylinders	5/8	25	Brass	1.25	1.20	1.04	~ 25
Full-Scale TOM	Cylinders	1-1/4	50	Brass	3.05	1.26	2.42	~ 100

are good to about 5, per cent. On top of this uncertainty, of course, are the errors in measuring the breakdown voltage of the gaps. These are felt to be, again, within 5 per cent for the more recent data. This is in part confirmed by the very good agreement obtained between the breakdown fields at low pressures between the WEB and TOM, data which were obtained in completely different set-ups with different monitors, etc.

RESULTS

PLATO Gap

The PLATO gap results show significant jitter and also the breakdown point on the charging waveform is sometimes a little obscure. However, by selection of the better data, the curve given in Figure 10d-2 has been obtained. It should be mentioned that in Ian's original work pressures up to 300 psig were used in various gaps and with various electrode materials, so the curve given only represents a small part of his work at that time in this area. An

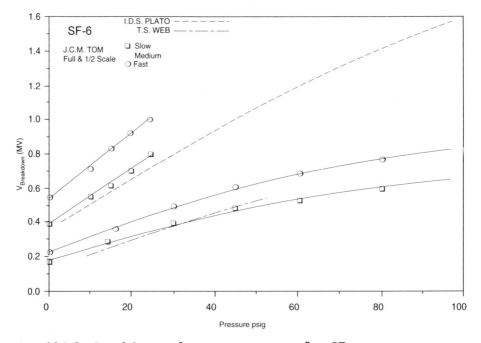

Fig. 10d-2 Breakdown voltage vs pressure for SF_6.

SWITCHING

additional point is that the values of t_{eff} used later are judged from the waveforms and may not be very accurate. However, as these numbers are taken to a low power, this factor is unlikely to have influenced the results.

WEB Gap

Tommy Storr's SF_6 data were obtained with the final electrode shapes of a short series of different forms and represent the values obtained after quite a lot of shots. Indeed, the initial curve obtained was some 10 per cent lower and some degree of conditioning occurred. The final curve given in Figure 10d-2 therefore represents a gap conditioned by maybe a hundred or more breakdowns. The coulombs passing through the gap are a bit difficult to define exactly (there being a high speed current pulse as well as a slower ringing) but are in the range of a few millicoulombs.

TOM Gaps

The design of the peaking capacitor made it comparatively easy to change the value of this. In addition, extra inductances could be placed between the Marx and the peaking capacitor. By means of these changes, the $\sqrt{LC}$ could be varied over a factor of about 4. Breakdown values were taken at 3 values of the rise time for the small gap and for the slowest and fastest waveforms for the large gap. The monitoring point was across the peaking capacitor and this measured the voltage on the high voltage electrode of the gap to earth. Fairly low resistances were placed across the output face of the generator, but because of the capacity of the gap, a small prepulse appeared on the output side of the gap. This varied between 2 per cent and 8 per cent, depending on the rate of rise of the pulse on the gap and on which gap it was. In addition to this, the field lines were distorted towards the high voltage electrode because of the presence of the resistors across the output face and the return earthy transmission line some 24" below the gap. This effect was estimated to be some 5 per cent and was included with the effect of the prepulse when the breakdown voltages were transformed into maximum field values on the electrodes.

The actual breakdown voltages were recorded on two 'scopes of different sensitivities and rise times and the slower 'scope corrected for this effect. In general the breakdown voltages derived from the two 'scopes were in good agreement, the average deviation between them being about 2 per cent, with no particular trend apparent between them. For the small gap working with air in it, only one of the 'scopes could be used and even this one tended to give rather small deflections, hence the air breakdown values for the small gap are less accurate than the other values.

Half way through the series, the polarity of the Marx was changed and a short run done with the medium rise time pulse and the half scale gap. The results showed no polarity effect apparent for the SF_6 data but did disclose a polarity effect for the air data. This effect was of the order of 10 per cent for 40 psig and decreased to about 3 per cent for 80 psig. The direction of the effect was that the positive electrode had the bigger strength. This at first sight was surprising; however, some fast edge plane work done recently shows that for small pressures the positive edge is stronger. If a differential relationship is inferred from the integral edge plane breakdown data, this does suggest that any streamers starting from positive whiskers will take longer to grow out from the electrode and that as pressurisation takes place, this effect should reverse. It should also be weakly gap length dependent.

Returning to SF_6 results, a weekend intervened in the two days of measurement and on the Monday the jitter of the SF_6 gap was seen to be much higher than it had been on the preceding working day. It was of the order of 6 per cent, whereas in general the jitter was 3 per cent or less, apart from an occasional drop out which was ignored in taking the data. A check was made that it was not the SF_6 flow rate and after perhaps 20 shots or so it seemed to disappear. The SF_6 data at the higher pressures for the medium rise time case half scale gap are maybe a little too low because of this effect. The runs were made alternately with air and with SF_6 and no very large number of shots was taken with each run, the total number in a run with one gas being of the order of 30, this number including a few initial shots after the gas had been changed which were not recorded. Figure 10d-2 gives the breakdown voltages (uncorrected for prepulse and field crunching) for SF_6 and Figure 10d-3 gives the breakdown data for air (again uncorrected) for the high voltage electrode being negative. The charge carried by the gaps was again in the range of a few millicoulombs.

In the case of the SF_6 measurements at atmospheric pressure, there were considerable signs of fizzle, there not being a clean transition to full conduction in the switch. This had also been observed as an effect on the rise time of the output pulse in other experiments. However, by the time the gap was pressurised to some 15 psig, the fizzle had largely disappeared. The transition to significant fizzle conditions was not a clear-cut one, but appeared to happen around the same pressure for both gaps for all rates of rise of the pulse, over the range examined.

BREAKDOWN FIELDS

Using the data in Table 10d-I and in Figures 10d-2 and 10d-3, the maximum breakdown fields for both SF_6 and air were calculated and these results are shown in Figures 10d-4 and 10d-5.

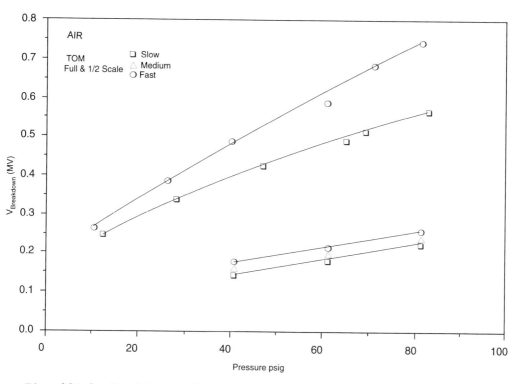

Fig. 10d-3 Breakdown voltage vs pressure for air.

The data for SF_6 from the half and full scale agree very nicely where the data overlaps and shows no signs of a gap length dependency. However, there is significant evidence of a time dependency of the breakdown field. The data from PLATO and WEB are in quite reasonable agreement with those TOM results obtained with the slow rate of rise of the pulse.

In the case of air (Figure 10d-5), there is again evidence of time dependency and also some gap length dependency as well.

Considering first the time dependency, I resort to the well tried fudge of a one-sixth power. I can plead very little rationale for this, except that the point and edge plane breakdown data fits this as well as any other power and Laird Bradley at Sandia, Albuquerque, has found a similar dependency for a pulse charged uniform field gap working in nitrogen in a particular regime. Table 10d-II lists the $\sqrt{LC}$ values for the systems and also the old t_{eff} (full time width at 63 per cent of peak) and the value that should perhaps more rigorously be used if the time dependency is one-sixth, that of the full time width at 89 per cent of peak ($t_{89\%}$). The effective time, of course, depends on the point on the

Table 10d-II
Breakdown Time Dependencies (All times in ns)

	$\sqrt{LC}$	$t_{63\%}$	$t_{89\%}$	(relative t)$^{1/6}$
PLATO	~ 220	~ 100	~ 40	1.26
WEB	100	75	25	1.17
TOM Slow	103	37	13	1.05
Medium	50	21	7	0.97
Fast	28	10	3.5	0.84

Base time for 63 per cent   27 ns *
for 89 per cent   9 ns *

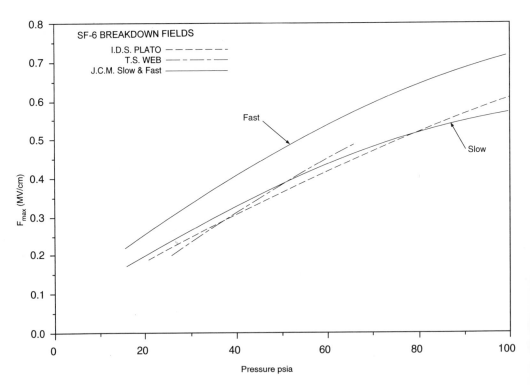

Fig. 10d-4 Breakdown field vs pressure for SF$_6$ gap.

*Editor's note: Slight change, ~ 10% from origional

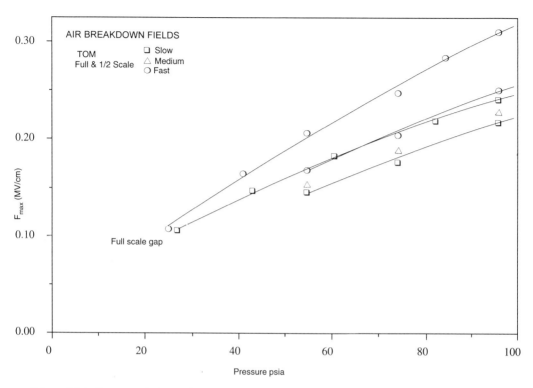

Fig. 10d-5 Breakdown field vs pressure for air gap.

rising waveform at which it was arranged in the tests that the gap would fire and is not directly proportional to $\sqrt{LC}$. Also given is a relative time to the one-sixth power. If the parameter being used is $t_{89\%}$, then this time is 10 ns; if the old t_{eff} is being used, the reference time is 27 ns.

The results for SF_6 are given in Figure 10d-6 and the data for the two TOM gap tests fall on essentially the same line, within a per cent or so. That for the WEB tests has a closely similar value at low pressures, but has a steeper slope. The PLATO data is quite similar at low pressures but is significantly higher than the TOM data and this would still be the case if the data were corrected to be the same at 20 psig. This is tempting to do for two possible reasons, one of which is that the time dependency may begin to disappear around the 100 ns time range, and the other is that the volt may have devalued some 10 per cent since 6 years ago. However, the differences are within any accuracy expected from the basic data and the uncertainties connected with the determination of the FEF factors, etc.

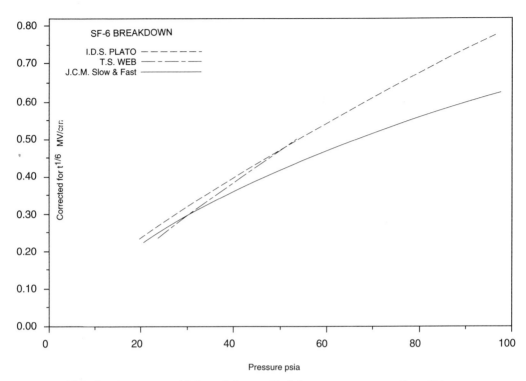

Fig. 10d-6 "Corrected" breakdown fields vs pressure for SF_6 gap.

The air results corrected, for a one-sixth time dependency, are shown in Figure 10d-7. The agreement after correction is not quite as good as in the SF_6 case, where it was excellent. Even so, the large gap data (which is rather more accurate than the half scale tests, as was mentioned earlier in the note) are brought into reasonable line. In addition there is a difference between the averaged data for the two gaps, which is essentially constant at a ratio of 1.20, the larger gap holding more field.

DISCURSIVE RAMBLE OVER THE RESULTS

It is amusing to note that when this series was mounted, while it was hoped to find a length dependency for the SF_6 breakdown field, none was found, whereas for the air case, where none was needed, it was found. A time dependency for SF_6 breakdown was found but it is a fairly weak one, and certainly not enough to account for the very high stresses mentioned by Ian. However, it is felt that Figure 10d-6 gives some clues as to how these could come about. Firstly, it is necessary to dispose of a previous explanation. This was advanced seven years or so ago, when the earliest

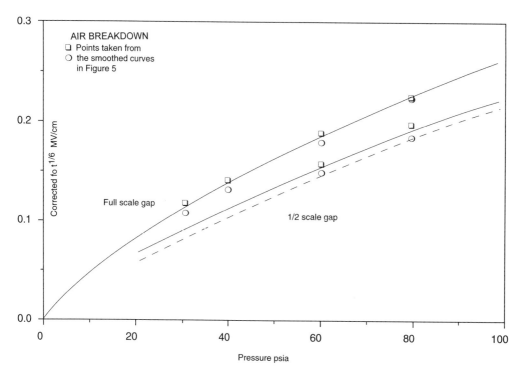

Fig. 10d-7 "Corrected" breakdown field vs pressure for air gap.

pressurised SF_6 measurements were made. I had predicted that fields of more than 1 MV/cm would be easily obtainable when pulse charged SF_6 gaps were used. This was based on consideration of the factors I believed to be responsible for the failure of DC charged gaps to have a more or less linear relation between gas density and breakdown voltage. All of these seemed to be slow acting effects and hence should not have time to occur in pulse charged gaps. When I failed to obtain the gradients I had so blandly predicted, I retreated behind the observation that as the pressure increased to around 100 psig, the jitter of the very small test gaps I was using became large, $\sigma \sim 10$ per cent or more, and at the same time the breakdown voltage/density curve became very decidedly nonlinear. This I claimed was what was causing the perverse behaviour of nature in failing to agree with my prognostications. Thus if one made many measurements, a few would be up on the more or less linear relation. Also if one could reduce the jitter, one would easily get to 1 MV/cm or more. Seven years later, that explanation looks as sick as the earlier one, since in these tests the TOM gap showed 3 per cent jitter or less up to the highest pressures used.

The explanation I now favour is the simple one of conditioning. Tommy's gap showed significant signs of conditioning and after a couple of hundred shots or so gave a much more linear dependency between breakdown field and density and, indeed, if extrapolated in a realistically slightly curved way, hits 1 MV/cm at about 100 psig. Table 10d-I lists the approximate stressed area and it is seen that the large TOM rail gap has ten times the area of the WEB gap. Thus to condition it to the extent that Tommy's gap was, it would have been necessary to fire a thousand and more shots. The size of defects one is trying to remove can be estimated very crudely as being comparable to the Townsend avalanche, which is of the order of 3×10^{-3} cm. Certainly in the large rail gaps used here, the existence of thousands of such projections in the highly stressed area is only too likely, as the electrodes were not polished or finely finished in any way. On this picture the jitter could be low after the first 10 or 20 shots and stay low as the conditioning slowly raised the mean breakdown field. For this assumed low σ conditioning to occur, firstly a fairly large area gap will be needed and in addition it would need to be rapidly pulse charged. If the TOM rail gap were very much more slowly charged, the channels instead of taking the shortest way across the gap, would be expected to begin to zigzag and to take longer paths; hence the jitter would be expected to increase considerably and the breakdown voltage at high pressures drop.

This picture of conditioning is supported by the PLATO results, where phosphor-bronze ball bearings were used. The stressed area was small and initially highly polished and the best curve lies quite a bit higher than the brass TOM results. However, in the PLATO tests the gap carried large currents and significant deconditioning would also be expected to take place. Certainly the scatter of results in the PLATO tests was much bigger than in the TOM tests. It is quite likely the charge carried in typical EMP generators is of about the right order to condition without creating too much in the way of secondary defects or whiskers. It should also be mentioned that the orientation of the gap might be important: fortunately most such gaps are likely to be mounted so that the debris falls off the electrodes.

While brass is a very good electrode material to use in DC charged pressurised high coulomb SF_6 gaps, there is no particular reason why it should be a good material for EMP output gaps. One would expect that a rather harder metal would be better, providing that it could take a good polish and that preferably it passivated when sparked in SF_6. The first property is easy to judge. The second can probably best be studied in intelligently guided tests. Unfortunately in the TOM tests the gaps were totally sealed ones, so they could not easily be opened and the electrodes removed and polished. Thus we were unable to check the above speculations in the couple of days we had available.

With regard to the air gaps, on a streamer transit picture, the larger gap would be expected to break at a higher field, but calculations suggest the effect is a few percent, not the 20 per cent found. (Similar calculations for pressurised SF_6 suggest tenths of a percent). Thus while the general gap length dependency of the breakdown field for pressurised air gaps and the small observed polarity effect are reasonable, the magnitude of the effect seems to be too large. However, further fiddling of the numbers might force agreement, but I have not had time to go into it properly yet.

ACKNOWLEDGMENTS

I should first absolve Messrs. Smith, Storr, Herbert, Hutchinson and Akers of all responsibility for most of the above. While they directly or indirectly provided much of the data, they are otherwise entirely blameless.

I would also like to acknowledge most warmly the very able and pleasant assistance of Mr. Graham Lovelock of AAEE, Boscombe Down, who, after suffering me for three months, still had enough uncurdled milk of human kindness left to slog through a hectic series of measurements with me in a conscientious and cheerful way. Without him, the measurements would not have been started, let alone completed.

Section 10e

FOUR ELEMENT LOW VOLTAGE IRRADIATED SPARK GAP*

J. C. Martin

INTRODUCTION

The higher the working voltage of a gap, the easier it is to trigger and the smaller the fraction of the DC volts the trigger pulse needs to be. The difficulty is usually to stop a high voltage gap working. For low voltage gaps (which the reader should be warned have always been a hang-up of mine) the difficulty is to trigger it quickly and reliably with a small trigger pulse. This difficulty is largely, if not completely, associated with obtaining electrons from the electrodes to start off the various breakdown processes.

Traditionally, field distortion gaps are useful for voltages of 20 kV and above and these derive their initiating electrons by field emission from whiskers on protruberances on the sharp edge of the trigger electrode. These gaps can be made to work at 10 kV, but need trigger pulses of greater than 15 kV and when used to pass quite small quantities of charge, may become erratic after a while. This is because either the thin trigger edge gets blunted, or because the whiskers get chemically rotted off and the rather low field on the trigger edge ceases to cause electron emission quickly enough.

---

* AFWL Switching Notes
  Note 26
  SSWA/JCM/778/514
  August 1977

The corona gap, which works down to a few kV, obtains the initiating electrons by producing a large field between a thin trigger strap and another electrode from which it is separated by one or two thou. (US mil.) of mylar. These initiating electrons avalanche across the plastic surface and irradiate or inject electrons into the main gaps which are between the trigger electrode and other electrodes. This gap, properly designed, works well but still needs a kV or two to trigger it, has a fairly large input capacity, and has the disadvantage for high current use that the discharge passes through the thin trigger strap.

Having a week or so to spare, a little while ago, I decided to indulge myself and look at other gap designs. Desirably these should have more or less uniform field electrodes, so that they can pass reasonable quantities of charge in use, have a wide operating range, need a small trigger pulse into a lower capacity, be fast, and have a low jitter. Because of the uniform field requirement they would have to be u.v. irradiated, and the other objectives suggested that three electrode gaps would not be optimal for operation at 10 kV or lower.

This note briefly summarises the result of a few days' hurried work, so the arrangements tested have not been well optimised. In addition, four element gaps have an extra degree of freedom, compared with three element ones, so only a few of the possible arrangements were looked at, and are results for only two are given below. Lastly, my notebook was even more inscrutable than usual and while this would not have mattered if I had been able to write up the results straight away, this was not possible. Returning to the notebook after a couple of months has been a sobering experience (to be compared with an archaeological dig) and while I am sure the results quoted were obtained with the mechanical configuration described, I am decidedly less certain that the particular values of the resistances given were actually in the circuit at the time of the tests; my apologies. Thus this note should be read as a description of a certain class of gaps, the results given as examples of what can be achieved, and the circuit values as possible examples of what actually were used.

MODE OF OPERATION OF GAP

Figure 10e-1 shows a schematic of the switch, which has four electrodes and three gaps. These gaps are numbered from the bottom, as shown. Also shown in Figure 10e-1 is the ratio of the DC breakdown voltages of the three gaps, 1, 2, 1 in this case, and a resistor chain not shown keeps the individual gap voltages at these ratios. This particular arrangement is typical of a gap requiring about 5 kV to trigger it quickly for 8 kV operation, but also having a wide operating range. The spacings (in mm) of the gaps

SWITCHING

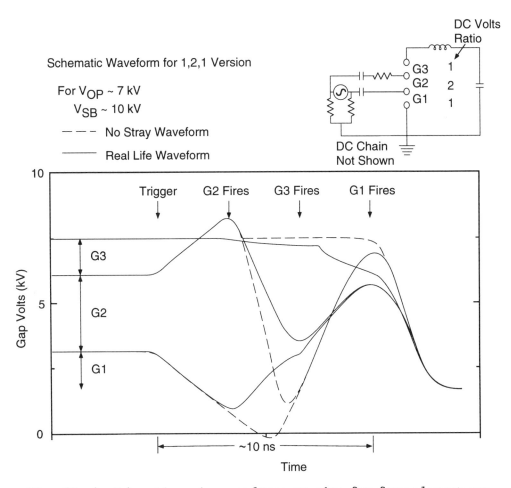

Fig. 10e-1 Schematic and gap voltage vs time for four element gap.

for breakdown at 10 kV DC are about 0.45, 1.2 and 0.45. The firing order of the gaps is G2, G3 and then G1.

I have an awful feeling that the nomenclature of this note is going to get out of control, but I will attempt to be consistent and follow the above pattern. Thus a fully specified switch would be: (1, 2, 1) (0.45, 1.2, 0.45) (3, 1, 2). This means that the bottom first gap has one-quarter of the DC volts on it, its spacing is 0.45 mm, and it is the third gap to close. In general, only the first set of numbers will be given in describing the switch, the others being included only occasionally. Hopefully all the circuits will be drawn for the switch so that gap No. 2 fires first, No. 3 next, and No. 1 last. This corresponds in the notation to 3, 1, 2 as the last set of three numbers. The perceptive reader will readily understand the potential pitfalls, but other methods of numbering seem to me to have even worse troubles.

Figure 10e-1 shows a distinctly idealised sketch of the voltages existing on the four electrodes (unnumbered; I've given up) as functions of time after the trigger pulse arrives. The circuit is arranged so that the volts of G1 and G3 go to zero and reverse, allowing a substantial pulsed voltage to appear on G2. Before either G1 or G3 can fire, G2 reaches its breakdown voltage and closes. Ideally the circuit should be arranged so that all the trigger pulse plus the DC voltage on G2 should now be applied to G3 (the dashed line). However in real life, where fast triggering is required, stray capacities, inductances, etc. mean that this does not happen and the electrode voltages follow the full line curves. G3 then fires and ideally the full DC volts appear across G1 (the dashed line again). However in real life, for quick operation, the potential of the electrodes follows the full lines again and less than the DC voltage appears promptly on G3.

In real life there are two further complications. The first and obvious one is that for a gap which can be triggered fast over a wide range of DC voltages, the standing voltage across each gap (and hence the total voltage applied to G2 and G3) alters as the DC voltage on the switch is changed. Thus the gap spacings should ideally be changed whenever the working voltage is altered. However, reader, take heart: in practice a gap self-breaking at 10 kV can be triggered quickly down to 4 kV with the above 1, 2, 1 arrangement, and even wider operation achieved with other DC breakdown ratios.

The second and more subtle way in which Figure 10e-1 is idealised is that it assumes the spark closure is well behaved and merely has an inductance and time-varying resistance, as is shown in Figure 10e-2A. In practice all gaps, I believe, have the breakdown characteristics shown in Figures 10e-2B and 2C. What happens is that an initiating electron (usually from an electrode) avalanches and produces a brief period of spatially non-uniform glow discharge. After a while the region where the initiating electron occurred gets warm, its pressure rises, and the gas expands. The combination of a gas density below ambient and glow discharge current necking in the region leads to it reaching plasma temperatures very much faster than would be the case if the glow discharge were uniform in cross-section across the gap and no hydrodynamics occurred in the gas. The front of the plasma spike then propagates across the gap and finally plasma channel closure occurs, the gap voltage falls, and the resistive phase formula can be applied.

The current passing before final plasma channel closure is known as "fizzle" in the spark gap world, and in the world of the

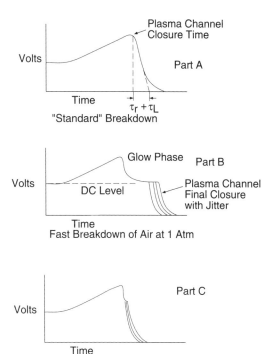

Fig. 10e-2 Pulse charged gap breakdown waveform.

gas laser as 'pumping'. (One man's meat is another man's poison.) The phenomenon is most clearly seen (for uniform field switches) for rapidly pulse charged atmospheric air dielectric gaps with large spacings (Ref. (1)), in particular where these are very well irradiated by u.v. Such gaps have a voltage waveform such as is shown in Figure 10e-2B. After reaching a field where the ionisation multiplication rate is very fast, the voltage falls to a level above but close to the DC breakdown level where the multiplication rate is low. After a period, which is longer the more intense and uniform the u.v. irradiation is, plasma channel closure takes place and the remaining gap voltage collapses. As is indicated, there is a jitter in the time of this final voltage collapse. For a uniform gap which is pressurised to a couple of atmospheres, the glow discharge phase still exists, but is much shorter, as is indicated in Figure 10e-2c. In addition, if the impedance of the pulse generating circuit is low, the glow discharge phase is shorter.

All the above applies to irradiation with u.v. which is only energetic enough to give rise to photo-electrons from the electrode. If u.v. of a wavelength small enough to provide a suitable level of ionisation in the gas is used, the gap volts can collapse very quickly, as was shown by Laird Bradley, but these rather special conditions are not likely to be relevant to this note.

In the present experiments the atmospheric air gaps were weakly irradiated by longish u.v. and were also very quickly pulse-charged by relatively high impedance circuits, so the gap breakdown behaviour sketched in Figure 10e-2B is very likely to occur on the few nanosecond timescale. This means that the pulsed voltage waveform that, say G3 sees after G2 fires may come in two bits. If the total of DC standing and pulse voltage is enough to fire the gap on the rapidly falling portion of the waveform, the jitter is very good. If, however, the gap closure occurs during the slowly falling portion of the wave, the jitter is very bad. If, however, the gap has to have most or nearly all the pulsed voltage, the jitter, while poorer than in the first case, is better than in the second. This I believe accounts for the occasional observation that lowering the DC standing voltage on the gap can improve its jitter.

Another way around the difficulty would be to pressurise the gap to two or three atmospheres, but firstly the spacings are already rather small, and secondly the gap loses its simplicity of construction, which is a practically nice feature of working in air. Reader, again take heart: the jitter of the gap was frequently sub-nanosecond in practice, without any care being taken. The above section is only included to account for surprising behaviour and to explain why the jitter is not in the tens to one hundred picosecond range, which calculationally it ought to be. It also gives hints as to how the design might be altered to achieve very small jitters indeed.

In designing a gap it is desirable to make the times for the three gaps to break down about the same; this seems to lead to the least jitter as well as the shortest operating time. Where a gap with a wide operating range is aimed at, then the above condition should be met for an operating voltage in the middle of the design range. For a low trigger voltage version, such as is next described, the condition should be achieved for an operating voltage of about 0.8 or 0.9 of the DC breakdown voltage. Using Figure 10e-4 (discussed briefly at the end of this section), it is relatively simple to do the initial gap design, although the final circuit components may have to be experimentally varied in order to get the best operating conditions. In addition, it is desirable to monitor the gap closure times to make sure they are approximately the values calculated.

Figure 10e-3A gives a sketch of a gap designed to operate with a near minimum trigger pulse. In this case it is a (5, 2, 3), (1.15, .35, .60), (2, 1, 3) gap. Figure 10e-3B gives a sketch of the waveforms, again showing how after firing G2 the pulse is used to help to over-volt G3, as before. After G3 fires most of the DC gap volts are applied to G1.

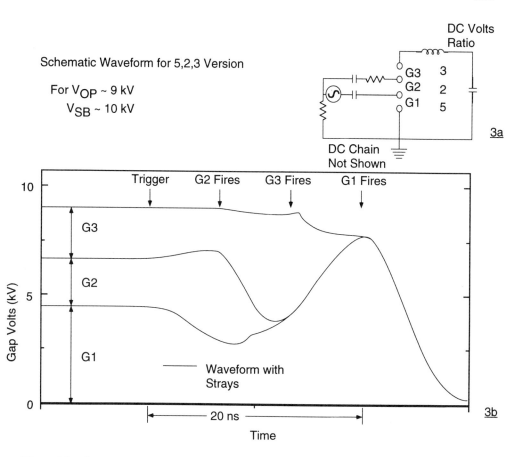

Fig. 10e-3 Gap voltages vs time for (5, 2, 3) four electrode gap.

The observant reader will have spotted that the ratio of the gap voltages is not much like the ratio of the spacings. This is because the DC uniform breakdown field for air at one atmosphere is given by

$$F = 24.5 + 6.7/d^{1/2} \text{ kV/cm}$$

where d is the spacing in centimetres. Thus for a 1 mm gap, the breakdown field is nearly double that of a large gap, even with uniform field conditions, and for smaller gaps it is even higher. This is shown in Figure 10e-4, which is taken from Reference (2). Reference (2) was written shortly after the period of experimentation with the gaps described in this note and provides the input data necessary to calculate their performance. The data given in Figure 10e-4 and Reference (2) is not of the highest accuracy, but is adequate for the purpose for which it was obtained.

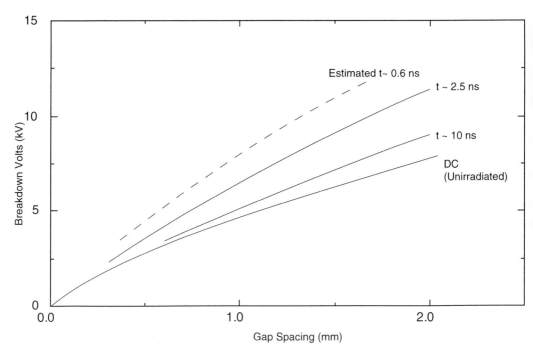

Fig. 10e-4 Irradiated pulse and DC breakdown voltage of atmospheric air gap.

In Figure 10e-4 it is seen that the DC breakdown field curve is far from linear and in addition the pulsed breakdown curves intersect it. Thus for a gap of about 0.5 mm spacing, a pulse voltage equal to the DC breakdown voltage with irradiation will cause it to break down with a t_{eff} of about 10 ns. (t_{eff} is the time width of the pulse at 89% of peak amplitude.) To make the gap break in a time of 2 1/2 ns only needs about a 3.5 kV pulse, while its DC breakdown level is about 2.7 kV, i.e., a pulse to DC ratio of about 1.3. For a centimetre gap with a DC breakdown of about 31 kV, a pulsed voltage of around 60 kV is needed to break it down in the same time, that is, a ratio of about 2.0. Thus small gaps need significantly less fractional pulsed volts to trigger them quickly. This is not so much that the pulse voltage requirement is down as that the DC breakdown voltage is high for small spacings. This phenomenon would not be expected to be observed for gases such as SF_6, where the data suggests that there is very little gap length dependency of the DC breakdown field.

SWITCHING

DESCRIPTION OF GAP AND UV IRRADIATION

It should be stated at the outset that the gap was just botched together and worked well enough that there was no cause to change it, and that it could be greatly improved. For instance, it is at least twice as long as it need be for operation at 10 kV and in addition its inductance could easily be halved from the present value of about 40 nH. In addition, the method of adjusting the gap was very inferior and this could and should have been improved. The DC breakdown of the gap was quite stable, except when dirt got into it, and some simple housing to exclude this would be useful.

With all these reservations, Figure 10e-5A gives a sketch of the gap. The two bent rod electrodes were made out of 1/8" brass, while the outer two electrodes were made of slightly thinner copper, suitably rounded. Under the gap was a reasonably low inductance strip line feed to a home-made condenser of about 1.5 nF capacity. Figure 10e-5B shows a cross-section of the gap with this feed. As the electrodes were rail ones and the maximum spacing about 1.2 mm for a rod radius of 1.6 mm, there was little field enhancement on them.

The pulsed irradiator was also highly unoptimised. The first version is shown in Figure 10e-5C and the surface flashover from it used to provide the u.v. irradiation of the main gap when it was about 3 cm away. The electrodes were mounted on a 3/4" OD perspex rod and a thin copper sheet acted as a backing plate or electrode. Two thou. mylar was placed between this and a tensioned 2 thou. thick, 2 mm wide, steel strap. The tensioning was provided by squashing a small slab of sponge rubber under the steel and away from the gap. This meant that if the steel expanded slightly, the tension in it was not lost. In the case of the irradiator shown in Figure 10e-5C there was no need to keep the capacity low or to make it work with the lowest possible trigger pulse: hence the backing plate was about 3 mm wide and the length of mylar over which the surface discharge took place was about 1 mm. Thus the capacity before the discharge started was some 3 pF and the pulse volts needed to flash it over in a few ns about 4 kV. The irradiator was driven from a 50 ohm cable unterminated.

Figure 10e-5D shows a sketch of the very low capacity u.v. irradiator which was designed to require the minimum trigger energy to operate. In this, 1 thou. mylar was used and the area in contact with the backing rod was about 1 mm^2. Thus the starting capacity was about 1 pF plus the lead strays. The mylar overlap was about 0.3 mm and the irradiator needed some 1.2 kV to flash it over in a few ns. The backing metallic rod had a diameter of about 3 mm and was half sunk into the perspex rod. The metal rod stopped just under the edge of the strap and was replaced by a perspex rod of the same diameter.

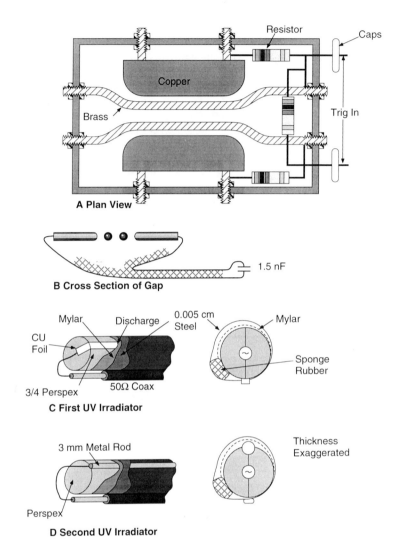

Fig 10e-5 Spark gap construction.

In operation, this irradiator was placed about 2 cm above the gap and positioned so that the uv could irradiate the cathode of G2 at its closest point. This did not seem to be very important, but appeared to have a marginal advantage.

SWITCHING

PERFORMANCE OF GAPS

Wide Operating Range Version

This had approximately the following characteristics:(1, 2, 1), (.45, 1.2, .45) (3, 1,2). Figure 10e-6 shows the believed circuit parameters, the trigger volts being provided by a 50 ohm cable; and also the pulsed trigger volts applied to the electrodes of G2 to fire it. The pulsed voltage required in the cable was roughly 0.7 of that shown in the figure. In the actual test set-up the trigger pulse had a rather poor rise time (~ 8 ns to 70% of peak), which is not of course necessary. Because of this, the circuit values were optimised for a rather slowly rising pulse. If these were re-optimised and an adequately fast rising trigger pulse provided, it is calculated that G2 could be fired in not much more than 2 ns. Because of this the observed delays have been reduced by the delay in firing G2 minus 2 ns; that is, the delays shown in Figure 10e-6 are based on G2 firing in about 2 ns.

As can be seen from the figure, operating at 0.8 of the DC breakdown voltage of 10 kV, the firing delay can be around 8 ns, using a trigger pulse rising very quickly to about 5 kV. The version shown here triggered down to a bit less than 4 kV. A slightly different version (1, 3, 1 probably) triggered down to 3 kV for a 10 kV self break, but with slightly longer delays at higher operating voltages.

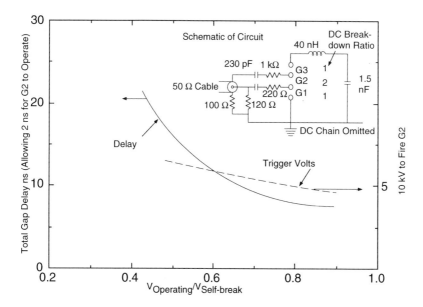

Fig. 10e-6 Wide operating range spark gap version (1, 2, 1),

Small Trigger Voltage Version

This had the following characteristics: (5, 2, 3), (1.15, .35, .60), (3, 1, 2). Figure 10e-7 gives the circuit values used and the pulsed trigger volts needed across G2 to fire it. In this case the pulse in the cable was about 0.6 of that shown in the figure. Once again the observed delays have been reduced to show what would have been the gap firing time if G2 fired after 2 ns. For operation at 0.9 of the 10 kV DC breakdown of the gap, the delay is about 17 ns and a 550 volt fast rising trigger pulse on G2 is needed to achieve this. If this is generated in a 50 ohm cable, the pulse voltage required in the cable is some 330 volts.

Table 10e-I gives the observed times for the gaps to fire, in nanoseconds. Δt_1 is the time for gap 1 to fire after gap 3 has fired, etc.

From Δt_2 the time the trigger pulse took to rise to the triggering voltage is subtracted and 2 ns added, to give the numbers in final column. No notice should be taken of fractions of a ns in Table 10e-I; the monitoring was not capable of resolving these, really. The actual numbers were obtained by averaging a number of individual measurements; hence the fractions.

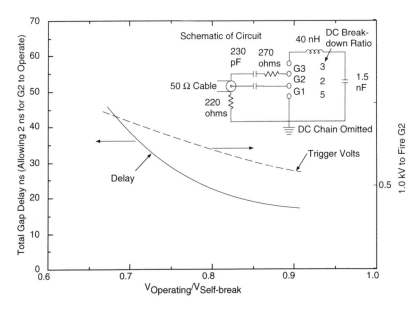

Fig. 10e-7 Small trigger voltage gap version (5, 2, 3), (1.15, .35, .60), (3, 1, 2).

Table 10e-I
Firing Delays for the Gaps
Self Break of Gap ~ 10 kV
Volts in kV; Times in ns

Operating Voltage 10 kV	Trigger Volts $t \to \infty$	G2 Firing Voltage	Δt_2	Δt_3	Δt_1	Gap Firing Time (Fast Trigger)
0.9	0.9	0.55	7	7.5	5	17
0.8	1.0	0.67	10	9.5	9	22
0.69	1.1	0.85	16	20	17	41

The total capacity across G2 in this version was about 6 1/2 pF, including the strays of the blocking capacitors. By building a half length version and tidying up the feeds, this capacity could easily be reduced to 3 pF.

Such a gap, at 0.9 of its self-break would need 550 volts into 3 pF to fire it quickly, whereas even the low capacity irradiator (shown in Figure 10e-5D) needed 1.3 kV into around 1.2 pF to operate it so as to close the surface discharge in a spark. It would work as a u.v. irradiator (without a faint closure spark appearing) at under 1 kV, but I would be reluctant to use it in such a mode, as the mylar might charge up and it could become irregular in performance. Thus the gap needs 1/2 microjoule to trigger it, while the u.v. irradiator needed about 1 microjoule, a most annoying outcome. In addition, the u.v. irradiator needed about 5 ns to provide sufficient u.v. to reduce the time jitter in operation of the gap to a low value (~ 1 ns). This was partly because of the slow rate of rise of the pulse supplied to the u.v. irradiator (it was the same pulse rise time as the trigger pulse). Thus with a faster pulse operating the uv irradiator gap probably there would be no need to delay the gap trigger pulse. Even if this were not so, only a couple of ns need be added to the times shown in Figures 10e-6 and 10e-7 to get adequate u.v. irradiation in the gap by the time it was needed.

In some applications it might be possible to provide a trigger pulse a few tens or hundreds of ns before the time the gap needed to be fired; however, from the point of simplicity this is obviously undesirable. Thus when I left the work I was contemplating improved u.v. irradiators, i.e., ones needing less volts and energy to make them operate more quickly.

When the trigger pulse was larger than the minimum required, multichannel operation was frequently observed. This was particularly true of the 1, 2, 1 gap, which when operated at around 0.7 to 0.9 of its self-break, would have a few channels in G2 and about 10 each in G1 and G3. As the capacity across the gap was only 1 1/2 nF, the multichannelling might not be so spectacular for larger capacitors and longer switched pulses.

While both gaps were fired many thousands of times, they were only tested with the small value of load capacitor, so questions as to electrode erosion, etc. were not addressed in the tests. However, because of the constancy of operation at peak currents of up to 2000 amps ringing on for quite long times, it is not expected that erosion will be a significant problem for most applications.

With regard to the jitter in gap firing time, there were operating regimes where the standard deviation was 0.3 ns or less, particularly with the 1, 2, 1 version. In general, the standard deviation was less than 1 ns, except for peculiar operating modes or where the trigger pulse amplitude was marginal. As was mentioned earlier, there were occasionally operating modes where the time jitter was relatively poor and where this improved considerably as the DC voltage on the gap was actualy reduced.

POSSIBLE VARIANTS

As has been mentioned earlier, there are various obvious improvements to be made to the first experimental gap, notably to build it about half the length over the electrodes and considerably less over the whole length. Where there is a trigger pulse available intermediate in amplitude between 1/2 and 5 kV, the spacings can be re-optimised to take advantage of this fact, to obtain an improved operating range and faster closure time than the 5, 2, 3 results.

I think the gap can be scaled down to work at 4 to 5 kV with reduced trigger volt requirements, but I have not shown this.

If there are trigger volts to spare and multichannelling can be confined with larger load capacitors, a very much lower inductance feed can be devised than is shown in Figure 10e-5B. Under these circumstances it may be possible to build a version with an inductance of significantly less than 10 nH, and switch currents of 5 kA or more rising in 5 ns or so. The latter estimate assumes that the full gap fizzle phase is short, which would probably--but not necessarily--be the case. Pressurising the gap to two or three atmospheres should certainly achieve this, or better.

Mr. Storr has devised an elegant version of the gap, where he has combined the u.v. irradiator and one of the electrodes of gap 2. This he has done by providing a corona gap version of this electrode, by making the current-carrying electrode a ring of 2 thou. metal separated by mylar from the rod underneath. The gaps are staggered out of a single plane, so all see the corona. This arrangement has the great advantage that no separate irradiator circuit is required: the trigger pulse provides the corona as well as over-volting G2. There is a minor disadvantage in that the gap will be somewhat limited in the coulombs it can pass, as the discharge current passes through the thin metal ring.

He has also built a small avalanche transistor circuit, so that the system operates with a 20 volt input. He has shown overall delays slightly shorter than those shown in Figure 10e-7, with very good jitter and the ability to run at modest rep. rates. Anyone interested should write to him and get details. (That should lumber him with writing it up.)

CONCLUSIONS

A four element u.v. irradiated gap of simple construction has been tested in a few of its many possible versions. With a suitably fast trigger pulse, one version can trigger down to 0.3 of its DC breakdown voltage, and at 0.9 of this fire in about 8 ns. The other version of interest requires only 0.55 kV to trigger a 10 kV DC breakdown gap operated at 9 kV in about 17 ns. Both versions have quite good jitter ($\sigma < 1$ ns) which can almost certainly be improved for a given set of operating conditions by tinkering. Low inductance versions of it are very likely to be constructable and with some air flow and short output pulse operation, versions of the gaps can probably be operated at 50 pps or more.

ACKNOWLEDGMENTS

It is, as always, a great pleasure to thank Vikki Horne, who has been typing out a series of notes for me over the past two weeks, this one being the last. I fear the strain of straightening out my syntax, pruning my punctuation, and doing major surgery on my spelling, for so long may have been too much for her. This is because as she handed me the typed script for part of this note, she snorted and said: "Archaeological dig, indeed," as she retired in good order. My apologies, Vikki, and my heartfelt thanks. I promise next time I'll use the word 'excavation'.

REFERENCES

1. "An Auto Irradiated Pulse Charged Divertor Gap", T. H. Storr and J. C. Martin. AFWL Switching Note 25, May 1977.

2. "High Speed Breakdown of Small Air Gaps in Both Uniform Field and Surface Tracking Conditions", J. C. Martin. AFWL Switching Note 24, April 1977. (Section 6e)

CHAPTER 11

BEAMS

Section 11a

PERFORMANCE OF THE TOM MARTIN CATHODE*

J.C. Martin

INTRODUCTION

The cathode arrangement known at AWRE as the Tom Martin (in fond acknowledgement of a very good friend and pioneer pulse powerer) is a hemisphere mounted on a long stalk of equal diameter. It has been used for a number of applications including large volume X-ray irradiation (Refs (1) and (2)) and as far as we are concerned as a machine performance check and part calibrator. It works at relatively large AK gaps, has a pretty constant impedance during 50 ns or so pulses and provides a cool electron beam under many conditions of operation. In the early days it was also important that it was tolerant of large prepulse voltages. For high-impedance high-voltage machines it is an excellent cathode that can be stuck in and fired many times while a machine is being run up and calibrated. It also can be used to check the machine performance over a long period to ensure there is no drift in its operation or serious change in its monitors.

Recently I had occasion to analyse the performance of the Tom Martin on MOGUL D and with the help of George Herbert and Pete Carpenter, went on to look at its operation on MOGUL C and SWARF. I also dug out a few other references on its performance on other machines and have included these in this short note. The main reason for writing the results up is that otherwise I will lose the data and at some later date have to do it all over again.

---

* H36
HWH/JCM/82/9
16 March 1982

GENERAL COMMENTS

Usually the Tom Martin cathode is used opposite a flat anode of substantial extent. This nearly always terminates the end of a cylindrical extension (known as the dust bin to us) which is mounted on the output face of the vacuum envelope or diode. Thus, geometrically, there are three dimensional variabilities, the radius of the stalk and hemisphere (a), the radius of the dust bin or return coax (b) and the AK distance between the hemisphere and the flat plate anode (d) (see Figure 11a-1).

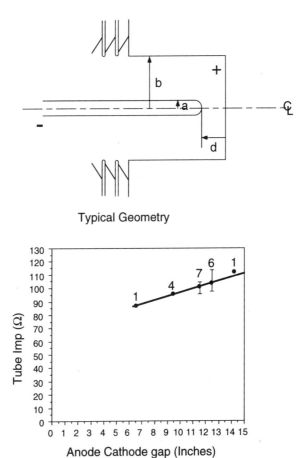

Fig. 11a-1 The Tom Martin cathode. Hermes II Data, tube impedance versus anode-cathode gap, for a 8 to 10 MV, 6 inch Diameter Cathode and ~ 60" O.D. dustbin.

For a given voltage of operation there are some obvious constraints on the values of these three dimensions. Firstly, the radius of the dust bin is usually dictated by the inner diameter of the tube and tends therefore to scale with the peak voltage of the machine. The choice of d is bounded on the lower end by the desire not to have the electron flow pinch so that the anode does not get damaged. If d is made too big the hemisphere and some of the stalk does not "light up" and so become reasonably uniformly covered in plasma. Also the radius of the hemisphere cannot be too small or it will start to emit during the prepulse. For applications where the maximum X-ray output is wanted there are two further constraints. The AK impedance should be around 3 times the generator impedance (typically ~ 90 ohms) although this criterion does not have to be exactly met. In addition it is desirable that the mean angle of the electrons at the anode be as small as possible, so that the maximum dose on axis may be produced (typically ~ 10° or less can be achieved). It is a tribute to the usefulness of the cathode that these requirements can be quickly and easily met and that then a given AK set up will work well over a wide range of voltages. In addition its impedance is pretty constant during the high voltage pulse.

With regard to prepulse, the cathode can be arranged to tolerate over 1/2 MV prepulse voltage, however modern machines tend to have rather smaller prepulses than this. Mike Goodman has the feeling that the Tom Martin likes some but not too much prepulse. However the data on MOGUL C (which had ~ 20 kV prepulse in its later versions) seems no different from the other machines (see below). However at relatively low peak fields in the cathode, a number of pulses may be needed to get the cathode down to a stable impedance operation. Unfortunately I do not have electrostatic plots to give me good estimates of the maximum field on the cathode but a *rough* value of this is the voltage divided by the radius of the hemisphere (ie. $F \equiv V/a$). At a field of about 1 MV/cm some 3 or 4 shots are needed to achieve the stable impedance with a new stalk and ball. At 1/2 MV/cm a fairly large number is needed to condition the cathode and stalk so as to obtain stable and θ symmetric operation.

When the conditions for maximum X-ray output are met (ie. ~ 90 ohm operation and small angle at the anode) the diameter of the electron beam at the anode is of the order of the AK spacing and sometimes may be slightly hollow.

MORE DETAILED ANALYSIS OF ITS PERFORMANCE

Despite the cathode's usefulness, there is not a lot of data about its operation over a wide range of parameters. In a way this is a consequence of its flexibility, it's easy to find good operating conditions, hence one settles for these. Thus it is necessary to use data from different machines in order to draw some hopefully more general conclusions. This is obviously a fairly dicey way of going about things and most of the statements made below are true of at least one setup and may be more general. In addition they only apply to "conditioned" cathodes.

Table 11a-I lists the machines for which I have some data. The data given is only approximate (in some cases I have had to take data off sketches of the machine). Also typical AK gaps are given and where the maximum voltage has only varied over a small range, the average of this is given.

The pulse length of all the machines is around 50 ns except for the long pulse Sandia one. For this, the time at which the impedance is taken is around 150 ns. The number of shots is not necessarily the number of times the machine is fired, it being the number of shots used in the analysis. The standard deviation per shot in the impedance is only rough. For MOGUL C and D the results analysed were spread over a long time, in batches of a few shots each. That for Hermes 2 is plausible for one series of shots. Thus the two sets of standard deviations are not necessarily comparable. Consideration will now be given to some of the results obtained.

Voltage Dependency of the Impedance

The MOGUL D results do not show any statistically significant variation of impedance with voltage over most of a range of two. In addition the impedance dependency on the geometry suggested below implies (if it is real) that there is little or no voltage dependency. The latter of course can only be true if (a) the conditioned cathode lights up during the pulse and (b) does not light up at all during the prepulse. In addition the gap must be large enough so that plasma closure during the pulse does not significantly close it. Ref (3) implies a cathode plasma closure velocity of about 2 cm/μsec for its conditions of operation. Also the anode should not go into plasma. This will normally only happen if the beam punches.

Variation of Impedance with AK Gap

The Hermes 2 note (Ref (1)) shows a gentle variation of impedance with AK gap. Figure 11a-1 reproduces the data. Unfortunately I have been unable to find any other data where the AK gap has been varied over a reasonable range, preferably at roughly constant volts.

Table 11a-I
Typical Operation Condition

	Diameter Stalk (2a) cm	Dust Bin Diameter (2b) cm	AK gap (d) cm	V_{max} (Average)	No Shots	θ Average	σ per shot of Z	Ref
SWARF	5	28	8	2.5	10	~ 10°	± 7%	
MOGUL C	3.8	50	10	4	7	11°	± 10%	
MODUL D	3.8	74	10	4 1\2-8	18	10°	± 7%	
HERMES 2	15	~ 150	15-35	9	19		± 4%	(1)
EROS	10	86	13	5		~ 12%		(2)
Sandia Long Pulse	7.6	~ 43	10	0.6				(3)
Phys. Int. Facility 2	10	~ 85	12.7	4				(4)

Variation of Impedance with Geometry

In the Hermes 2 note reference is made to some calculations of Jack Boers which suggested that Z = 30 ln b/a (Ref (1)). In order to try to take account of any slow variation of impedance with AK gap I decided to use Figure 11a-1 and convert all the data to a scaled gap of d/a = 4.3. This value was mid way between the main body of data of MOGUL C and D and the rest. This treatment obviously assumes that Figure 11a-1 applies approximately to all the set ups. However the maximum change of the observed impedance was only 20% and most of the data was increased or decreased by less than 10%.

Figure 11a-2 shows the impedances (scaled to a d/a of 4.3) plotted against ln b/a. The somewhat odd symbol against the long pulse Sandia machine reflects my attempt to correct this point back to about 50 ns on a fast rising pulse.

The line drawn in Figure 11a-2 is an eyeball one and the standard deviation per point off it is some ± 8%. This is remarkably low since there are many machines involved and any errors in calibration etc. will add to the intrinsic error. The numbers in the brackets beside the machine names are the peak voltages given in Table 11a-I. There appears to be no obvious trends with voltage

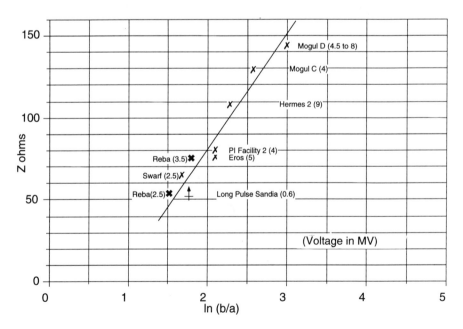

Fig. 11a-2 Conditioned Tom Martin Impedances at d/a = 4.3.

and indeed an attempt to correct the data with a Child Langmuir space charge limited voltage dependency led to a very bad scatter. While most of the data points had adequate fields on the stalk to make it reasonable that they will be plasma covered, the long pulse data is surprising. In this case the maximum field on the stalk away from the cathode is only some 90 kV/cm and this is a very low field at which to get emission unless special care is taken to ensure it.

CONCLUSIONS AND ACKNOWLEDGEMENTS

Some of the operational characteristics of the Tom Martin cathode are documented. In particular its conditioned impedance appears to depend almost solely on ln b/a with only a weak dependency on d/a and little or no voltage dependency.

I would like to thank the Foulness SWARF team for their help in obtaining some of the data and also the H Area pulse power group for accumulating other data. What I have done with it is nobody's fault but my own.

REFERENCES

1. "Summary of the Hermes Flash Xray Program", T.H. Martin, K.R. Prestwich, D.L. Johnson, Sandia Labs, 5C-RR-69-421 Oct 1969.
2. "EROS Pulsed Xray Facility", E. Thornton, UKAEA AWRE Report NRN 10/71, April 1971.
3. "Pulsed Nanosecond High Energy Electron Beam Accelerator", T.H. Martin and R.S. Clark, Informal note, undated.
4. "Design Review 3 Aurora Vol 2", PISR-127-4, July 1969.

Section 11b

ELECTRON BEAM DIAGNOSTICS USING X-RAYS*

D.W. Forster
M. Goodman
G. Herbert
J.C. Martin
T. Storr

INTRODUCTION

The X-ray spatial distribution produced by an electron beam hitting a high Z target has been used as a diagnostic for many years. Quite early in the development of very short pulse X-ray systems we used it to provide indications of the spot diameter and, a little later, the mean angle of incidence of the electrons at the anode. Physics International in particular have used the whole X-ray field to check electron flux and mean angle as a function of radius across the cathode, using low Z material to map the front surface dose and high Z to map the intermediate field, during the Aurora programme. Cornell have employed both framing and streak camera techniques to give a time-resolved pattern of the electron flux at the anode. Sandia Laboratory, Albuquerque, have used polar diagram diagnostics on Hermes II and have also used the spectral content of the X-ray flux to obtain time-resolved data on the energy of the original electron beam in the diode.

While many of the remarks made below apply to the high voltage range, the aim of this note is to extend the simpler of these techniques to the low voltage range and to point out their advantages and some of their limitations. It is intended to be a user's note

---

* SSWA/JCM/714/162
AFWL Radiation Production Notes
Note 10
April 1971
Editor's Note: Several figures in this section have been renumbered to allow sequencing with the text.

and will concentrate on the practical issues involved; consequently it is redolent with simplified treatments and adequate approximations which ease the reduction of the data. It is also parochial in that it lays out the parameters we use and how we obtain them on a routine basis. It is hoped that the justifications offered will explain our choice but, as with all simplistic parameters, many can, in principle, be used.

The section below deals with the beam parameters which can be measured and in this section some of the techniques possible will be outlined, although we have not yet found it necessary to use them all.

The following section covers the experimental data for the X-ray generation from an axial beam of electrons and the way this is treated to analyse experimental results.

The next section deals with some of the many practical aspects and attempts to assess the accuracy obtainable.

The last section deals with some experimental results obtained at AWRE as illustrations of some of the simpler applications.

This note is not intended to be the last word on the subject and, as such, the accuracies aimed at are of the order of 10 per cent rather than 1 per cent. However, in view of the fact that the mean angle of the electrons in a high flux beam can be obtained easily, such an accuracy is a very considerable improvement on previous techniques, while at the same time being quick and convenient.

OUTLINE OF DIAGNOSTIC TECHNIQUES TO BE CONSIDERED

The parameters which can be measured are the current density across a plane, the mean angle of arrival of the electrons as a function of the radius and also the energy of the electrons, all of the measurements being time dependent if required. While separate experiments may sometimes be required, several of the measurements can usually be done at the same time, at the cost of more recording channels. The great advantages of the techniques are that they can be employed at flux levels which will vaporise any material and also that the measurements are not restricted to the plane of the anode, but can be employed at the end of any transported beam.

The following list of experimental techniques covers and comments very briefly on some possible approaches.

Spatial Distribution

In general, a high Z target is used for these measurements, so that a flat polar diagram is obtained. For time integrated results a pin hole camera can be employed, using a series of films within it to cover a large range of intensities. Because radiographic film has a high contrast ($\gamma = 3$ typically) and variations of density of a percent or two can be detected, rather high accuracies can be obtained for the resulting time averaged electron flux. The resolution can frequently be 10^3 to 10^4 elements if sufficient intensity is available to use small pin holes, which can be conveniently made in heavy metal. A lower resolution technique is to cover the high Z converter with dosimeters, of which TLDs are a very compact, cheap and convenient example. If these are required to respond to only small areas of the beam, they can be set in a high Z egg box structure, but in general it is adequate just to strap them on the rear face of the converter. For time dependent measurements an image intensifier camera may be used, either in the framing mode or in a streak mode. A thin slab of plastic phosphor is mounted on the back of the converter and light from this recorded. Unfortunately, as the voltage on the diode changes, large variations of the light output result. For a constant diode impedance the light will go roughly as the fourth power of the voltage and hence the camera has to cover as large a range of light intensities as possible. This may mean a number of separate image converter cameras, or possibly a camera with a series of images side by side with different optical attenuators in each of them; this, however, reduces resolution rather seriously.

In obtaining the electron flux from the data, the assumption has to be made that the energy of the electrons across the plane is the same and sometimes, as in the case of a diode pinched beam, it is conceivable that this is not so to an adequate accuracy. However, a subsidiary experiment using techniques given below can show if this is the case and provide a measure of the departure from a uniform instantaneous electron energy.

When using any spatial resolution techniques at voltages of several million volts the angle of the electrons at the converter becomes of paramount importance, since, even with a high Z target, the cone of X-rays is small and may largely miss the detector from some area of the anode. This polar diagram vignetting can be of considerable importance in interpreting current distributions from an X-ray pinhole radiograph. This effect can also be of importance where a pinhole camera is added to other measurements in a low voltage shot using a low Z target, where again the polar diagram can be quite peaky. The dosimeter plot across the rear face of the converter is much less susceptible to this difficulty and can be very useful where the electron mean angles become significant compared with the polar diagram width. The corresponding time

dependent measurement can be made in these conditions by using
phosphor blocks in egg boxes on the rear of the converter and connecting these to photo-diodes or photo multipliers by shielded
light pipes, or maybe it would be better not to do the experiment
at all under the circumstances.

Another potential difficulty may arise in a drifted beam,
where an original monoenergetic beam develops a spread of energies.
This can conceivably come about from two-stream instabilities, or
because of electrons being slowed at the front and/or being over-
taken by faster ones from later in the pulse. In these circumstances the higher energy electrons produce most of the X-ray emission and the results are then biased towards the distribution of
these. A subsidiary experiment of the kind outlined below can warn
when this is happening.

Polar Diagram Measurement of the Mean Electron Angle θ

For low Z materials the polar diagram is comparatively sharp,
even for low energy electrons. For instance, for carbon, for a
1 MeV electron beam the half width at half height of the polar diagram occurs at about 27° and even for 0.2 MeV the angle is only
some 57°. Thus mean angles as small as 10° can be measured with
reasonable accuracy.

The polar diagram is ideally measured in the far field of the
X-ray pattern at a distance remote from the source. A fairly useful terminology, taken from rather longer wave EM usage is: near
field - adjacent to the X-ray source and within a fraction of its
diameter; intermediate field - that which is neither near or far;
and far field - where the source dimensions are small compared with
the distance. The near field measurements have been covered above;
the far field can be analysed to give θ_{avg}; the intermediate field
contains information relating to both the spatial distribution and
the angular distribution of the incident electron flux at the converter. Ideally, measurements should be made all over the field
and then unfolded, but life is a bit too short and hence the selection of one or the other region to simplify the data reduction.
For polar diagram experiments there is pressure, because of dose
limitations, to work as close to the edge of the intermediate field
as possible and then a spot size correction is taken off, a matter
dealt with reluctantly in the next section. However, where the
radius of the source is less than 10 per cent of the distance to
the dosimeter array, the correction is small and reasonably well
known.

To obtain a time average mean angle of the electrons, an array
of dosimeters is set up at a constant distance from the source in
one plane or, with asymmetric sources, two orthogonal planes. By

comparison with the polar diagram to be expected from axial electrons the mean angle can be determined, as is discussed later. In principle, distributions of electrons more complicated than a simple mean can be derived from the polar diagram, but experimental uncertainties render this a difficult and unrewarding task, except in a few special cases.

One of these is in the case of a pinched beam, where it can be estimated what fraction of the X-rays come from the pinched phase; then two mean angles may be derived with reasonable accuracy. However, time dependent measurements, with or without blanking off of sections of the converter, can lead to better estimates. As dosimeters we use lithium fluoride TLDs and find these cheap, convenient and reliable.

In some beam transport experiments, stray dose may come from the outer regions of the converter or return conductors. These X-rays can be shielded out with lead absorbers, or the TLDs mounted in lead telescopes so that they only see the intended converter area. As is mentioned later, we use lightly shielded TLDs to reduce the response to stray background scatter X-rays, but, if necessary, background TLDs shielded from the main source can be used to subtract out this component.

For time dependent polar diagrams a plastic phosphor photodiode or photo-multiplier combination works well, the response of which can be separately calibrated or normalised with an adjacent integrating dosimeter.

The experimentally determined polar diagram is corrected for absorption in the converter, as is further explained in the next section, before being used to obtain a mean electron angle. However, a point worth dealing with at this stage is the definition of the mean angle and its relevance. As with all parameters which attempt to categorise a complex real situation with a single number, cases can easily be constructed where it is meaningless or misleading. However, as with the rise time of a voltage pulse, it is extremely useful to have such a parameter, accepting that it is only a rough approximation to real life. The real battle occurs when it comes to selecting the parameter. One possible choice is to assume that all the electrons are travelling at one angle and use this to fit the experimental pattern. This approach, in our experience, rapidly leads to highly non-observed polar diagrams for large electron angles. The parameter we use is to assume that the electron flux uniformly fills a cone out to some angle θ_{max}. The relation between $\bar{\theta}$ and θ_{max} is not exactly constant but $\bar{\theta} = 0.65 \theta_{max}$ is within a couple of percent as θ_{max} goes from 0° to 90°. Experimentally, this parameter appears to fit most of the data we have. In addition, considerations relating to electron trajectories in diodes and beams tend to suggest a fairly well smeared out distribution of electron angles. A sharp cut off at θ_{max} is practically very unlikely, but an angle above which there are few electrons is quite reasonable, if only in some cases because 90° is a limit for a progressively minded beam. Incidentally it is worth pointing out that a mean angle of 57° corresponds to a cone filled out to 90°.

Anode Cathode Voltage

The X-ray dose rate for paraxial electrons is given by a relation of the form $dR/dt = k\, i\, V^n$ where $n = 2.8$ over a wide range for all materials and k is a function of the material. The values for these are discussed in the next section. If i is measured as well as dR/dt, then V is obtainable. Indeed, because of the high power of V a 10 per cent accuracy in the dose gives V to about 4 per cent. For a beam of electrons which is not axial, a polar diagram measurement made at the same time allows a correction to be made to allow for this. This correction can be made in two ways. If the polar diagram leads to a unique value for θ_{avg} this can be used to correct the dose on axis to what it would have been if the electrons were paraxial. A theoretically sounder method is to integrate the total flux of X-rays and then use this integral radiated X-ray energy to give the incident energy of the electrons. Unfortunately, a lot of the X-ray energy is radiated at large angles, even when the polar diagram is fairly peaky, and this may mean making measurements on the beam side of the converter. Absorption corrections in the target can also become important and tiresome; so, while this is sounder in theory, in practice the simpler procedure can be practically as accurate and a lot less trouble.

Experimentally a mean anode cathode voltage can be obtained with a dosimeter array and an approximate knowledge of the diode impedance, or a measurement of the mean current. However, if the converter is made part of a Faraday cup current monitor and if the dose on axis is measured by means of a phosphor photo-diode combination, the voltage can be obtained as a function of time. An example of this is given in a later section, where the mean angle of the electrons was small and hence the time-dependency of it was unimportant. Where this is not the case, a time-dependent polar diagram measurement is also required, so a Faraday cup converter and a phosphor photo-diode array is then required.

Where the electron flux is not too intense, the carbon block which doubles as an electron stopper in the Faraday cup and an X-ray converter can also function as a calorimeter.

The main difficulty with this measurement is the accuracy of the constant k and the determination of the absolute dose rate.

However, accuracies of ± 20 per cent in k are probably on with existing data, corresponding to an accuracy of ± 7 per cent in V. However, if the technique becomes popular, this constant can obviously be measured with entirely acceptable accuracy, although care will have to be taken in defining the experiments to make the new measurements of immediate practical use.

All Singing, All Dancing Measurement

Set ups can be devised which measure just about everything that is worth measuring, at least as far as the diode is concerned. Such a set up would involve an anode divided into, say, 3 equal sectors by sheets of lead orthogonal to the converter and containing the axis of symmetry of the beam. Within each 120° sector, lead shields are placed over the converter, letting X-radiation out from 3 annular zones, one in each sector. The converter is in the form of a 3 sector carbon Faraday cup and measures the current to each carbon zone from which the radiation is being allowed to reach its corresponding photo-diode dosimeter array. There are 3 of these, each of which measures half of the polar diagram from one sector zone. From such a set up the values of θ and the electron energy may be obtained as a function of time and of radius, providing the beam is axially symmetric. Yet more complicated set ups can be visualized, but it is doubtful whether there is any practical application for them - a comment which may well apply to the last proposal.

The approaches outlined above are in fact all fairly easy to set up and, as has been mentioned earlier, yield data for beams of extremely high fluence. While attention has been drawn to a lot of possible snags and difficulties, in practice they are easy to use and reasonably unambiguous in interpretation. Used in conjunction with standard tube voltage measuring techniques, they give real confidence in the internal consistency of diode characteristics, as well as providing values of θ simply, for diodes and beams under most conceivable experimental conditions such as magnetically confined ones.

EXPERIMENTAL DETERMINATIONS OF POLAR DIAGRAM AND DOSE ON AXIS

As input for the analysis of X-ray diagnostics, the polar diagram for axial monoenergetic electrons for various target materials is needed, as is the dose on axis. The two papers containing most of the available data are: Buechner et al., "Thick Target X-Ray Production in the Range from 1250 to 2350 Kilovolts" - Phys. Rev. 74, No. 10, Nov. 15, 1948; and Rester and Dance, "Thick Target Bremsstrahlung Produced by Electron Bombardment of Targets of Be, Sn and Au in the Energy Range 0.2 to 2.8 MeV" - Jnl. App. Phys.

Vol. 41, No. 6, May 1970. Comparison of these shows that the polar diagram data is in essential agreement and fits rather well the Universal Polar Diagram of Limited Applicability, even down to 0.2 MeV. However, the data for dose on axis is in a considerably less satisfactory state and while the dependency on Z is very similar in each paper, the absolute agreement is poor.

A few general points will be made before proceeding to the comparison of the results. Buechner and his colleagues rotated the target, and the detector, which was mounted at right angles to it. Thus the self-absorption of the target changed only a little and they quote results where this has been removed. Rester and his partner fixed the target at right angles to the beam and moved the detector around it. Their results are quoted for external radiation uncorrected for target absorption. One main aim of their paper was to measure the spectra of the X-rays and these are compared with ETRAN 15, consequently they used a NaI anticoincidence counter, from which the integrated energy flux was subsequently calculated.

Dealing first with the polar diagram results, a point which immediately arises is whether the data required should take a form corrected for target self-absorption, or as externally measured. If the data is quoted for external radiation this will clearly depend on the target thickness used and also the solid angle subtended by the detector. Moreover, as the target is viewed from the front face or the back face there will be a big difference in absorption, since in the latter case the X-rays have to travel through much less absorbing material. Thus no form of universal curve will result as the transition from front face measurement to back face will occur at different points up the curve for different energies and Zs. Consequently we have elected to follow Buechner and remove the target absorption for the polar diagram results, otherwise ideally an ETRAN run would be needed to compare polar diagram results for each case. The consequence of this decision is that for any polar diagram experimentally obtained the target absorption must be allowed for, but for low Z materials this is a small correction, even out to 75°. This is, of course, the case of most interest for determining θ. However, even for high Z materials the correction is not large; for instance, even for 0.2 MeV at an angle of 60° it amounts to about 1.25 with an electron range thick target. Thus the correction for target and other backing material absorption is left to the experimenter, although guidance is given on it later.

The data of Rester on polar diagrams is given on p. 2690 of the journal, but before the polar diagram can be extracted one must take into account the fact that the backwards X-ray flux comes from about one-sixth of the way through the target, while the forward flux passes through more like five-sixths on average. This effect is only of importance for the higher Z targets, namely tin and gold. Using the experimentally determined spectra given earlier in the paper, the absorption was removed at the small and large angles. A smoothed curve was used to join these points. The same was done for the low Z targets and from these the polar diagrams in the absence of target absorption were obtained. These were then compared with the "universal" curve originally obtained from Buechner (SSWA/JCM/711/149). Again, the Z dependency factor of $(Z/74)^{0.30}$ was used and the data were plotted against $U\theta$ where $U \equiv V + 1/2$ and V is the electron energy in MeV. Figure 11b-1 reproduces the "universal" polar diagram and Figures 11b-2, 11b-3 and 11b-4 give the Rester results obtained, as outlined above. Also shown in Figures 11b-2, 11b-3 and 11b-4 is the universal curve for an electron range thick target.

Even for 0.2 MeV the agreement is quite reasonable. There is a suggestion that for Be the curve is fatter than for the other materials, but the original points were taken from a very small scale graph and thus could be reading errors. However, at large values of $U\theta$ the low Z materials consistently lie beneath the universal curve. This had been noted for the Buechner data, but the scatter there was sufficient not to make the effect certain. The values for Au for 1 MeV lie above the curve significantly, but there the absorption correction may not have been quite correctly applied. Considering the original curve was partly based on data (high Z only) up to 27 MeV, the universal curve does a good job over a rather large energy range for this class of material.

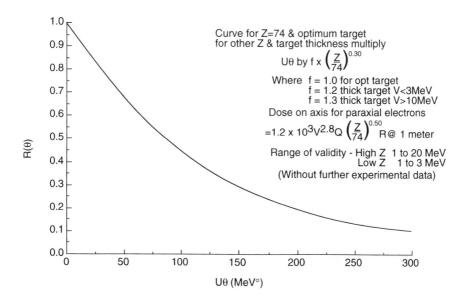

Fig. 11b-1 "Universal" polar diagram and dose on axis relation.

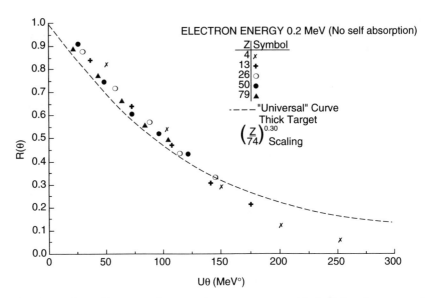

Fig. 11b-2 Polar diagram for various targets (0.2 MeV electrons).

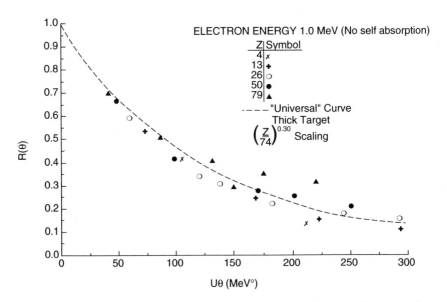

Fig. 11b-3 Polar diagram for various targets (1 MeV electrons).

BEAMS

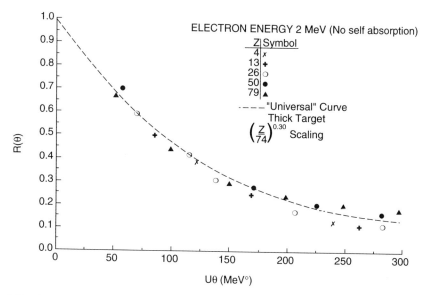

Fig. 11b-4 Polar diagram for various targets (2 MeV electrons).

With-in the originally stated accuracy of ± 10 per cent the curve fits the new data: however, for the case of particular interest, that is carbon, the polar diagram ratios below about 0.25 should be shifted to the left a bit. Figure 11b-5 gives our best estimate of the polar diagram curve for an electron range thick target of this material. Also included are the results of an ETRAN calculation, kindly supplied by Dr. Peter Fieldhouse and Mr David Large of CNR, for 2 MeV. The calculations were done with and without absorption and these gave essentially the same results as expected.

We now turn to the question of the dose on axis from axial electrons. The Buechner data had been analysed to give a $(Z/74)^{0.50}$ dependency. Figure 11b-6 gives the smoothed on axis data.

Also included is the data of Rester as given and also corrected for target absorption. (Note that the 0.2 MeV data has been multiplied by 10). Dodging the question of absolute yield for the moment, the two sets of data have been normalised at 2 MeV. The data is in very reasonable agreement as regards Z dependency and also shows that the slope is reasonably constant for $1.25<V<2.35$, but for low values of V the slope rises. The slope of the lines is slightly higher than 0.50 because these curves are for zero absorption. When the intrinsic absorption of the optimum high Z target is included, as is shown later, the slope is a bit lower and this is the one quoted on Figure 11b-1.

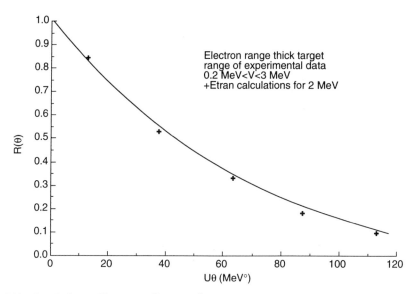

Fig. 11b-5 Polar diagram for carbon.

In order to compare the absolute output on axis, the data in both papers have been converted to Roentgen at 1 metre per coulomb. The conversion for Buechner is simple, but the data from Rester is a bit more complicated. This has been analysed firstly for a cut off of 0.05 MeV. This value is taken because below this value most detectors become rather sensitive to low energy X-rays, and because any scattered field is usually in this range, it is desirable to cut off the low energy response. This point is dealt with further in the next section. The data is then converted, using as a factor 1 R = 2 x 10^9 MeV/cm^2. This is a reasonable average for X-rays from 0.06 to 2 MeV. In addition, of course, the correction for target absorption is employed to make the data comparable. When this is done, the dose levels are very different, Rester's data being 1.62 times as large as Buechner's on average. This is troubling but, being unable to suggest that either set of data is greatly better than the other, some sort of average has to be taken. It is true that Rester's experiments are much more recent but the dose is more derived than that of Buechner's. When two sets of data are this far apart, it is a good question as to what average should be employed: we have elected to use a geometric mean, multiplying the results of Buechner by 1.27 and that of Rester by 0.79. Table 11b-I gives the data so treated for 0.2, 1 MeV and 2 MeV. The Buechner data has been extrapolated to give the data for 1 MeV for comparison with Rester's. In addition, Rester's data has been interpolated to given values for the Z's used by Buechner.

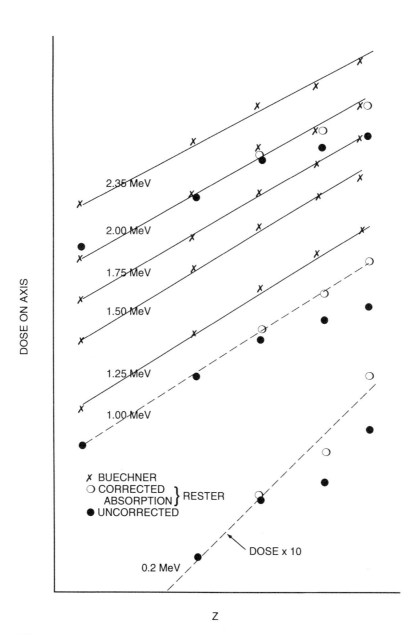

Fig. 11b-6 On axis dose for various targets and electron energies.

Table 11b-I

Dose on Axis for Axial Electrons
(in Roentgen at 1 Metre per Coulomb)

Z → E MeV		4	13	26	47	74
0.2	Rester	2.5	9.5	18.5	27	44
1.0	Buechner	276	575	1010	1430	1810
	Rester	314	630	1020	1340	1730
	Average	295	600	1015	1380	1770
2.0	Buechner	2240	4250	7050	8000	10400
	Rester	2520	4340	6600	8000	10200
	Average	2380	4300	6800	8000	10300

To obtain the dose on axis from real targets it is necessary to include the absorption. As was mentioned in the original polar diagram note, it is possible to have an "optimum": target for high Z materials where a thin layer of converter is backed by a low Z material. For high voltage electrons (V > 10 MeV) this two layer optimum thickness is one-third of the radiation length and for tungsten this thickness is 2 grams/cm^2. For low Z materials this is impracticable and unnecessary because self-absorption in the converter is very much lower, so the target is made an electron range thick. It may be worth while saying a few words about electron ranges at this point. The universally quoted range tables are for aluminium, with the statement that the range is largely independent of Z. The range most often quoted is the practical range which for 2 MeV electrons in aluminium is 1.0 gm/cm^2. However a significant number of electrons get beyond the practical range and the extrapolated range, which is what is needed essentially to stop the electrons, is about 1.2 gm/cm^2. However, for high Z materials the extrapolated range is more like 1.5 gm/cm^2 at 2 MeV, thus a fair degree of confusion can arise as to what is an electron range thick target. These elaborations are not relevant to the polar diagram data because the width is only a weak function of the thickness of the target. However, for the dose on axis the question is of significant importance. However, this sensitivity only applies for high Z targets and is avoided if the data is quoted for the optimum two layer target. This is also of considerable practical interest since this is the target which gives the largest dose on axis, which is normally the name of the game when high Z convertors are being employed.

Figure 11b-7 shows some data kindly supplied again by Dr. Peter Fieldhouse and Mr. David Large from an ETRAN run for 2 MeV electrons. It gives the generation of dose as the electrons move through the target (no absorption curve) and also the transmitted X-ray flux (with absorption curve). These show the low self-absorption of a carbon target and also that the optimum thickness for a high Z target is again about one-third or the extrapolated electron range. Calculations suggest that the optimum target for a high Z material continues to be about one-third range thick and that the output level is about that appertaining to the 2 MeV case, or a little lower. This applies where a cut off of 0.05 MeV is again used.

Using the data in Table 11b-I, the output for a full range carbon target and for an optimum tungsten target can now be obtained and these values are given in Table 11b-II.

These results are shown in Figure 11b-8 where the dose per coulomb is plotted against V. For the high Z target, Tom Martin of Sandia has shown, in SC-DR-69-240, that up to about 30 MV the dose

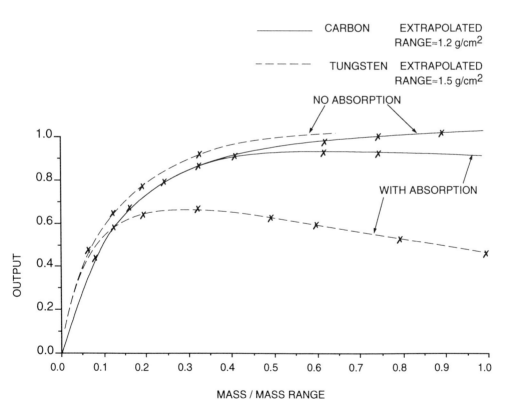

Fig. 11b-7 Normalized dose generation on axis (2 MeV electrons).

Table 11b-II
Dose on Axis for Paraxial Electrons
(in Roentgen at 1 metre per Coulomb)

Z	E = 0.2 MeV	1.0	2.0
W optimum target	28	1.17×10^3	6.9×10^4
C full range target	3.7	3.5×10^2	2.6×10^3

goes at $V^{2.8}$ and the intercept at 1 MeV agrees with that given by Tom Martin for the optimum target. However, for voltages rather less than 1 MeV the dose from a high Z target goes as a lower power. The carbon data, over the range it is available, goes as $V^{2.8}$. Because of the change of power of the on axis dose for tungsten for voltages below 1 MV, the relation for the Z dependency of the dose given before as $(Z/74)^{0.50}$ does not apply. However, above 1 MV the relation is approximately true. If the carbon curve continues to rise as the 2.8 power of the voltage, the Z dependency will continue to hold. A rather rocky justification for the averaging used is the fact that the high voltage data for high Z optimum targets joins smoothly with the low voltage data. However, despite this consideration it must be reckoned that the dose relationship shown in Fig. 11b-8 is uncertain in absolute terms. An error of ± 15 per cent in the dose axis is probable and it could well be higher.

This completes the analysis of the input data needed to use the X-ray diagnostic techniques. In general, the agreement is reasonable but the dose on axis badly needs supporting by better data.

One of the incidental points which has not been covered so far is the dose produced by beams which are non-orthogonal, in particular for carbon. For high Z targets a large fraction of the electrons can be back scattered at small angles of incidence. However, in order to scatter out of the target, the electrons have to undergo a number of small angle scatterings, or a single large angle scatter. In both cases their energy is significantly reduced and because of the powerful law of X-ray production efficiency, an electron leaving with even half its energy will have generated most of the X-rays that it can. Thus the generation of X-rays should not be seriously affected when the beam is incident on the target at considerable angles. The argument is very much stronger for low Z targets when a smaller fraction is backed scattered. These contentions are supported by Buechner's data, where the target was rotated so that the electrons were sometimes incident on it at

BEAMS

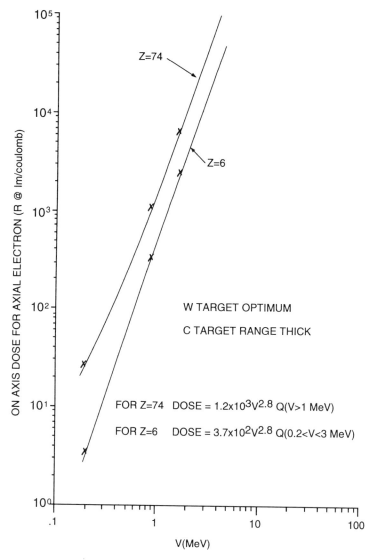

Fig. 11b-8 On axis dose for axial electrons vs voltage.

small angles to its surface. The polar diagrams given by Buechner are smooth through this region, even for high Z materials. Thus the fact that electrons are non-orthogonal should not change the efficiency of generation of X-rays significantly.

Figure 11b-9 gives the normalised spectrum of X-rays for 0.2 MeV to 3 MeV for carbon at 0°. However, the spectrum holds pretty well out to 45° or more. This curve is of use for absorption calculations and is taken from Rester's paper.

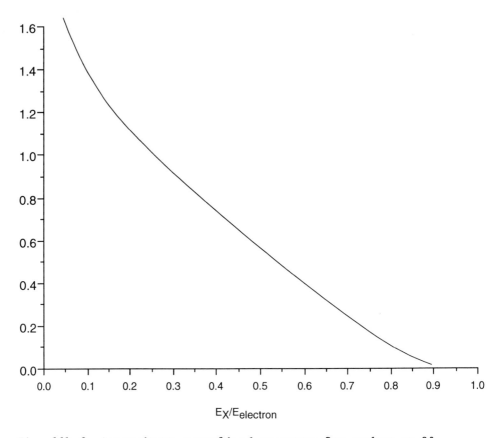

Fig. 11b-9 Approximate normalized spectrum for carbon at 0°.

The above section deals with the required input data: what follows is an outline of how we use this data to derive the required beam parameters, in particular $\bar{\theta}$. A few words ahead of the self absorption correction discussion may serve to justify some of the rather crude approximations we employ Basically, the data points have an error of the order of ± 4 per cent when sensible numbers of TLDs are employed. To achieve even this accuracy may mean the deployment of 20 to 30 TLDs. Some error in the input polar diagram must also be assumed to be present and these basic limitations imply that over-refinement in the treatment of the data is unnecessary. In addition, the basic assumptions in defining the parameter selected to categorise the mean angle also obviously have limitations, since while Nature is kind and tends to provide an approximation to a fully filled cone, she cannot be relied upon for slavish fulfilment of this requirement.

With these rather transparent justifications for our laziness, we reveal all.

Self Absorption Correction

Firstly the correction to the polar diagram for converter self absorption will be discussed. What is required here is the difference in X-ray path length between the on axis X-rays and those at any other angle. The X-rays are assumed to be produced in about 0.2 of the electron range through the target on average. The extra path length is then easily calculated for the target and any other backing material. For carbon and other low Z elements, the absorption is rather small and while it is a little difficult to decide what cross-section should be used, it is fortunate that it does not matter much. The complication is that for low Z materials the interaction is mainly Compton scattering and hence a build-up field of softer secondary X-rays is developed. For infinite media the build-up factors are well known but for thin slabs these calculations are not applicable. A second problem is that the build up field is largely isotropic and thus tends to reduce the peakiness of the polar diagram by X-rays scattered into the large angle TLDs. This effect is mitigated by the use of copper shields around the TLDs, to some extent, but is still present. Figure 11b-10 shows the absorption factor for carbon (and other light elements) as a function of thickness and original electron energy and was derived from Figure 11b-9 and the published X-ray total cross-sections. Again, a cut off at 0.05 MeV has been used. The absorption factor used here, and later, is the reciprocal of the actual absorption and is the factor by which a data point should be multiplied to remove the absorption effect approximately. As can be seen, for 1 MeV and a target of the order of the electron range thick, the correction factor is 1.05. For a point at 60° this is also the factor which would account for the extra absorption, if the full cross-section were to be employed. However, as has been stated, a lower factor than this applies and effectively we use two-thirds of the total cross-section. This fudge factor was derived from some very crude calculations of the build up factor and its effect on the polar diagram shape. Thus, in this case the 60° data would be raised by only 1.03. For high Z targets the situation is easier, since most of the absorption is photo-electric and here the full cross-section is used to obtain the attenuation factor. In the experimental data given later the error bars for the points at 45° and 60° have been increased to allow for the plausible range of the attenuation correction factor.

For the dose on axis the curves give the value for an electron thick target for additional absorbers and the situation is different because the scattered X-rays are essentially lost where a relatively peaky polar diagram occurs and, typically, a cross-section equal to the total cross-section is used in such cases. Again, for high Z additional absorbers the full absorption factor would be used.

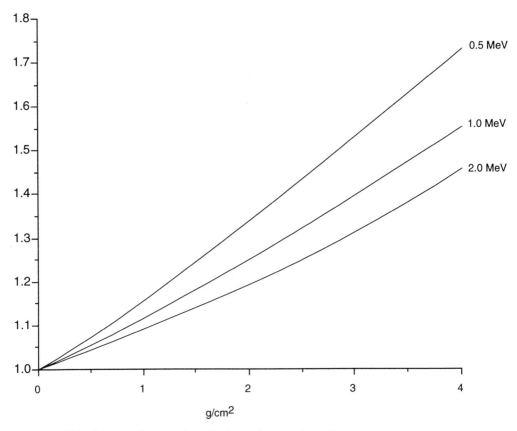

Fig. 11b-10 Carbon multiplying absorption factor.

Determination of $\bar{\theta}$

Using the experimental data corrected to a zero absorption target, the normalised polar diagram is plotted against θ. On the same graph the axial electron polar diagram is plotted (using Figure 11b-5 if the target is carbon). This raises the question as to what V_{eff} should be employed. It is not difficult to derive a series of polar diagrams for the real time varying electron energy, but if the anode cathode impedance is constant a value of V_{eff} equal to 0.84 V_{max} is a good approximation.

The question now arises as to what the polar diagram is for various values of $\bar{\theta}$ for the incoming electrons. A series of calculations for the fully filled cone and for other reasonable distributions has shown that the following approximation is an adequate one.

BEAMS

For the 0.75 level add 0.9 $\bar{\theta}$

For the 0.50 level add 1.0 $\bar{\theta}$

For the 0.25 level add 1.1 $\bar{\theta}$

Using this approximation, curves are easily constructed from the $\bar{\theta} = 0$ curve for various values of $\bar{\theta}$. Figure 11b-11 gives an example of the results and shows that the experimental data fits a mean angle of a little over 10° rather well.

If the above scheme offends people's sensibilities, Tom Martin, in SC-RR-69-241, gives a computer code for predicting the X-ray polar diagram. The polar diagram used in his report is slightly different from the one advocated in this note and in the

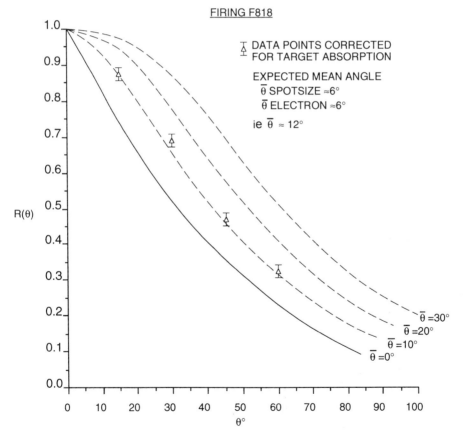

Fig. 11b-11 Full length razor blade with anode cathode gap of 1.5 cm with carbon block target.

examples calculated explicitly the spot size is also made a function of the voltage, but this is not inherent in the treatment. However, because of the factors mentioned earlier, it is believed that the approximation used in our approach is sufficiently good, especially for low values of $\bar{\theta}$.

Spot Size Correction

In general, the dosimeter array should be at a distance at least 10 times the spot radius and preferably more. Where this applies, the spot size contribution to the polar diagram flattening is quite small and roughly equals $\theta \sim 57\ r/d$ degrees, where r is the effective radius of the spot and d the distance to the dosimeter array. The effective radius is two-thirds the outer radius for a uniform spot and equal to the mean radius for an annular spot. Once again, providing the correction is small, it does not matter exactly how you get it. This spot size correction is then taken away from the observed $\bar{\theta}$ to give the mean angle for the electrons incident on the target.

Dose on Axis Measurements

Firstly any absorption over and above that in the optimum target (high Z) or electron range target (low Z) must be allowed for, as is indicated above. Then a correction has to be applied for the effect of the mean angle of incidence of the electrons. This correction involves the real angle and hence cannot be applied directly to the universal curve. This has first to be reinterpreted to give the polar diagram against laboratory angle for the material and electron energy used and then weighted by a $\sin\theta$ factor and averaged. Figure 11b-12 gives as an example the drop in on-axis dose for a carbon target as a function of $\bar{\theta}$ and energy of the electrons. The observed dose is divided by the appropriate factor to yield what would have been dose on axis for axial electrons.

As was mentioned earlier, the other way to treat the problem is to integrate over all solid angles and obtain the total radiated flux, but for the reasons mentioned before we prefer the above treatment in cases where the polar diagram yields a reasonably unique value for $\bar{\theta}$.

For most of the cases we have been interested in, this factor is small and even when it becomes large, uncertainties in it only enter the resulting derived electron beam energy as the one-third power, approximately. Thus it does not contribute significantly to the uncertainty of the final answer, certainly not when compared with the basic uncertainty in X-ray generation efficiency covered in the first part of this section.

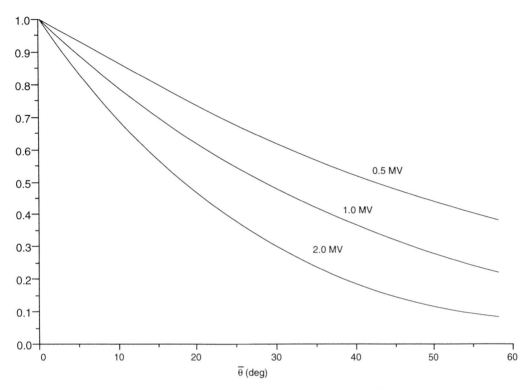

Fig. 11b-12 Relative dose on axis for differnt $\bar{\theta}$ for an electron range thick carbon target.

In conclusion, we would like to mention that the particular shape of the universal curve, coupled with the fact that basically an integration of it is being made to allow for the practically existing range of electron angles, means that for a widely varying smooth range of assumed electron distributions the curves look surprisingly similar. Also the half point of the polar diagram shifts by about the mean angle of the distribution. This fact enables the rather approximate treatments to give quite good answers: at the same time it implies that details of the actual electron distribution cannot be extracted from an experimental polar diagram without great accuracy in the measured data and in the input data. This point is briefly returned to in the experimental results section.

PRACTICAL ASPECTS AND ACCURACY

In this section the practical nitty gritty will be covered from an essentially parochial viewpoint. While the approaches we have employed are obviously not by any means the only ones that can be used, they are of course the only ones we can comment on with confidence. The aspects to be covered will be in roughly the order used in the diagnostic outline section, to which reference should be made for some of the practical aspects.

Spatial Flux Distribution Measurements

There are three main sources of error in taking pinhole radiographs, namely pinhole vignetting, polar diagram vignetting, and background on the film. The first one of these is fairly obvious but if care is not taken, a fall off of density towards the edge of the field of view can be interpreted as lack of flux rather than a too large ratio of length to diameter of the pinhole. It pays quite a lot to use the densest material in which to drill the pinhole and to have a series of inserts with different diameter holes in them. In the past we have also used a rectangular array of pinholes of different diameters and with different absorbers over repeats of the same hole. This is quite useful for covering a large range of intensities and also ensuring that the X-rays doing the photographic darkening have approximately the expected spectrum. In general we use a fraction of a mm of copper filtering to make sure that X-rays from the tail of the pulse or scattered X-rays do not have much effect. However, to prevent flux penetration through the front of a large multiple pinhole camera can be quite difficult and we now mostly use a high resolution camera and a low resolution one side by side.

Polar diagram vignetting was dealt with in the diagnostic outline section and again can lead, unless care is taken, to an under-estimation of the spot diameter.

Even when these two effects have been eliminated or allowed for, it is still easy to make a mistake as to the spot size. This is because of the very high contrast of X-ray film. For instance, a beam where the diode pinches during the pulse will produce a very intense black image; but this is surrounded by a diffuse halo of X-rays which can easily contain as much, if not more, integrated flux. We therefore use 3 films in each of the cameras of varying sensitivity to check this point. The poor resolution camera can also pick up low area density records from say the anode plane or the drift cones and enable crude estimates of the fraction of the electrons hitting these to be made. Allowance of course has to be made for the increased efficiency of production of the guide cone, if this is made of copper, compared with the main beam hitting carbon, and for any other extra absorption in the path of the X-rays. Mention should also be made of the use of a large aperture camera at right angles to the diode or beam axis for the same purpose of pinpointing stray stalk emission or beam losses.

The third factor is background. In general quite a modest amount of lead shielding around the sides of the camera will prevent background scattered X-rays getting in, but a check of this can be made by placing small lead absorbers up against the front of the film pack and midway between the film pack and the pinhole. While relatively modest thicknesses of lead prevent fogging from the back or sides, much larger quantities are required on the face of the camera, especially when small pinholes are used. In general we avoid using extremely small pinholes for this reason, otherwise absorptions through the front of 10^3 to 10^4 may be necessary. Such absorption factors can be obtained but the spectrum hardens considerably and hence becomes more penetrating. This difficulty particularly applies to systems working at 2 or 3 MeV. When a medium size pinhole is used, allowance for the effect of the penumbra can quite easily be made when the spot has sharp edges.

An additional source of error in deriving the edge of a spot applies where a central emitter is used, such as a small ball or a razor blade. In addition to any polar diagram vignetting, there is a cosine effect of flux arriving obliquely at the target, which again is amplified by the high contrast of the film. However, despite all these effects, quite accurate measurements can be made visually and, with film response calibration, quantitatively.

With regard to dose scans across the face of the target with TLDs, these give quite good relative results even when only placed against the back face of the target. However, some degree of local shielding with copper is desirable and of course the target must be at least an extrapolated range thick, so that electrons cannot get directly to the dosimeter material.

One point that is worth making is that the integral of the flux across the face in the near field should of course equal the integral of the flux in the far field. Where this does not happen (which in our experience is surprisingly frequently) two obvious explanations may be worth looking at. The first is that a significant fraction of the beam is hitting regions outside the nominal target area, with a low flux density. This of course can miss the array on the rear of the target but affect the polar diagram TLD array. A second possibility is that the super-linearity correction for the TLD may not be accurate and since the dose levels may well be 10^2 to 10^3 different, this will affect the close in readings much more than those of the polar diagram array.

Polar Diagram Measurements

The first and most obvious point here is that all the readings are referred to the reading on axis. If this data point is measured incorrectly due to variability in the TLDs, the whole pattern is thrown out. Consequently the number of TLDs at zero degrees should be higher than at the other angles. We find that unless special care is taken, the standard deviation for a single TLD is in the region of $\pm$ 5 per cent, providing the dose is in the range 0.5 to 500 R. Below this range the background correction on our TLD reader becomes significant and above it the super-linearity correction comes in. This is not to say that readings cannot be made above this range, but if they can be avoided it helps to increase the accuracy a bit. We deploy a symmetrical TLD array and if 2 TLDs are used at each angle and 4 at the zero angle position, the mean error for each angle (averaging the two minor image readings) becomes about 2 1/2 per cent and the error in a point normalised against the on axis dose is $\pm$ 3 1/2 per cent. To obtain much better than this requires a large number of TLDs. However, this accuracy is of the order of that of the polar diagram curves themselves and hence a reasonable balance between errors has been struck.

The second point concerns the shielding of the TLDs. We use copper pipe of 1 mm wall thickness and 6 mm bore. This arrangement dates from our radiographic days when we investigated an "air wall" arrangement and found that for X-rays from about 4 MV electrons the absorption in the copper was balanced by the extra dose arising from the extra knock-on electrons getting through the polythene powder container and coming from the copper. Figure 11b-13 gives the attenuation correction factor for 1 mm thick copper tube and also for a tube of half the thickness. The latter might well be preferable if measurements were being made around or under 0.5 MeV electron energy. Again these numbers are calculated for a cut-off in the original spectrum of 0.05 MeV, in conformity with the dose calculations. We have not made comparisons between the copper clad TLDs and the bare ones, but we feel that the cleanliness of the polar diagrams we obtain (such as in Figure 11b-11) is partly because of this light local shielding. The effect of the copper shielding is to help to wipe out secondary scatter from the dustbin, etc. and also it avoids the region where corrections for the absorption of the LiF itself compared with an air dosimeter become necessary. Because we use an intermediate Z material of small thickness, the absorption is well known, being mainly photo-electric in the region where it is removing the low energy tail of X-rays. Another factor that helps is that any low energy X-rays from large currents flowing late on in the diode at low voltages are wiped out. Polar diagrams from single razor blades have been taken with the standard holders wrapped in additional lead shielding and these gave essentially the same curve as in Figure 11b-11; they also showed that the spectral hardness was approximately what would be expected from the diode volts.

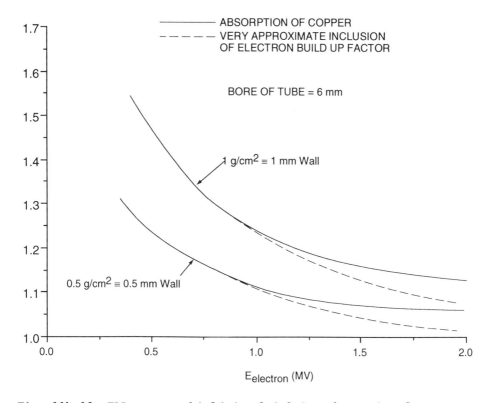

Fig. 11b-13 TLD copper shielded multiplying absorption factor.

Time resolved measurements made with photodiodes are easy and reproducible. We use a 2" diameter ITT photodiode, type W114 and a block of plastic phosphor NE102 of about 2" diameter and 3" long, white painted on the outside. This is shielded on the outside by lead and on its face by about 0.5 mm of copper. The photodiode-phosphor combination has very constant characteristics and directly drives the oscilloscope deflection plates, since it can produce up to 3 amps linearly with about 2000 volts on the diode. The saturation current of the diode is over 7 amps and the turn over fairly sharp. Such a combination could readily be absolutely calibrated but we mount TLDs near the phosphor in order to measure the integrated dose, whenever this is necessary.

Beam Energy Measurements

The electron energy measurements basically require the absolute measurement of dose at a metre as well as the determination of the current in the diode, or, if a beam is used, with a Faraday cup. Some time ago we went through a lengthy comparison with other people's dose measurements, including Physics International and

Harry Diamond Laboratories, and eventually felt we were within 5 per cent of the true dose. However, something like a man-year of effort was needed to attain this state (which aged the gentleman concerned by at least twice this amount) and while we think we are still about as accurate as this, this is more an act of faith than a statement of scientific certainty. What we did find, and have found since, is that other laboratories (neither of those mentioned, we hasten to add) can be up to 70 per cent in error and hence the measurement of dose at a metre is not something that can be tossed off in an afternoon, but is a fairly difficult activity needing continuous effort after it has been first achieved. Once again it is desirable to lop off the very low energy end of the spectrum, but where you do the lopping will be important if the technique is to be pushed down under 0.5 MeV. Also for very low X-ray energies the thickness of the phosphor will have to be reduced (or allowed for) as the phosphor then essentially absorbs all the X-rays, in the very low energy region. This is incidentally yet another reason for having a cut-off at some level, so that the phosphor is approximating an air-wall chamber.

Thus the rather unexpected result of investigating the determination of the absolute rate of production of X-rays is that the time dependency of the X-ray flux is rather easy; the uncertainty lies in measuring its absolute level.

EXPERIMENTAL RESULTS

A representative series of experimental results will be covered in this section, partly as illustrations of the way the various calculations are performed, and partly as examples of the sort of data and consistency we obtain.

In the rather daunting Figure 11b-14 the results from three particular shots are shown as plots of the experimentally obtained polar diagrams after correction for target absorption. The experimental set up was essentially the same in all three shots, a single 3.9 cm long razor blade being the cathode. A carbon block formed the anode whose mass was 0.75 gram/cm^2 approximately. The peak volts were about 1 MV but the current differed significantly between the shots. The TLD array had to be installed within a cylinder behind the anode (the dustbin) and so only a limited radius could be employed and this was 13.5 cm. Paired TLDs were used at each point and the angles used were $\pm$ 0°, 15°, 30°, 45°, and 60°. Two arrays at right angles to each other were also deployed, one along the blade and the other at right angles to it. As it turned out, this was unnecessary, because of a compensation which took place, so that there were no statistically significant differences between the two orthogonal polar diagrams obtained. The probable explanation for this was that while along the blade the spot size

BEAMS

was significantly bigger, the mean electron angle was low, while at right angles to the blade the spot size was smaller but the mean angle up. As such, the data points plotted in Figure 11b-14 are the average of all 4 similar angle positions.

Taking Firing 750 first, this is representative of a well-behaved constant impedance shot. The anode cathode distance was 1.5 cm and the blade stuck out from the flat cathode plate 0.8 cm. The anode spot was well defined at 5.6 cm by 2.4 cm and in the form of a rectangle with rounded ends. The mean angle was about 10° or a little larger, about equally due to spot size and angular divergence of the beam at the anode. This point is gone into more deeply with respect to Figure 11b-10 and 11b-15.

Firing 742 had an anode cathode spacing of 0.75 cm and as such, exceeded the critical current for pinching during the pulse. As can be seen, the mean angle is now a little bigger than 20°, say

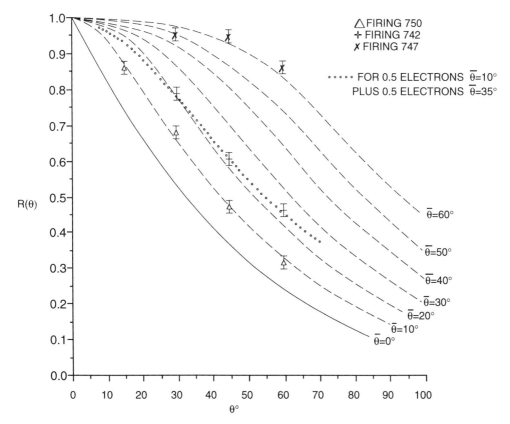

Fig. 11b-14 Polar diagrams corrected for target absorption showing effect of impedance collapse.

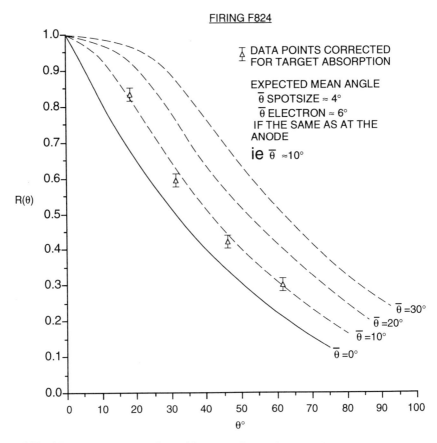

Fig. 11b-15 Average angle effects of drifting the beam 34 cm.

22°. Because the spot size will be rather smaller during the latter part of the pulse, the time-averaged mean electron angle has increased from something over 6° to more like 18°. As an illustration of the relative insensitivity of the polar diagram plot to details of the assumed distribution, Figure 11b-14 also shows one line calculated as follows. It is assumed that for half the pulse the mean angle (spot size plus electron mean angle) is 10° and for the other half a value of 35° is taken. As can be seen, this assumption fits the data as well if not slightly better. However, one could not distinguish between the two possibilities on the basis of the experimental points. On the basis of other reasoning, the second possibility is a bit more likely. The way to distinguish between the two possibilities is, of course, to make time resolved polar diagram measurements.

Firing 747 again had an anode cathode gap of 1.5 cm but in this case significant current flowed during the prepulse, the pre-pluse gap broke down and applied a 50 kV prepulse, which in its turn largely collapsed before the main pulse arrived. The current was several times larger than in a normal firing and it is expected that plasma filled the anode cathode region early in the main pulse, if not before. A strong diode pinch resulted and this is rather dramatically confirmed by the polar diagram measurements, which show a mean angle of about 60°, corresponding to a fully filled cone of 90° angle.

These three shots show the full range of angles that we have observed, although we are fairly certain we can produce beams with a mean angle of a degree or so, but whether the polar diagram input data is accurate enough to show such an angle is rather debateable.

The next example is Firing 818 which again is a single full length razor blade with an anode cathode gap of 1.5 cm. the anode was again an 0.75 gram/cm^2 carbon block but backed this time with 0.6 cm of lucite (Perspex for our English readers). The polar diagram obtained is shown in Figure 11b-14. The calculated spot size angle is 6° and the estimated mean angle of the electrons at the anode is again 6°. The experimental points almost unbelievably fit a 12° total mean angle, which must be partly fortuitous. We should perhaps reassure the dubious reader that the new curve for carbon given in Figure 11b-5 was derived from the Rester paper completely independently and the experimental points are entirely unadjusted apart from a very small target absorption correction for the 45° and 60° angles.

Figure 11b-16 gives a welter of information. The voltage and current curves are derived from our normal monitors, the voltage curve having been corrected for the L di/dt as is usual. Also shown is the smoothed impedance curve with a rather satisfying constant impedance of about 50 ohms during the first pulse, which is highly characteristic of a single full length blade. There are two pulses because the 2 ohm generator is very lightly loaded. The climbing current during the second pulse is almost certainly associated with current emission from the cathode stalk. This is because some shots show that the razor blade on its own keeps a good impedance out to 300 ns; Figure 11b-17 shows an example of such a shot. The carbon anode was also a calorimeter which recorded 1.3 kJ while the energy obtained by integrating the first pulse plus a small fraction of the second is 1.55 kJ. This difference is likely to be caused by calibration errors of the calorimeter but could be due to some 1 1/2 kA flowing other than from the razor blade cathode. Usually the system is operated at 100 kA or so; consequently we would not normally bother about such small background stray current flow.

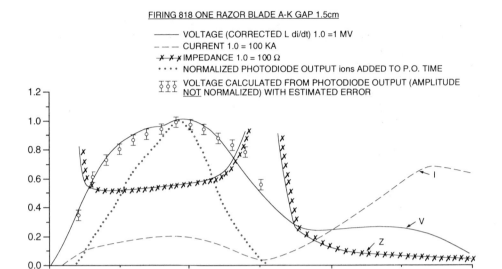

Fig. 11b-16 Parameter comparison showing voltage calculated from photo diode output.

Also given in Figure 11b-16 is the photodiode pulse, normalised in amplitude to unity at peak and adjusted to fit the peak with regard to the zero time of the sweep. The current and voltage waveforms are accurately correlated in time but the photodiode record is not tied in with this system. From the polar diagram TLD's, and others at 42 cm and 1 metre, the dose at a metre was measured to be 0.46 R with an estimated error of $\pm$ 10 per cent. The peak current was 19 kA and the effective pulse width (measured wither from the photodiode record at 50 per cent height, or the voltage wave form at 84 per cent of peak) is 70 ns. Thus Q_{eff} = 7 x 10^{-8} x 1.9 x 10^4 = 1.2 x 10^{-3} coulomb.

From Figure 11b-8

Dose = 3.7 x $10^2 V_{max}^{2.8} Q_{eff}$ = 0.46,

giving V_{max} = 1.00 MV.

Using the measured currents, the photodiode record is used to give the electron energy as a function of time and the peak voltage used is that derived above. The points so obtained are also shown in Figure 11b-16. The error bars on these points are obtained as follows. Dose at 1 metre, $\pm$ 10 per cent input data, $\pm$ 15 per cent

BEAMS

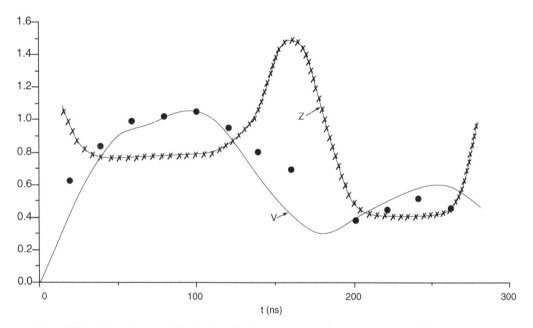

Fig. 11b-17 Photo diode derived output voltage compared to measured voltage (6 razor blades).

current, ± 7 per cent, giving an overall error in $V^{2.8}$ of ± 20 per cent and hence an error in the voltage of ± 7 per cent. The agreement obtained for the peak voltage is obviously a little fortuitously good, again, but the agreement as regards waveform shape is quite acceptable.

Summarising the results of this shot, we would conclude that our monitoring is in reasonable good shape and that indeed, as we had calculated, the razor blade cathode gives a cold beam with $\bar{\theta}$ ~ 6°.

The next shot (F 824) to be described, while it has some minor lacunae in the interpretation, also has some interesting conclusions which, again, did not come as a surprise.

The set up was once more a single full length razor blade in the centre of the cathode backing plate, with an anode cathode spacing of 1.5 cm. Beyond a double 0.3 mil. mylar window was a drift region of 34 cm length. The outer return conductor for this was made up from bits to hand and is not what we would have designed. It took the form of a copper cone, entrance diameter 26 cm, coning down to 13 cm diameter over 15 cm length. This was followed by a parallel section 19 cm long made out of 5 mil. melinex covered internally by 1 mil. aluminium. At the exit window there was a carbon target-cum-calorimeter. Unfortunately this did not take the form of a Faraday cup current monitor because this was a shot sandwiched in another programme. One of the current monitors measured the anode-cathode current and another the current just behind the diode face. The drift region pressure was about 1 torr. The pinhole camera showed a clear X-ray pattern on the target and no detectable X-ray emission elsewhere. The X-ray spot was slightly distorted from the original rectangle, with rounded ends, but was largely rectangular with dimensions 5.2 x 2.6 cm approximately and with the long axis in the original razor blade direction, as would be expected. The polar diagram plot for this shot is shown in Figure 11b-15. As the TLD array was now outside the dustbin, the radius could be increased to 20 cm, reducing the spot size angle to 4°. If the input data of the section on experimental determination of polar diagram and dose on axis is to be believed to the required accuracy, the mean electron angle after drifting remains at 6° or a fraction less. However, it can certainly be stated that it has not increased after drifting 34 cm. This is in line with previous measurements made on MOGUL and collectively remembered, that drifting did not significantly increase the mean angle of the beam. In both cases the ν/γ was about 0.4, although for MOGUL the mean voltage was like 4 MV and the present result is much more accurate. Thus we detect no transfer of energy to the perpendicular direction. The other possibility for two stream instability is that the beam energy is modulated in the direction of its propagation: this was investigated in the second part of the measurements, although here the proof that it did not happen significantly is not quite so direct.

Figure 11b-18 gives the anode cathode voltage, current, and impedance, as measured by our standard monitors. Because the pressure in the tube was rather high and also because the false work is more than a little battered, the tube flashed around 70 ns. This is readily apparent from the current monitor beyond the tube vacuum interface. One of the effects that has been observed as a consequence of this is that more current apparently quickly flows in the anode cathode monitor loop. Whether this is very fast plasma jetted from the tube interface, or the effects of a bar of much greater current (~ 1/2 megamp) advancing towards the anode cathode monitor, is not clear; it is perhaps connected with both effects. Anyway, it is considered that the fall in impedance after 70 ns is caused by the tube flash over and is not connected with the current flowing from the razor blade, which invariably has a constant impedance

BEAMS

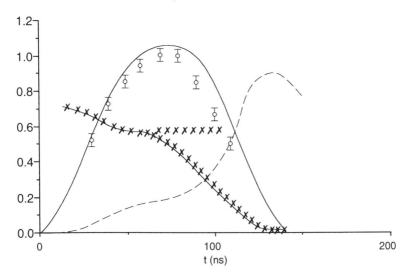

Fig. 11b-18 Photo derived output voltage compared to measured voltage (one razor blade).

during this phase. Thus it is assumed that the cathode impedance stays constant at the flat level it had reached before 70 ns, as shown by the lightly dotted line. This is confirmed by the calorimeter reading, which was 900 joules. Multiplying by the factor previously found, this corresponds to about 1.1 kilojoules, while the expected beam energy was calculated to be about 1.15 kilojoules, using the reduced pulse length shown by the photodiode record. The excess current from the interface flashover, if it really is a current, would not enter the drift region - a fact confirmed by the lack of any extra X-ray on the radiograph.

The dose at one metre was measured to be 0.28 R as an average of a number of determinations at different distances. Using the current deduced from a constant anode-cathode impedance of 58 ohms and this dose, the electron beam energy was calculated as before and is also shown on Figure 11b-18. As before, the zero time of the photodiode record has been slightly shifted and the amplitude has not been normalised, but is obtained from the radiation measurements. It should be noted here that if the full anode cathode current waveform had been used, the points beyond 70 ns would fall more sharply because of the increased current reading.

A point should also be made about the L di/dt correction applied to the observed voltage waveform for times after 70 ns. As the monitor is mounted in the tube face, its record is held up by the large current starting to flow across the interface. The flashover was approximately opposite to the voltage monitor and hence there is a considerable inductance between the two. This makes determination of the L di/dt correction uncertain after the insulator breaks down, so that the anode cathode voltage beyond this time is somewhat uncertain; the curve given is our best estimate. It might be supposed that the tube flashover was an unmitigated disaster, but in fact it was not. Two previous attempts to make the same measurement had failed, partly because extra emission had started from the cathode stalk during the second pulse, and on each occasion some of this had got into the drift region and messed up the polar diagram. So while it has rendered the interpretation a little less direct, it enabled a rather clear polar diagram to be obtained, and also where assumptions have to be made they do not disturb the major conclusion about a longitudinal variation in energy due to two-stream instability.

As is shown in Figure 11b-18, the voltage waveform, deduced from the absolute photodiode record, lies, if anything, below the voltage monitor waveform. If the larger current values had been used, it would be even lower for times after 70 ns. Thus there is almost certainly no excess of dose on this shot. If a rather naive picture is taken that the beam splits into two equal populations, one with $V + \delta V$ and the other with $V - \delta V$, then the dose increase to be expected is $V(1 + 2.5(\delta V/V)^2)$. If any dose increase is taken as less than 10 per cent, and in fact the observed dose is some 15 percent less than would be expected, then $\delta V/V$ is less than 0.2. The only way any significant spread of energy in the transported beam could be allowed is to reduce the mean energy of the beam as well. However, this is unlikely because of the calorimeter reading. Thus while we would be reluctant to claim that this firing disposes of any significant two-stream instability effects, it certainly does not support such effects and does suggest a technique of proving whether or not they occur. In such an experiment, the carbon block would be a Faraday cup and so measure the relevant current directly and accurately.

The last experimental example is not a very good one but it contains a couple of features of interest. On a number of shots we have had two bumps on the photodiode record, the second about 20 percent of the first. Firing 819 (Figure 11b-17) was such a shot, although from several other aspects it was a rather unsatisfactory one on which to do an X-ray voltage measurement. The set up in this experiment was 6 half razor blades set out on the curve of the cathode plate. Beyond two 0.3 mil. aluminised mylar windows was a drift region with a series of 6 different guiding wire arrays. The pressure in the drift region was 0.17 torr and at the far end was a carbon target. Around the edge of this was copper foil and the copper wire supports. The average diameter of the array of beamlets was 20 cm but the pattern of each beamlet was, to say the least of it, complicated. Part of the beamlets hit the copper surround as well as the guiding wire supports. The energy transfer was good and, as before, anode cathode voltage and current measurements were made in the normal manner.

Figure 11b-17 shows the peak voltage and also the derived smoothed diode impedance. Also shown in the electron energy measurement deduced from the two humped photodiode record. In this case both time and amplitude have been normalised at the peak. This was because of the big spot size and uncertainty as to what fraction of the beam hit copper rather than carbon. The normalised voltage derived from the photodiode record is not in good agreement in this case, but does follow in general the ordinary voltage waveform. The early high reading at 20 ns is probably because more of the beams were hitting the copper surround at the beginning of the pulse and the same explanation may well apply around 150 ns. The fact that the second derived voltage pulse is lower than it should be may well be because of small extra currents from the cathode stalk. Not altogether a convincing example but one which does display almost certainly a genuine double photodiode pulse.

The second point of interest in this firing is the way the impedance changes with time. The starting plateau, at about 16 ohms, is just about what we expect. Normally a half razor blade has an impedance of 100 ohms at 1.5 cm and there were 6 of them and although these were at 1.8 cm from the anode foil they were in a diverging field. Thus the impedance per blade of 95 ohms, approximately, would be reasonable. The impedance climbs between the pulses and although the peak value of 30 ohms is rather dependent on the exact phasing of the V and i records, it is in the expected street. When the impedance falls again during the second pulse it is about 8 ohms. As the plasma blobs expand the spacing decreases, and a small but significant droop would be expected in the impedance during the first pulse. However, we generally see a very flat impedance in this phase, or frequently a slightly rising one. this could be if the front of the plasma were developing a resistance in the region where the electrons were running away. However, when the current is drastically reduced between the pulses, this resistance may disappear. Anyway, taking the impedance change from 40 ns to about 220 ns, this implies a change of spacing of about 9 mm and hence a velocity of the order of 5 cm per microsecond for the plasma front, under these condition of current density. This deduction is distinctly speculative but not out of line with expectations.

CONCLUSIONS

The various simple techniques outlined have a range of application for low voltage electron beam diagnostics. In particular they enable the mean electron angle and the electron energy to be determined as a function of time, almost regardless of the flux level. Certainly they can be used at much higher intensities than are necessary to vaporise any material. They are relatively simple to use and yield unambiguous results. Their range of applicability is from 0.5 MeV upwards and in some instances it may be possible to push the lower limit down to 0.2 MeV. They are a very useful adjunct to the normal monitoring techniques in diode and beam work but they should be used in conjunction with these, not as any sort of replacement. The results using them that we have obtained in the past agree with various other measurements and calculations and hence give considerable confidence that these are correct.

Various pitfalls await the unwary experimenter, but it is hoped that by now we have uncovered the more obvious ones. There is no guarantee that this technique washes whiter than all others, but with luck it will remove some of the stains from the experimental log book.

CHAPTER 12

HIGH VOLTAGE DESIGN CONSIDERATIONS

Section 12a

MEASUREMENT OF THE CONDUCTIVITY OF COPPER SULPHATE SOLUTION*

J.C. Martin

Much of the confusion in copper sulphate solution resistivity measurement derives, it seems to me, from surface insulating films and/or electrolytic action at the electrodes. The behaviour of the Wayne Kerr bridge is particularly revealing in this respect, since if a measurement of resistivity is made with it of a liquid resistor which has been made up some time, a large value of the capacity is also read, while at the same time the resistance comes out too high. This is interpreted as the effect of partial capacity in series with a resistor. If a current of a few tens of milliamps per cm of electrode is passed for ten or twenty seconds, new electrode surfaces are formed in part and the resistance falls and the capacity reading in the bridge becomes very small, one or two units on the dial. In one case this treatment caused black flakes to come off the electrodes and float about in the solution. In other cases, copper is presumably laid down on the resistive film while sulphate ions are free to attack the film at the other electrode. volt-ohmmeter readings also become lower and agree with the bridge measurement, although care must be taken that a small cell voltage (because of the different ion concentrations at the electrodes) does not strongly influence the ohmmeter readings. This can be done by measuring the resistance both ways round. The act of reversing the clearing voltage a few times does not seem to interfere with clearing the insulating film and does not leave any significant cell voltage afterwards. Another way round a residual voltage is to short out the resistor for a few tens of seconds.

---

* AFWL Energy Storage and Dissipation Notes
  Note 4
  SSWA/JCM/667/67
  22 July 66

If a battery for clearing the electrodes is not to hand, the volt-ohmmeter can be used to perform this function mostly. As such it is switched to the ÷ 100 range and then left until the reading finishes falling. It is this reading (or lower) which is the correct one. Even when the reading has to be taken on the x 1 range, staying on the lower range for a minute will reduce the reading obtained and bring it closer to the truth.

With a cell with fairly old and used electrodes, the reading obtained on the x 1 range of the volt-ohmmeter can be three or four times as high as the real one, and the reading on the lower range can come close to the proper value. After passing a reasonable current from a 9 volt battery the volt-ohmmeter reading fell further and came into alignment with that given by the Wayne Kerr, which showed only a very small value of capacity during the measurement. These values were compared with those obtained by using a pulsed bridge technique which put 600 volts across the resistor and agreement got to within $\pm$ 2% between the three values. The mean value of these determinations was also within this error of the resistivity calculated for the solution from standard tables from the concentration of copper sulphate.

A check was made on the resistivity test sometimes employed, using the volt-ohmmeter metal ore leads. These are held close together and put in the solution. Once again the more consistent values are obtained on the lowest range of the ohmmeter that can be used. The multiplying factor needed to get the resistivity of the solution is about 5 for the values obtained with the ÷ 100 and 3 1/2 for the readings on the x 1 range. However, the error on these measurements can be quite large (up to $\pm$ 20%) and may well depend on the condition of the ore clips. However it is a quick way of obtaining a rough measurement of an unknown solution. Since there is a test cell now available, it should be easy to make a more accurate measurement quickly and easily.

As an additional aid to obtain solutions of known resistivity, a standard solution has been made and a dilution graph attached to it (Fig. 12a-1). This gives the volume of deionized water to be added to unit volume of standard solution to give any required resistivity. The solution chosen, 250 grammes of hydrated copper sulphate crystals with 750 grammes of water, is rather concentrated and if allowed to cool well below 17°C., some of the sulphate may crystallise out. That this has not happened should be checked before the stock solution is used. In the case of very weak solutions, the dilution should be done in two or three stages and the quality of the deionized water also checked to make sure that it will not reduce the resistance required by adding a significant quantity of other ions. The accuracy obtainable is probably of the order of 3 or 4% when using this dilution technique.

HIGH VOLTAGE DESIGN CONSIDERATIONS

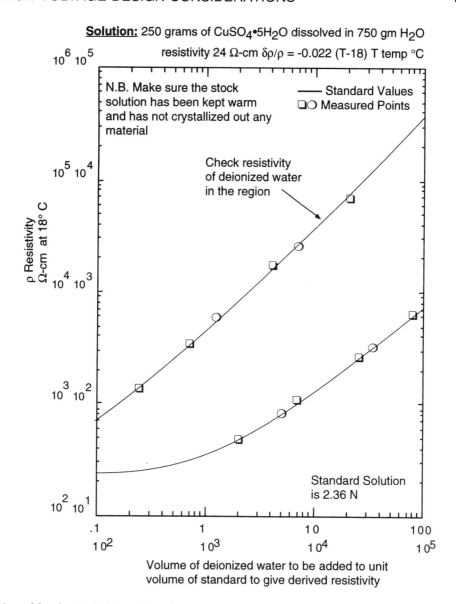

Fig. 12a-1 Dilution Graph

Section 12b

ELECTROSTATIC GRADING STRUCTURES*

J.C. Martin

This note briefly summarises some data obtained in a couple of days' work with electrolytic tanks on the optimum profiles for field grading electrodes. Symmetrical cases only were considered and Figure 12b-1 shows the nomenclature used for both the 1-D and 2-D cases. In each case the thickness of the slabs forming the two electrodes is chosen and the question is asked for a given separation of these, what the optimum profile is so that a minimum field enhancement exists around the edges. For the 2-D case a second question is what is the field enhancement in the centre of the electrodes. The electrolytic tank is a particularly powerful tool for this work, since even in the limited work noted here, some 20 configurations had to be optimised, each configuration requiring four or five attempts at the profile before a satisfactory one was obtained. Allowing for calibration, rechecks, etc., the actual time to determine these was about four hours.

The accuracy aimed at was only 10 per cent or so, since the finished electrodes were not intended to be of great accuracy. As such, no extrapolation techniques were used to obtain the field on the electrodes. However, in general the radius of curvature of the electrodes was considerably greater than ten times the half separation of the voltage probes used to measure the field. A second and probably more serious error is in determining the field at the centre of the disc electrodes with the optimum profile. These

---
* First Printed as SSWA/JCM/706/66
  AFWL High Voltage Notes
  Note 1
  26 Jun 1970

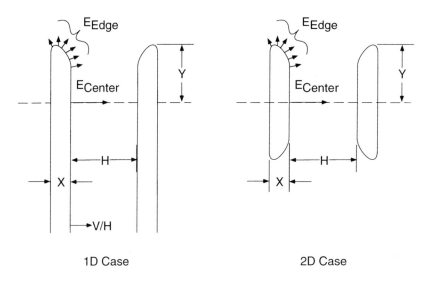

Fig. 12b-1 Electrode field grading.

measurements had to be made in a wedge electrolytic tank and the central field corresponds with the region where surface tension has important effects. However, the scale of the electrodes was such that this field varied only slowly in this region and it is felt that the curves that give this data are probably within 20 per cent of reality, and these are included to give some measure of this parameter. The more important parameter, the field enhancement factor (FEF), is defined as maximum field around the electrode edge divided by the mean field between the electrodes, i.e.

$FEF = E_{max} / (V/H)$.

This factor of course does not involve the field at the centre of the disc and also is the one which is by far the most useful practically.

Apart from the approximate nature of the profile to be expected in large, cheap electrodes, the electrostatic solution to the problem is only the first step in obtaining this profile when pulse voltage breakdown is considered. Some of the additional factors which have to be included when approaching the true optimum profile and field enhancement factor are

HIGH VOLTAGE DESIGN CONSIDERATIONS

(a) the area term;

(b) any streamer transit time, which means the shorter parts will break first;

(c) any radius of curvature effects - this effect in air can mean that at the sharp upper edge, up to 30 per cent more or so field can be tolerated in reasonably small electrodes;

(d) roughness and the statistical probability of distribution of sites of initiating electrons for gases.

However, most of these effects tend to mean that the electrostatic first order profile is on the safe side and usually it is not seriously affected by any of them. However, to an accuracy of better than 10 per cent they should be considered, hence there is little point in refining the first order solution beyond this point for pulse work.

I am aware that this work has almost certainly been done much better and published but the search to find it would have taken more time than the experimental approach, also the sensitivity of the FEF to incorrect profiles was not felt likely to be covered.

The curves enable the FEF and profile to be obtained for the symmetrical case and of course for a single electrode over a ground plane. In the case, which is of some interest, of a generator sitting some way above an earth plane with its bottom earthed, the top and bottom electrodes should have different thicknesses, the top one of course being the thicker. If the equipotential which is reasonably flat in the central region can be found, with an electrolytic tank or by field sketching, the problem can then be split into two parts and an adequate solution then obtained.

Figure 12b-2 gives the FEF for the optimum profiles. The 1-D case is covered by the line $Y/X = \infty$. It is clear that even for this case a uniform field gap cannot be made with $H/X > 3$.

Figure 12b-3 gives the factor $E_{edge\ max}/E_{centre}$ for different values of H/X and Y/X.

Figure 12b-4 plots the FEF factor at the <u>centre</u> of the disc for different H/X and Y/X. These are the curves which should be treated with rather more caution than the previous ones. As an example of the gain to be had from using these profiles rather than say, large spheres, a case is chosen where the FEF factor of 1.5 is taken as a required input. For a ratio Y/X of 2.5, this is achieved with H/X of 9. Hence for the unit spacing ($H = 1$) $X = 0.11$ and the diameter of disc = $2Y = 0.55$. The same FEF for equal

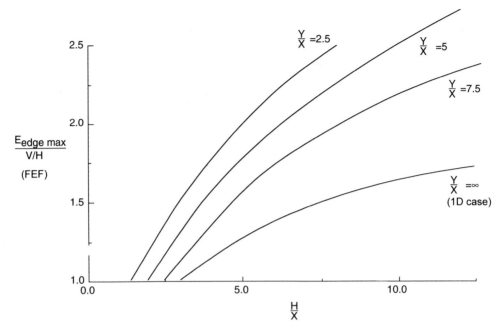

Fig. 12b-2 Field enhancement factors for optimum profiles.

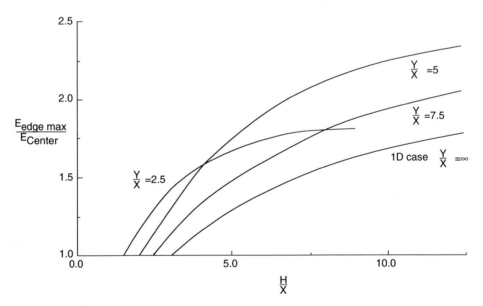

Fig. 12b-3 $E_{edge\ max}/E_{centre}$ for different H/X and Y/X.

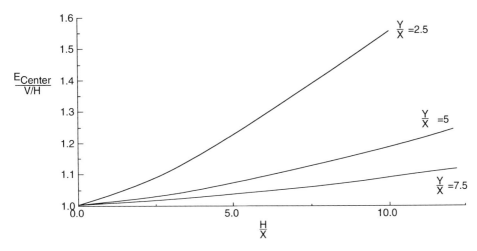

Fig. 12b-4 Field enhancement factors at centre of disc for various H/Y and Y/X.

spheres with unit spacing requires a diameter of 1.4. Thus compared with spherical electrodes, the optimum electrodes in this case are only 0.4 the diameter and 0.08 the thickness.

When the various profiles were compared, it was found that over the range of parameters investigated these changed little and to a first approximation depended mainly on the factor H/Y. Figure 12b-5 shows a couple of profiles for H/Y < 0.5 and H/Y > 1.5. In a couple of typical cases the sensitivity of the FEF to the profile was approximated by substituting the wrong one. The average increase of the FEF given in Figure 12b-2 was 7 per cent and hence over most of the range covered either profile could be used within the stated accuracy. However, certain sorts of departure from the profile are quite important, in particular bumps with fairly large derivatives of the slope: thus it is moderately important to keep the profile smooth. In making quite large electrodes with a file, the correct profile can be obtained without checking after only a little experience and a bumpiness which is acceptable is easily obtained. This judgment of bumpiness is difficult to quantify but was of course directly experienced in optimising the profiles in the electrolytic tank.

With regard to making the electrodes cheaply and quickly, the following approaches have been found to be useful by the members of the SSWA pulse group at AWRE.

For oil systems, plywood worked over with a file and covered in copper foil stuck down with an impact adhesive was found to be useful. Corners (the 2-D case) were formed by cutting the few thou. (mil.) copper and soldering the joints down.

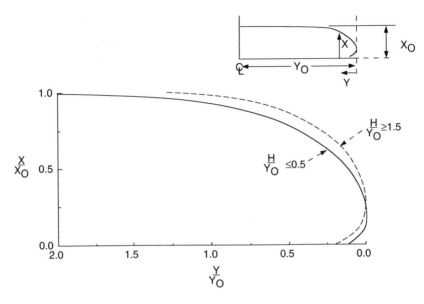

Fig. 12b-5 Examples of electrode profiles.

For water, balsa wood can be quickly worked and covered with thin lead sheet which can be gently hammered and worked onto the shaped wood backing. No serious deleterious effect on the water resistivity was observed.

For electrodes for use in air and other gases, polyurethane foam of density between 2 and 6 lbs per cubic foot can be quickly worked and is nice and light. This is then covered with about 2 thou. aluminium foil. Double-sided sticky tape is then stuck to the flat foil before it is formed over the contour. The creases and folds are then pushed into the foam by smoothing with a wooden stick (the handle of a hammer of modest size works fine). The finished metal surface is quite smooth and such defects as there are in it are pushed inwards and small smooth dents are not important. A large electrode of dimensions 3 feet by 2 feet by 4 inches can be made in about two and a half hours. All the approaches mentioned above are largely self-healing, in the sense that a discharge between the electrodes pushes the metal into the underlying material. In general this leaves a smooth contoured dent which does not lead to subsequent discharges at the same site.

In concluding it should be re-emphasised that the data produced is not of great accuracy, but is adequate for the design of cheap, large field shapers. In particular it is warned that the data in Figure 12b-4 is less accurate than the rest. While the data it gives can be derived from Figures 12b-2 and 12b-3, the smoothing was done independently for the three graphs and is reflected in differences of two or three per cent between the internally related graphs. As such, this is a technological user note rather than one in the purest traditions of Science, something for which I make absolutely no apologies.

Section 12c

SOME COMMENTS ON SHORT PULSE 10 TERAWATT DIODES*

J.C. Martin

INTRODUCTION

An area of considerable interest at the present time is generators capable of delivering 10 terawatts at voltages of one to a few megavolts. For some applications, the pulse can only usefully be 10 or 20 nanoseconds long. This implies the rise time of the pulse at the load must be short. Generators can be built for such requirements, using multichannel switching techniques and fast charging of the high speed section or sections. However, the diode presents significant difficulties and this note covers some aspects of the design of this and makes estimates of possible rise times. It should be emphazised that in a properly designed system the cathode interacts strongly with the diode design and this in turn interacts with the pulser design. Thus the design of the diode should not be considered in isolation. However, in this note this real life difficulty is partly sidestepped by taking a cathode of a given radius and assuming the inductance of the current flowing at or within this radius is small. With regard to the interaction of the diode/ cathode assembly back on the generator some comments on this will be made towards the end of the note.

One of the objectives of this note is to show which parameters are the most important in the overall inductance of the diode. It turns out that liquid breakdown in the feed and vacuum breakdown of the surfaces within the diode are the most important factors and that more work is needed in these areas. However, using small extrapolations of existing data, it appears that diodes can be built with quite reasonable rise times, providing cathodes with the assumed properties can be made to operate.

---

* SSWA/JCM/748/790

SIMPLIFIED TREATMENT

Figure 12c-1 gives a simplified picture of the tube and its feed from the pulse generator. The cathode is taken to have a radius r_c. This is fed by a vacuum feed (taken as parallel sided in this simplification) which extends out to a radius r_f. The diode multistage insulator is at a mean radius r_d and beyond this a matched tapered liquid or solid insulated line begins at radius r_o. The inductance between r_o and r_f is taken as the tube inductance, while that between r_f and r_c is the vacuum feed inductance. The inductances of these two are given by

$$L_{tube} = 2A/r_d \text{ nH}$$

$$L_{feed} = 2d \ln (r_d/r_c) \text{ nH} \quad \text{where A is cross-section of tube.}$$

In real life designs $r_d \sim r_f \equiv r$

Hence total inductance = $L = 2A/r + 2d \ln (r/r_c)$ nH.

This has a minimum when $r = \dfrac{A}{d}$.

Putting $\dfrac{A}{dr_c} = n$ and for $n \geq 1$ the optimum inductance is

$$L/2d = 1 + \ln (n) \text{ nH/cm.}$$

For r_c greater than A/d the lowest inductance is where the cathode is located as close to the vacuum interface as possible and then for $n \leq 1$.

$$L/2d \simeq n \text{ nH/cm,}$$

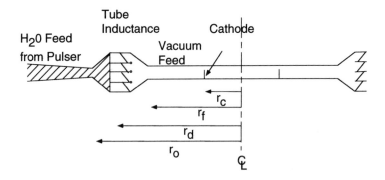

Fig. 12c-1 Schematic of diode.

HIGH VOLTAGE DESIGN CONSIDERATIONS

assuming as before that the dimensions of the tube are small compared with the radius of the cathode. Figure 12c-2 shows a graph of L/2d against n.

If the voltage of the diode is doubled, so is d, and A increases by four, approximately. If the radius of the cathode is kept constant and n is less than one, the inductance is increased by four. Thus for a constant delivered watts the rise time is constant. However, if n is greater than one, the increase in inductance as the voltage is doubled is less than a factor of four and the rise time at constant watts decreases as the voltage increases.

The next sections consider the practical values of the fields that can be applied to the various parts of the transmission system.

LIQUID INSULATED PULSE FEED FROM GENERATOR TO DIODE INSULATOR SURFACE

As an example of a practical feed, water will be taken as the insulating medium on the line. For short pulses water potentially has as large a Poynting vector as mylar and is considerably better than transformer oil. The feed will be tapered to keep a constant

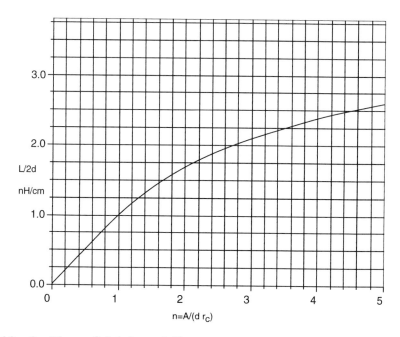

Fig. 12-c2   Plot of L/zd vs A/dr_c.

impedance until near the diode insulator surfaces. Since the tube cannot support anything like the field that the liquid insulated line can, the metal will have to flare apart close to the insulator sections. However, metal flux excluders can be introduced into this region, leaving only a small volume close to the tube as a non-matched line. Figure 12c-3 indicates how this could be arranged. However, this can only be completely done if the water feed can stand the necessary field at r_0.

To calculate the breakdown voltage of the tapered line, an effective area is required. This is taken to be the area out to where the field has dropped to about 92% of the peak value. This comes from the (area)$^{1/10}$ dependency of water breakdown. Thus the effective area is taken to be about $0.6\ r_0^2$.

The old AWRE breakdown relation can then be used to obtain the breakdown field for, say, a 20 ns pulse. However, some later work with largish areas (~ 3000 cm^2) of polished stainless steel at AWRE showed that breakdown fields some 40% higher could be supported. Work at Maxwell Laboratories by Dick Miller also showed that carefully polished and conditioned stainless steel surfaces could stand fields considerably above those given by the old relationship. Based on these data, it is assumed that with considerable care, the feed lines can <u>work</u> at 1.3 times the old breakdown field.

Using the data given later for the optimum radius of the tube where the radius of the cathode is less than this, Table 12c-I can be obtained.

It is seen that for V ≥ 2 MV the feed line can be matched down to the optimum radius. However, for 1 MV the insulator surfaces would have to be located out at a radius of about 70 cm, if this condition were to be met, unless even higher water breakdown fields than already assumed could be obtained by some means. Of course the 1 MV diode insulators could be located out at this large radius but in the event that only relatively small cathode radii could be used, a large inductance would exist in the vacuum feed.

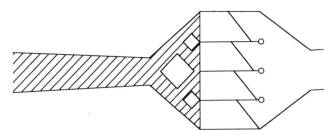

Fig. 12c-3   Flux excluders.

HIGH VOLTAGE DESIGN CONSIDERATIONS

Table 12c-I
Diode Parameters

Peak Volts (MV)	r_0 (cm)	Eff. Area (cm²)	F (MV/cm)	Separation (cm)	Impedance (Ω)
1	23	320	.82	1.2	.35
2	36	780	.77	2.6	.45
3	48	1400	.71	4.2	.56

There are two reasons why only relatively small cathode radii may be useable. Firstly there seems to be a finite time during which the diode pinch is establishing itself. The larger the radius of the cathode, the longer this time. While there may be ways of decreasing this pinch time, it is uncertain if it can be lowered far enough to allow 60 cm radius cathodes to be used. Secondly, if the aim of the system is to produce a tightly pinched multimegamp per cm² beam, then the inevitable small imperfections in the system will limit the degree of convergence practically achievable. Once again, the smaller the cathode, the less convergence needed to give the required current density. Thus too large a cathode radius may lead to lowering of the pinched current density.

Thus the conclusion of this section is that for voltages above 2 MV the water feed can be matched down to the necessary outer diode radius, providing a modest increase over the old AWRE breakdown data can be obtained by surface treatment.

FLASH OVER FIELD OF THE INSULATOR

"Fast Pulse Vacuum Flashover", SSWA/JCM/713/157, March 1971, shows that the average of series of determinations of the flash over of 45° perspex cones gives the relation:

$$FA^{1/10} t^{1/6} = 175$$

where F is in MV/cm, A in cm² and t in μsecs.

It is shown that this relation agrees rather well with test results for the best graded multistage tubes. For typical areas of insulators of 2500 cm² and times of applied pulse (measured at 89% of peak) of 10 ns, the factor $A^{1/10} t^{1/6} \sim 1$. thus, for a very well graded tube the flash over gradient of the insulators would be about 175 kV/cm where a 45° angle is used.

A possible design of a low inductance tube would be as is detailed in Figure 12c-4. This contains four slightly unusual features which will now be briefly discussed.

Firstly, the perspex "cone angle" is now 60°. In simple cone experiments there is a large increase of the field towards the positive electrode, due to the dielectric discontinuity. The electrodes are shaped to decrease the field at the positive electrode.

Secondly, in a simple cone test the field is reduced at the negative electrode. In this design, the negative side of the metal spacers are shaped so that the field is kept more nearly constant as this is approached.

Thirdly, the metal spacers are not rounded uniformly, but are shaped to reduce the field on those surfaces which can emit electrons, allowing the field to rise on those parts which cannot.

Fourthly, the metal spacers are thick enough in places to take the O rings but are thin under the central stack of insulators.

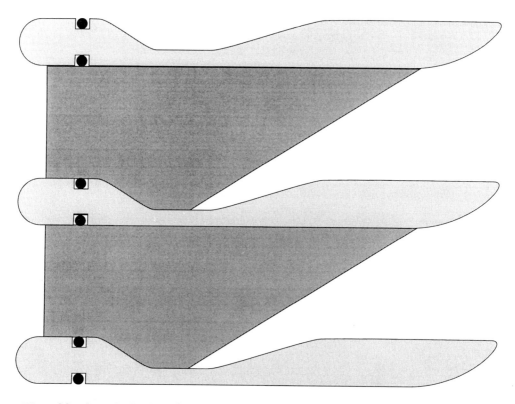

Fig. 12c-4 Diode insulator sections.

HIGH VOLTAGE DESIGN CONSIDERATIONS

This enables the thickness of the metal spacers to stay at about 10% of the total height of the diode even when the stages are 1 cm thick or less.

There are two separate reasons why insulator sections less than 1 cm thick are desirable. The thinner the insulators, the narrower these can be and hence the smaller the inductance of this part of the diode. In addition there is evidence that insulators or cones shorter than some critical avalanche length stand a higher field than given by the relation above. This was indicated by experiments in which small sections of a poor angle were introduced into a 45° cone. More importantly, experiments by O. Milton at Sandia Laboratories, Albuquerque, show that the field climbs as the cone height decreases. The effect is not large but is probably of the order of 10% for insulator stages a bit less than 1 cm high. This increase is taken to cancel out the effect of having to have a thin but finite thickness of metal spacer in the insulator stack. Thus for a very well graded 45° tube with a 20 ns type pulse, the gradient up the tube might be 175 kV/cm of tube. However, because a 60° reasonably well graded insulator surface is used, it is estimated that the flash over gradient might be 220 kV/cm. In real life tubes can only be used at 75 percent of their average flash over gradient. Thus a gradient of 160 kV/cm may be useable for the size of tube to be considered below.

The above obviously badly needs checking practically, but I feel that some working gradient near to this is in fact achievable in a very well graded system, for the short pulse application.

VACUUM INSULATED FEED FROM INSULATORS TO CATHODE

Some years ago a series of tests was performed with IT on 7 inch diameter metal blanks, in order to determine what surface treatments reduced, or at least delayed, significant cathode current. These were largely aimed at finding the highest field that would be supported by the surface without a significant (~ 20 ohms) impedance developing in times like 100 ns. It is intended that Mike Goodman and Dave Forster will include a section in the report on IT dealing with these rather ad hoc experiments. However, possibly doing some violence to the results, the following points emerged with more or less certainty.

Solid layers on the cathode could lead to a very dramatic drop in impedance, with the current being carried in a few very small diameter beams. Where liquid layers were used to cover the vertical cathode blanks, it was very important that a uniform film be produced and the layers must not "orange peel" or become highly non-uniform, leaving areas only thinly covered, or not covered at all. Some ten liquids were tested: of these, the best were unpolymerised araldite (epoxy) and a silicon based vacuum oil, Midland Silicone MS 704. The MS 704 was a bit more erratic, but slightly better than the liquid araldite and was mostly used by us. However, because it applied more easily and was more consistent, the data for araldite has been used in the analysis below.

It appeared that for pre-pulse fields up to at least 30 kV/cm there was no effect on the subsequent breakdown history. For a 1 1/2 cm gap, the nature of the anode material did not appear to have much effect. For 0.8 cm there was evidence that carbon was better than steel.

There did not appear to be a strong dependency on the nature of the cathode material between aluminium, steel and stainless steel. The steel was perhaps 10% lower in field, but had a poorer finish than the other two materials.

Considering some 20 tests, it appeared that the standard deviation of the field at which significant current flowed around 100 ns was ± 15%. This implies something like an $(area)^{1/10}$ dependency.

In addition, the time dependency appeared to be some $t^{1/4}$ on rather slender evidence.

Thus, for liquid araldite coatings, the following relation was not a bad fit:

$Ft^{1/4} A^{1/10} = 0.55$

F in MV/cm   t in μsecs   A in cm^2

This relation happens (quite possibly by accident) to agree with some old data of George Herbert's for thinned araldite on ball bearings, which gave 2.1 MV/cm for a time like 10 ns for an area of about 0.2 cm^2.

It must be emphasised that this relationship is only provided in desperation, since there is no better data to hand that I know of. It should not be taken as much more than an educated guess. The field given by the relation is that which gives rise to a current corresponding to an impedance of about 5000 ohms/cm^2 after a time t measured at about the 84% level. After a further period of t the impedance drops by most of another order of magnitude.

As was mentioned earlier, MS 704 gave higher fields on occasions and in addition some of the araldite data was almost certainly influenced by lack of uniformity of the layer. Thus it is to be expected that better polishing of the surface and more uniform application of the liquid will lead to better results. It is also

HIGH VOLTAGE DESIGN CONSIDERATIONS

unlikely that we tested the best liquid; thus, for the purposes of this note it is assumed that the constant can be raised to 0.7 rather than 0.55. Then for an effective pulse length of 10 ns the relation becomes

$$F\ A^{1/10} = 2.2$$

As with the liquid insulated feed, it pays to taper the vacuum insulated line slightly. This reduces the effective stressed area at the cost of a slightly increased spacing of the outer regions of the disc feed. The total inductance is then slightly lower than for the case of a parallel sided disc feed.

A second and more general point is that provided the onset of significant current from the feed surfaces can be delayed until the current has risen in the cathode, the disc feed may become magnetically cut off. The electrons emitted from the negative metal surface then return to it and the plasma blobs should only expand at about 2 cm/μsec. However, the self-magnetic field is unlikely to affect fast jets of dense plasma coming from the negative surface, so if these are formed (as in the hosepipe instability) the desired effect will not happen. The whole set of phenomena is complicated, involving the cathode turn on time, the nature of the breaks in the liquid film, and why these appear to be different from that involved in thin films of solid dielectric, and the density gradients in the resulting plasma blobs. Hence, while magnetic fields may help in the vacuum insulated feed, it is difficult to assess by how much, without experimentation. In addition, where liquid layers are used on the negative feed surface, even if the time for which the field has to be supported was reduced by a factor of two or so, there would not be a very significant increase in field that could be used because of the quarter power dependency of the field on the time.

Consulting the results given in the later sections and allowing for tapering of the vacuum insulated feed where this has an appreciable extent, the effective stressed area is about 1000 cm^2 for voltages between 1 and 3 MV. For the case where $r_c > r_{opt}$, the cathode is located 5 cm inside the inner diode radius. Thus the factor $A^{1/10} \simeq 2$ and the field allowed will be taken as 1.1 V/cm for the size of tube considered in this note.

INSULATOR STACK AREA

Figure 12c-5 shows a cross-section of the diode for the 3 MV short pulse case. On the water insulated feed side, 3 flux excluders have been inserted which match the impedance down to close to the tube. On the vacuum side of the insulator sections flux excluding shapes have also been introduced so as to reduce the

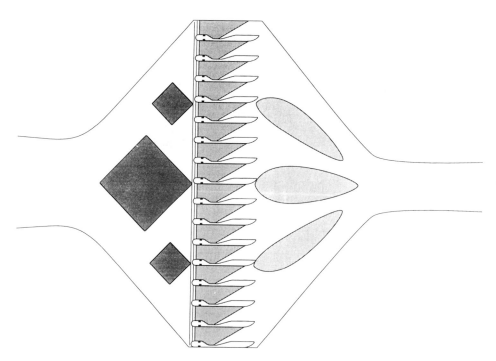

Fig. 12c-5 Scaled cross section of 19 cm high 3 MV insulator stack.

inductance in this region. The field over these is taken as about a half of that used in the vacuum feed. The flux excluders on both sides (but primarily on the water side) are located so as to ensure equal capacity division of the voltage gradient up the tube. There is a secondary function of the flux excluders on the vacuum side and that is to prevent direct access of u.v. from the cathode onto the tube insulators. This is because there is some evidence that hard u.v. irradiation of the insulator surfaces can lead to flashover because of photo emission currents changing the surface potentials. In a real life case the self magnetic field of the current flowing in the A-K gap will give rise to large magnetic fields at the insulator surfaces. There is some evidence that such magnetic fields can help to compensate for the drop in flashover voltage in single stage tubes, where a fall in this has been observed, due perhaps to u.v. irradiation of the insulator surface. Indeed, it is possible that a large magnetic field may actually help to raise the flashover voltage of a tube above that measured open circuit.

However, to date this has not been observed, as far as I know.*
However, the combination of u.v. shielding and large self magnetic
field should ensure that the tube maintains its open circuit flashover voltage even when large currents are flowing within it.

As can be seen from Figure 12c-5, no expense has been spared
with regard to the cost of the tube. Likewise, tiresome engineering details like tie-rods, structural stability of the insulator
stack, etc. have been ignored.

Table 12c-II gives the overall height of the insulator stack,
the number of stages, and the area from which flux has not been excluded. Also given is the fraction of this area in the insulator
stack region.

Also given in Table 12c-II are the approximate values of $r_0 - r_f$, the distance between the outer radius of the vacuum insulated
disc feed and the radius at which the water feed is fully matched.

Table 12c-II shows that for voltage greater than 2 MV, the
tube insulator sections give rise to less than half the inductance
of this portion of the total.

Table 12c-II
Parameters for Insulator Stack (Fig. 12c-5)
TUBE AREA A cm^2

Volts (MV)	Height (cm)	No. of Stages	Area A (cm^2)	% in Insulators	$\Delta r \equiv r_0 - r_f$
1	6	8	21	55	6
2	12	12	66	45	8
3	18	16	131	39	10

---

* Editor's Note: Magnetic flashover inhibition (MFI) has now been shown to work. First observed on Hydra at Sandia National Laboratories, it was then successfully utilized on Proto II.

DIODE INDUCTANCE

Using the values of A given in Table 12c-II, the optimum radius of the tube can now be calculated. In these calculations the fact that the centre of the insulator area is $\Delta r/2$ further out than the outer radius of the vacuum feed is allowed for. Table 12c-III gives the resulting optimum radius as a function of voltage. Also given is the inductance for a range of values of r_c. Where r_c is greater than the optimum feed radius, this is made 5 cm outside the cathode radius. Also given in the table is the e-folding rise time of such a diode fed with a fast rising pulse with a matching impedance of zero inductance at the cathode radius.

If such a very fast rising pulse were applied to the diode, a large inductive field would appear at the tube and the insulators, or the feed would break down. In addition to this, the impedance in the anode cathode gap will take some time to establish itself, so that the assumption of a constant impedance is not a very reasonable one. From some other calculations it appears that the rise time of the generator should be about 0.8 of the rise times given in Table 12c-III. In addition, the likely time varying history of cathode impedance combined with this generator rise time will mean that the voltage on the insulator stack will not exceed the matched output generator volts by more than 10% for a short time. These rough calculations also suggest that the 10% to 90% rise time across the anode cathode gap under these conditions will be about twice the e-folding rise times given in Table 12c-III.

It should be mentioned that in all the above, the limitation pointed out in the simplified treatment section, on the impedance

Table 12c-III
Diode Circuit Parameters with Optimum Feed Radius

V (MV)	r_f opt (cm)	L (nH)			τ rise (ns)		
		r_c 10 cm	r_c 20 cm	r_c 30 cm	r_c 10 cm	r_c 20 cm	r_c 30 cm
1	17	3.05	1.90	1.35	15.2	9.5	6.8
2	28	7.8	5.3	4.0	9.75	6.6	5.0
3	3	13.3	9.6	7.4	7.4	5.3	4.1

HIGH VOLTAGE DESIGN CONSIDERATIONS

of the water disc feed for voltages under 2 MV has been ignored. In fact, as Table 12c-III shows, in order to obtain a 10 to 90 rise time of, say, 12 ns, a cathode radius of about 35 cm would be needed, ignoring any questions of water breakdown. Allowing for the effect of the mismatch in the water feed, a cathode radius of about 45 cm is necessary to obtain a 12 ns rise time for an impedance of 0.1 ohm.

Thus imposing a condition that the 10 to 90 rise time should be 12 ns or less, the minimum values for the radius of the cathode can be obtained for the various voltages. These are given in Table 12c-IV.

Purely as an illustrative example (to which no significance should be attached), the cathode radius can be calculated if the parapotential impedance is assumed for the pinched cathode. In order to do so, the spacing of the anode cathode is needed and I have guessed a gap of 0.3 cm for 1 MV and assumed this goes as the square root of the voltage. This assumption gives the value of s in Table 12c-IV and the corresponding cathode radii. With these arbitrary assumptions, it would appear that operation of a pinched system at voltages greater than 2 MV gives quite reasonably small cathode radii for powers of 10^{13} watts.

With regard to the values of r_{opt} given above, the diode inductance is only a slowly varying function of this parameter near its optimum value and consequently the rise times are hardly increased if the tube radius is made significantly smaller than it. Thus, practically, for a 3 MV diode its diameter need be no more than 2 feet. This last comment only applies to those cases where the water insulated line can be matched down to the diode.

Table 12c-IV
Minimum Cathode Radius for 10 to 90 Risetime of $\leq$ 12 ns
0% 90% rise time $\leq$ 12 ns

V (MV)	Min r_c (cm)	Assumed AK Spacing s (cm)	r_c Parapotential (cm)
1	> 45	0.3	70
2	> 23	0.43	25
3	> 15	0.51	13

HIGHER POWER DIODES

A detailed discussion of these is not merited until more information about allowable cathode radii has been obtained experimentally. However, on the data given above, it is reasonable that diodes delivering 1.5×10^{13} watts can be designed for 3 MV systems.

In order to increase the power levels again, it will be necessary to work on the breakdown fields of all the component parts, but most importantly on the vacuum one. For instance, doubling the vacuum breakdown field would enable diodes carrying about 2×10^{13} watts to be designed at 3 MV while still using the water and insulator flashover fields assumed above. The radius of such a diode would be around 100 cm and it would have some pretty serious engineering problems associated with it. However, a pair of such diodes would deliver about 0.8 MJ in about 20 ns.

CONCLUSIONS

Assuming significant but probably achievable increases in working stresses of the various components, diodes delivering 10^{13} watts at 2 MV and above can be designed, using standard techniques. Such diodes are physically quite small and can produce acceptable rise times with quite small cathode radii. Pushing the vacuum breakdown fields up by a further factor of two enables paper designs to be done which will deliver some 0.4 MJ per diode in a 20 ns pulse. Around this level everything becomes rather difficult, to put it mildly, and hence without more work on what is practicable in the way of cathode radii, etc., it is probably not worth doing further work at this stage. The most important area for additional research would appear to be the short pulse vacuum breakdown of large areas of metal surfaces, particularly in the presence of large self-generated magnetic fields.

Section 12d

PULSE CHARGED LINE FOR LASER PUMPING*

J.C. Martin

INTRODUCTION

The aim of this short series of experiments was to show the feasibility of a generator providing a current of about 300 kA with a maximum rate of rise of about 8×10^{13} amps per second, by means of a cheap, simple system. In addition, it was desirable that the generator should use gas switches, so that the rate of firing could be a few a minute. The above requirements meant that the approach used by John Shipman of NRL (which uses triggered solid dielectric switches) would not be satisfactory. John Shipman's elegant system provides a short circuit current of 500 kA and a slightly faster rise to the current pulse, but it was felt that even though the performance of the new system was going to be poorer, its speed of operation and cheapness would compensate for this.

There was not time in the 6 weeks available to build a full system (whose width of lines would be about 1.4 metres) so a 60 cm wide strip of it was built, mainly to test the multichannel operation of the start and pulse sharpening gaps and to check out other features of the construction of the lines. Two people were engaged on the building and testing of the sub system and the work was terminated a little prematurely by one of them (JCM) getting plastered (the right leg). However, it was felt that demonstration of all the main points had been achieved and that with only a modest amount of further development the full system could be made to work.

---

* SSWA/JCM/732/373
  February 17, 1973

OUTLINE OF SYSTEM

The basic generator is a strip line Blumlein circuit. The Blumlein circuit is normally shown with 2 equal impedance lines (Z) and when charged to V_0 and switched with an ideal gap, provides an output voltage of V_0 into the matching impedance of 2Z. The duration of the pulse is the two-way pulse transit time. However, there is no need for the lines to be of equal impedance: they can be unequal and in this case it is desirable that they should be. Neither need they be of equal length and again, in this application it is functionally desirable to make the unswitched line significantly shorter.

Figure 12d-1 shows the schematic of the Blumlein employed and the equivalent circuit. Switched with an ideal gap, the first pulse duration is now that of the two-way transit of the shorter line. Thus, if the load R approximates to a short-circuit, the current provided will reach a value of $2V_0/Z_1+Z_2$, with ideal switching. In practice, if the start switch has a rather slow rise, the pulse voltage across R will have a poor rise and also not reach the ideal switch voltage value in a finite length line. The second point is not as serious as the first from a laser application point of view.

The rise time of the start gap was expected to be 30 ns or so, far too slow to provide good laser pumping. However, this difficulty was to be circumvented by providing a pulse sharpening gap to

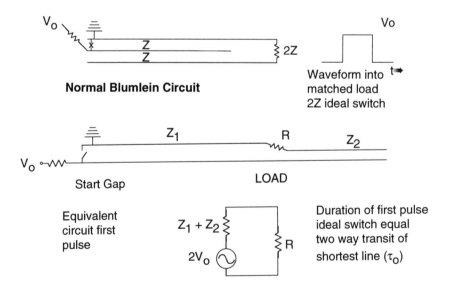

Fig. 12d-1 Blumlein circuits.

the left of the load. This gap would be much more quickly charged than the start gap and hence would be more compact and could be operated at higher fields and with many more channels than the very much more slowly charged start gap. However, the use of a pulsed sharpening gap throws away the first 20 ns or so of the rise of the pulse before it closes, providing the fast rising pulse to the load. However, the gas in the lasing cell only provides an output for a few ns, so the length of the right hand part of the Blumlein needs only be some 10 ns in electrical length, while the left hand line needs to be some 30 ns.

Figure 12d-2 shows the circuit with the pulse sharpening gap in it and also gives representative wave forms.

A major objective of the experiments described here was to show that the pulse sharpening gap would go into proper plasma conduction channels quickly, since recent Russian work has shown that uniform field unpressurised gaps had a fairly long (~ 10 ns) phase when current is carried in broad columns of ionised gas before the thin plasma channels form. This gives a poor rise to the output pulse. This phenomenon had been seen in edge plane gaps charged in times like 100 ns, when significant current occurred before the rapid voltage collapse phase took place ('fizzle'). It was known that this phase got much shorter as the gas was pressurised: however, it was not certain that in terms of a few ns the phenomenon had disappeared with pressurised edge plane gaps. Indeed, with carbon dioxide it was found that the fizzle phase was significant at quite high gas pressures.

CHOICE OF START GAP

While it was clear that the sharpening gap should be a pressurised edge plane gap, there were two choices as to what the start gap should be. The two distinct systems would have been a DC charged Blumlein system with a triggered uniform field rail gap, or a pulse charged line with an edge plane gap. For the DC approach, the uniform field gap would still have to be operated multichannel and this would have required a fast trigger pulse and, even with this, multi-channel operation of a DC gap at 70 kV has not been demonstrated with an adequate number of channels, to my knowledge. However, edge plane gaps have been operated with an adequate number of channels. (Multichannel Gaps, J.C. Martin, SSWA/JCM/703/27, referred to in future as 'The MC Note'). The use of edge plane gaps, however, meant that the Blumlein would have to be pulse-charged quickly and this would require the addition of a low inductance small bank to the system. A secondary advantage of this approach was that tracking problems would be much eased by pulse charging. A DC charged 70 kV line can be run in air, but very considerable care is needed to prevent tracking around the line and

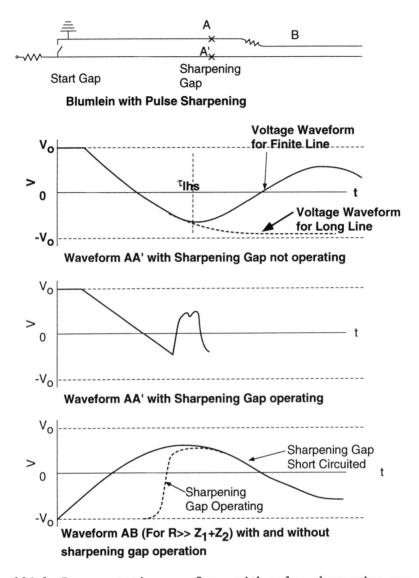

Fig. 12d-2 Representative waveforms with pulse sharpening gap.

the start gap. However, a pulse charged line can easily be made and indeed during all the tests no tracking around the line was observed at any time, despite the edge grading getting pretty grotty at times.

Thus, to enable an edge plane start gap to be used and to ease tracking problems, a pulse charged Blumlein approach was selected. In retrospect I think the decision was a good one. However, a DC

HIGH VOLTAGE DESIGN CONSIDERATIONS

charged system with a multichannel uniform triggered gap would be simpler (once the development work had been done) and almost certainly is feasible, but would require significantly more development work, in my opinion.

Thus the final schematic of the system is as shown in Figure 12d-3.

The inductances L_1, and L_2, are selected so that the two roughly equal capacities of the two parts of the Blumlein charge at the same rate, preventing a significant prepulse appearing across the pulse sharpening gap.

The 60 cm wide system will now be described and the results obtained with it given. The full system will then be outlined and some comments given about its testing and expected operation.

60 CM WIDE LINE

Bank

The small bank consisted of two AWRE-made 100 kV condensers in parallel with a rail gap at one end and the pulse output at the other - see Figure 12d-4. The capacitors are low inductance and have a value of 70 nF each. The tabs of each winding in them are brought out on one face and form a low inductance line when the return conductor is placed close to this face. The spark gap is a pressurised rail gap in a 2" OD perspex cylinder with 0.5 cm wall thickness. The rods which form the two electrodes are 5/8" OD brass rod, 6" long, and are spaced 0.60 cm apart in this gap (which is not optimum but just one that happened to be available). The ends of the brass rods are shaped as shown in Figure 12d-5 and rounded in all dimensions, so that the field decreases away from the maximum value it has in the central portion of the rail gap. This shaping is not critical and after carving off the brass bits indicated can be done with a coarse file in about half an hour for

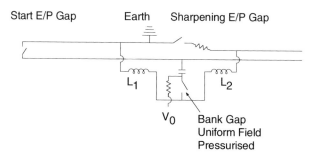

Fig. 12d-3 Pulse charged Blumlein.

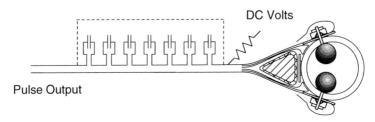

Fig. 12d-4 Schematic of condenser and gap.

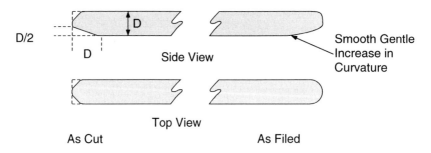

Fig. 12d-5 Uniform rail gap electrode shaping.

a pair of electrodes. No very perfect finish is required, but a rough polish with one grade of sandpaper is used after a smooth contour has been filed.

The low inductance connections between the spark gap and the capacitor are shown in Figure 12d-6. There is a flux excluding prism which also stabilises the voltage at the mid plane of the gap at V/2. Where the gap is charged 0, V this metal insert is allowed to float at V/2. Where plus and minus charging is used (which is the preferred arrangement, but was not used in the 60 cm tests), the metal flux excluder is earthed via a few kilohm resistor chain. The fact that the surface voltage gradient is divided into two half voltage gains because surface tracking goes at V^2 or higher. There are also ~ 1/16" flat sheet perspex guards simplexed onto the perspex tube and the mylar insulation of the main line is twin stuck against the spark gap body. Additional sheets of mylar are included around the main feeds. The flux excluder (which lowers the inductance of the gap significantly) is made by wrapping ~ 4 thou copper around a wooden triangular cross-section prism. This in turn is covered with double-sided sticky tape holding a mylar wrap down and the smallest face is then twin stuck to mid plane thin perspex sheet. The construction and insulation external to the gap is very similar to that of the start gap, which will be sent to the builders of the 140 cm line.

HIGH VOLTAGE DESIGN CONSIDERATIONS

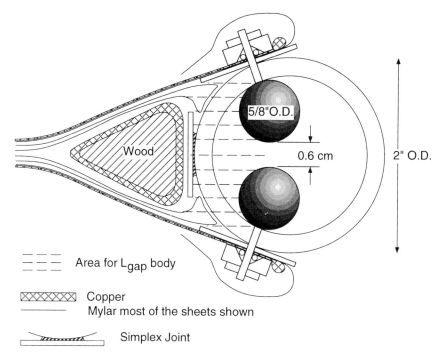

Fig. 12d-6 Bank gap.

The DC breakdown of the gap can be calculated to a couple of percent, using the treatment given in DC Breakdown Voltage of Non Uniform Gaps in Air, SSWA/JCM/706/67.

The calculated and experimental breakdown values for air are given below (Table 12d-I). These values apply after the gap has been fired a few times initially to condition off projections and whiskers. The breakdown channels should be distributed over the

Table 12d-I
DC Breakdown of the Rail Gap

$P_{absolute}$ (atmospheres of air)	$V_{breakdown}$ Calculated (kV)	$V_{breakdown}$ Observed (kV)
1	21	20
2	37.5	37
3	54.5	52.5

central region of the rods and not occur frequently (if at all) at the curved ends, if their shaping has been done correctly.

It might be as well to insert a word of explanation about all the calculations in this note. These are essentially designer's calculations and are only good to 10 per cent, or even poorer. In other words, these are designed to act as good estimates before manufacture and as a check that operation is as intended, as the system is made. The only real test of the system is that it performs as required at the end of the day: however, it is useful to check numerically that everything is roughly going according to plan at each stage, so rather crude calculations of expected performance are made before testing each stage and it is ensured that these agree tolerably, but not slavishly, well before proceeding to the next one.

The calculation of the inductance of the bank and switch up to the output strip transmission line is given below, as an example of how it is treated. Again, only crude approximations are used because this is all that is justifiable without a great deal of work which would be a waste of time and uneconomical.

INDUCTANCE OF SPARK GAP

Spark Channel

The plasma column expands at the rate of $\sim 3 \times 10^5$ cm/sec and hence at ~ 50 ns the channel radius is $\sim .015$ cm and the radius of the rod is $\sim .4$ cm. Hence the inductance of the plasma channel at the time of interest is given by

$$L \sim 2 \ell \ln b/a$$

where ℓ is channel length, b rod radius and a channel radius.

So $\quad L_{spark} \sim 2 \times 0.6 \times \ln 30$

$\quad\quad\quad\quad \sim 4$ nH.

The rail feeds to the channel have an inductance along them of about 4 nH/cm. The channel is fed from two directions and assuming the rails (width w) are fed uniformly along their length, the inductance $L_{feed} \simeq \dfrac{4 \times w}{12}$ nH

i.e. $L_{feed} \sim 5$ nH.

There is then the fact that the rails are fed only in three places (4 cm apart) by short connectors through the perspex spark

HIGH VOLTAGE DESIGN CONSIDERATIONS

gap wall. Each of these stubs has an inductance of about 2 x 0.7 x 2 ~ 3 nH and hence L_{stubs} ~ 2 nH. The copper feeds widen as they leave the spark gap to lower the inductance, but taking them as still being ~ 15 cm wide, the inductance of the area inside and just outside of the spark gap shaded in Figure 12d-6 is given by

$$L_{gap\ body} \sim \frac{12.6\ \text{Area}}{W}\ \text{nH}$$

i.e. $L_{gap\ body} \sim \frac{12.6 \times 14}{15} \sim 12$ nH.

Thus the spark gap inductance ~ 23 nH.

In addition to this there is inductance of the short feed between the spark gap and the condenser ~ 2 nH, giving an inductance up to the condenser of ~ 25 nH.

The capacitors each have an inductance of ~ 25 nH, giving 12 nH in parallel. There is ~ 3-4 nH in the low inductance beyond the capacitors, giving a value all up of ~ 40 nH. (Which, by experience, is usually a little bit on the low side.)

The measured value was obtained by firing the bank into a short circuit at ~ 25 kV and was found to be about 45 nH and also gave an internal resistance of 0.2 ohm approximately from the damping. The capacitor internal resistance depends a bit on frequency but is of the order 0.4 ohm each from previous determinations, in good agreement with the above.

Thus the bank basic characteristics are as given in Figure 12d-7. The bank should not be rung repeatedly at high volts into a short without any extra damping, as this reduces the capacitor life: however, at 70 kV it would give 90 kiloamps approximately into a short. The bank is simply fired by just reducing the gap pressure by venting the compressed air. The air flow is arranged so that the fresh air is introduced at the bottom of the gap and exhausted at the top of the cylinder and the rate of change in general for gaps should be about once every shot, but this is not at all critical and can be much lower for this gap which can be over-pressurised. The gap body can take well over 100 psig and is tested to this value after construction, while wrapped in waste cloth, in case it were to blow up. It is normally run at 30 per

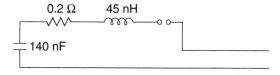

Fig. 12d-7 Bank characteristics.

cent over the breakdown pressure and is automatically flushed when the pressure is lowered to fire it.

THE LINES

The lines shown diagrammatically in Figure 12d-2 were made 250 cm long from the start gap to the pulse sharpening gap and 110 cm beyond to the end of the system. Initially the insulation of the LHS line was a total of 2 x 1/32" polythene and 4 x 2 thou mylar. The RHS was insulated by 4 x 2 thou and 3 x 5 thou mylar. The impedances of these (if perfectly assembled) would have been about 0.76 and 0.20 ohms respectively for a 60 cm width, giving a total Blumlein output impedance of 0.96 ohm. However, life is not quite that simple, because in a practical assembly there are thin films of air between the dielectric sheets. This has two effects: firstly these air films may increase the impedance somewhat (up to 10 per cent), and secondly (and more importantly) these air films break down during the pulse charging of the lines, increasing the capacity of the lines viewed as a "lumped constant" capacity. This introduces a $(dC/dt)^{-1}$ term into the charging circuit of the low inductance bank, causing increased damping and reducing the ringing gain from the bank to the lines. Thus it is desirable to assemble the lines in such a way that these air films are as thin as it is practically possible to achieve.

The method used to achieve a good assembly is to put the dielectric layers between fairly thin copper sheets (3 to 4 thou) and then push them together with a steady DC pressure. This is achieved by placing the lines between two 1/2" plywood sheets which are faced on the inside with spongy 1/4" sheets of rubber. This rubber needs to be pretty compressible (~ 70-80 per cent air) and when localised loads are placed on top of the sandwich, the plywood acts as a stiff member, flexing only a little under them, and the spongy rubber applies essentially uniform pressure all over the copper dielectric line. In actual practice, 6" concrete cubes were available to load the line (weighing about 20 lb each) and some 25 of these were distributed over the top of the sandwich, giving a pressure of about 10^{-2} of an atmosphere to assemble the lines.

After the line has been fired a number of times, charge separation occurs inside the line and this leads to additional electrostatic assembly forces which help to compact the line. As an example of what can be achieved, the capacity of the LHS was calculated to be 15 1/2 nF with no air present and the measured capacity was 12.6 nF after a fair number of firings. Thus, in addition to the 70 thou or so insulation, there was only some 6 thou of air. For the RHS the calculated capacity (allowing for extra insulation under the load and near the peaking gap) was about 26 nF, while the measured capacity was 17 1/2 nF. This indicates that in addition

HIGH VOLTAGE DESIGN CONSIDERATIONS

to 21 thou of mylar there was some 3 thou of air, on the average. There are as many interfaces on the RHS as on the LHS, but the mylar here is thinner and lies flatter than the dielectric on the LHS.

Figure 12d-8 gives a sketch of the cross-section of the line near the edge of the copper, showing the details of the edge control as well as those of the line construction.

A few notes about the construction of the line follow. The sponge rubber is Evo-stuck to the plywood (Evostik is an impact adhesive). A layer of 2 thou mylar is then stuck to the rubber face. The copper line is stuck to this mylar with two lengths of 3" Twin stick double-sided tape. The edge tracking control blotting paper is then stuck outside the edge of the copper with Twinstick. The dielectric layers are then placed in position and the top assembly then placed above and weighted down.

EDGE CONTROL

As the line is pulse charged, very big stresses are generated at the edge of the copper lines and in the absence of tricks to smear out these in a controlled way, flash round may occur. The tricks used in this line are threefold. Firstly the copper is bent back on itself (1/2" or so), giving a smooth edge of twice the thickness of the main copper, In addition, the space between the mylar sheets stuck to the sponge rubber immediately adjacent to the copper edge is filled by blotting paper or filter paper for a width of about 1 1/2". This edging can be conveniently made by sticking lengths of the paper to 3" twinstick and cutting this up the

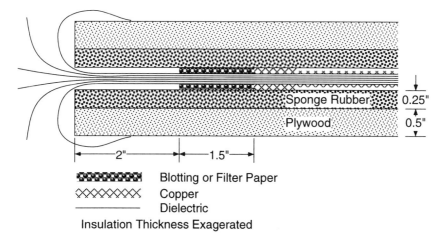

Fig. 12d-8 Edge grading of line.

middle. It is very desirable to use a porous paper for the edge control and not to fill up the space with solid material, otherwise the necessary grading at the edge will not occur by using the corona and edge puncture of the main dielectric may well take place. The third trick is to fold the top and bottom 2 thou mylar up and down over the plywood/rubber, so that any tracking that does get as far as the edge of the wood is forced to move away from the final edge of the dielectric, breaking the field here into a series of disconnected regions, each holding only a fraction of the voltage.

GAIN FROM THE BANK TO THE LINE

In theory, it is a simple matter to calculate the ringing gain of the bank/line when the latter is viewed as a "lumped constant" capacity. Figure 12d-9 gives the relation for a loss-less circuit and also a reasonable approximate gain for a circuit with a series and parallel resistance, valid when the damping is not too large. The difficulty in applying this relation is in defining what R_p should be. Any real resistances can obviously be allowed for, but there are two additional effects which decrease R_p. One is the air breakdown within the lines, mentioned above, for which an estimation can be made. Certainly the value of C_1 to be used is near the theoretical or no-air one, as the air must break down at operating voltages on the lines. This is because the field in the air reaches values of 1 MV/cm or higher. The air when it breaks may not become totally conducting when looked at from the point of view of a pulse travelling up the line, but during the relatively slow

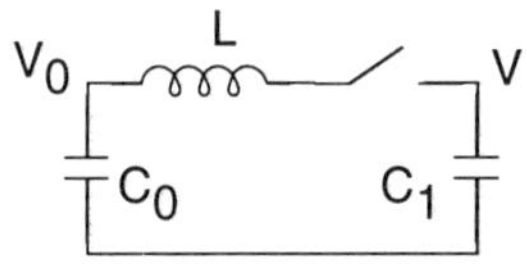

$$\frac{V_{max}}{V_0} = \frac{2C_0}{C_0+C_1}$$

$$\text{Time Constant} = \sqrt{LC_{eff}}$$

$$C_{eff} = \frac{C_0 C_1}{C_0+C_1}$$

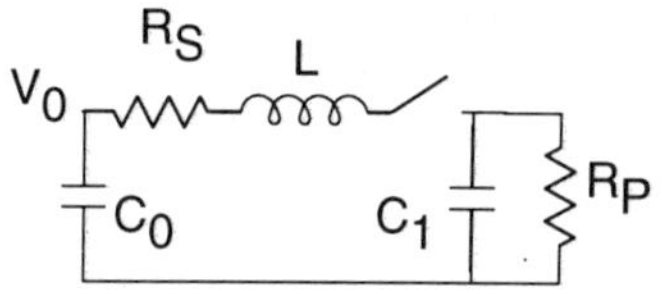

Approx Gain (1st Peak)

$$\frac{V_{max}}{V_0} = \frac{C_0 C_1}{C_0+C_1} \left(e^{-(\pi/2)\frac{R_s}{\sqrt{L/C_{eff}}}} + e^{-(\pi/2)\frac{\sqrt{L/C_{eff}}}{R_p}} \right)$$

Fig. 12d-9 Line ringing gain.

HIGH VOLTAGE DESIGN CONSIDERATIONS

charging phase, the capacity must rise to close to the theoretical value. Another dC/dt term arrives from corona moving out from the edges of the line. This effect is minimised by the edge control techniques mentioned above (sealing the edges of the copper completely will just lead to main dielectric breakdown, a phenomenon exploited in solid dielectric switches), but it still exists and it, too, causes a reduction in R_p.

An estimation of the ringing gain to be expected will be given later in this note, as an example.

After the line had been used for a while to do various tests and to make some edge plane breakdown measurements, its impedance was reduced by removing one of the 1/32" polythene sheets from the LHS, this giving an impedance (complete air breakdown) of 0.42 ohm and a generator internal impedance of 0.62 ohm, with air breakdown, and 0.72 ohm if the air did not break down completely. To cover the two possibilities, the line impedance would be taken as 0.65 ohm which probably closely corresponds to the impedance of the second version of the line.

There is an effect due to the fast rate of charging of the line. Typically, the $\sqrt{LC}$ of the bank charging is of the order of 40 ns. The "effective" $\sqrt{LC}$ of the LHS line, the longest, is about 8 ns, thus there is a significant voltage drop down the line, if it is fed at the sharpening gap, and the voltage at the start gap may be ~ 10 per cent lower, since it fires before peak volts. However, this is only of real importance in measuring the breakdown voltage of the gap; the pulse travelling up the line is, of course, the voltage across the start gap when this fires. This effect is minimised by taking a little care about the location of the feed point of the lines from the bank. The start gap is arranged to fire near, but before, peak volts on the charging waveform, in order to achieve multichannel operation, and it is the time of this breakdown on the waveform which decides the actual difference between start gap volts and feed point volts.

START GAP

The actual start gap will be sent to those intending to build the large systems, so that an extensive description of the gap and its insulation can be avoided.* However, Figure 12d-10 gives a sketch of the gap. The perspex body is 1 1/2" OD, with 3mm wall thickness, and the "plane" electrode is made from 1/2" brass rod. The edge is made from 1/32" thick strip, which is 1.1 cm wide. The gap between the edge and the rod is 0.70 cm and is uniform to about .005 cm. The length of the rod and edge are about 8" and the uniform length some 16 cm, the ends of both rod and edge being curved

---

* This offer has expired.

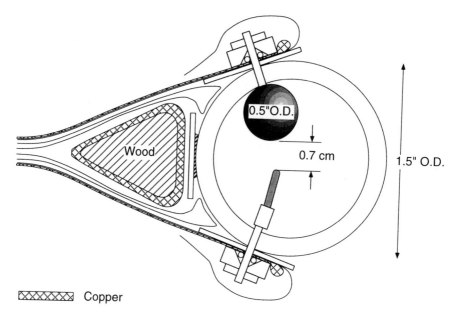

Fig. 12d-10   Start edge/plane gap.

back as in the bank gap, but done rather crudely in this case. The brass strip is sharpened, rather poorly, by filing at an angle of about 20° and with a final edge radius of about 0.01 cm. The overall length of the gap is about 15", including the demountable pressure caps at either end.

The two electrodes are fed in four places each along their length, brass fittings on the outside of the gap containing small O-rings and also pressing down onto the 3 thou copper of the output feeds to make electrical contact, as well as gas sealing. The inside of these stubs are fitted to the edge by slotting and then soft soldering. This method is fundamentally sound, but hard soldering should have been used instead of soft, as the stubs could be torn off by the over-enthusiastic use of the spanner.

The line tapers from the 20 cm or so of the switch out to the 60 cm of the line over a distance of some 30 cm and the insulation thickness is reduced over this region to a little under a half of what it is in the main line, in order to keep a roughly constant impedance. The insulation in this region is quite tricky and the use of Twinstick has been minimised, because it has an appreciable thickness in its own right.

In retrospect, the start gap was made too narrow and it would have been better to have electrodes some 35-40 cm wide. It was the first small pressurised E/P gap I had made and I was not sure how

HIGH VOLTAGE DESIGN CONSIDERATIONS

difficult it was going to be to fabricate it, when made by largely unskilled labour. As it turned out, it proved quite easy to make and a very fair uniformity of gap was obtained even when made by an amateur in a couple of days. The consequence of making it too small in length was that even when it operated multichannel (which it did with delightful regularity), its inductance was quite big and the resulting pulse travelling up the line had a poor rise (e-folding ~ 25 ns at best).

THE SHARPENING GAP

Figure 12d-11 gives a sketch of the gap. The electrodes are 55 cm long and the overall length of the gap about 85 cm. The "plane" electrode was made from 1/4" OD brass rod, while the edge electrode was made of strip 0.6 cm wide. The edge of this was filed down to a 15° angle and the final radius was about .005 cm. The gap was 0.61 cm, but the variation in this was up to 0.02 cm. This came about from over-confidence, leading to failure to drill the holes in the perspex tube at regular enough intervals. Despite this big variation in gap width, up to 40 channels were observed, although there was a definite series of bald patches where the gap was bigger than average. There should be no difficulty in holding the spacing constant to .005 cm when the gap is made properly. It may pay to thicken up the strip slightly to, say, 50 thou, to get extra stiffness.

Both electrodes are fed every 2" by stubs through the tube wall and make contact electrically to a square section brass bar which has O ring cones drilled in it at each stub point. The pressure scaling was quite satisfactory up to 80 psig, except when a stub was pulled off the edge strip by over-enthusiastic use of the spanner and/or poor soldering.

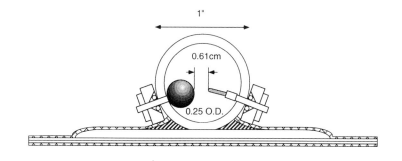

⊠⊠⊠⊠⊠⊠ Copper

Fig. 12d-11 Sharpening edge/plane gap.

The bottom of the gap is sealed to a 3" wide, approximately 2 mm thick, perspex sheet by some careful simplexing. The bottom of this perspex sheet is twin-stuck to the insulation of the line and at up to voltages of 90 kV across the gap, no tracking at all was observed. The pulse volts are on for an effective time of only a few ns, of course, which helps enormously in easing tracking at this point.

In the real application the start gap will be inclined across the strip line at an angle to the voltage front, but in these tests it was at right angles to the long length of the lines. The length of the sharpening gap in the full system will be some 200 cm and while it can be made in one piece I am sure, there is no reason why the electrodes cannot be made in, say, 60 cm lengths, or even three gaps can be made and simplexed end to end without degrading its performance significantly, providing the edge plane spacing is maintained constant throughout, should a single gap prove difficult to make.

This completes the description of the major items in the system. It is worth giving a few general remarks about the accuracy with which everything is made. In general, the assembly can be very poorly toleranced. Very little in the system need be carefully made and great care does not have to be taken over any part of it, with the exception of the gaps. Here there are three regions where some care has to be exercised. Firstly the insulation wrapping to prevent tracking should be done with some care. Secondly the gaps have to be pressure tight. Thirdly the gap in the edge plane should be held constant to better than 1 per cent. This will mean over a considerable length the standard deviation of the gap will be some ± 0.2 per cent. The absolute value of the gap is not too important, since this can be corrected for by changing its operating pressure, but it should be within, say, 10 per cent of the ones built for the 60 cm line tests.

In addition to the fact that the system can be fairly crudely made (except where noted), the numbers given in this note may not be exact and errors may have crept in on occasions; again, this is not of major importance. So the calculations given, while in the correct ball park, may not be exact to a few per cent; even so, they are probably better than the theory justifies.

ARRANGEMENTS OF EXPERIMENTS AND MONITORING

Initially, as was mentioned above, the bank characteristics were measured and checked against theory. The next set of experiments were carried out with the LHS line only in place, but with the start gap in position. This was over-pressurised so that it

HIGH VOLTAGE DESIGN CONSIDERATIONS

did not break. Some 30 ohms were placed across the line at the end remote from the start gap, in the form of 3 parallel chains of 8 resistors each. A signal was tapped off one resistor in one chain to monitor the waveforms. The ringing gain was then obtained and the operation of the start gap investigated for a range of gases at different pressures and for each polarity on the sharp edge. The ringing period of the line was investigated and also the rise time of the pulse from the start gap.

The sharpening gap was then built and placed in position and the line on the RHS of the system finished. The charging rates of the two bits of line were then balanced to give as small a prepulse as possible. The main monitor at this time was a chain across the LHS line just before the sharpening gap. This consisted of a chain of ten 10 Ω resistors tapping off one or two into the 100 Ω cable of the 'scope. To measure the out-of-balance signal, the 'scope cable was placed straight across the sharpening gap. In addition to these monitoring points, the load had been represented by 100 10 Ω resistors in parallel across the width of the line, giving a nominal 0.1 Ω series resistor. These were mounted in the line in a low inductance way and acted as a current monitor. The resistors used in this monitor were not the same as those used elsewhere in the monitoring, being physically rather smaller.

The system earth was arranged to be at the top line, just before the sharpening gap. In order to reduce earth currents, a 600 Ω approx. resistor was placed in the earth line from the bank, comprising 8 large 1 watt resistors, 2 in parallel, 4 in series. There was also, of course, a resistor in the charging lead from the power pack (2 megohms). Thus the system was DC earthed in two places (three when the safety dumps were down on the power pack) but had only one earth (that of the signal cable) from a pulsed point of view. Even this was not a direct earth for the medium-high frequencies, because of another pick up suppression technique we employ. This consists of winding some 50 yards of mains lead on a cardboard or plastic drum to form an inductance of the order of a millihenry. This acts as a choke for the high frequencies (of impedance of the order of a kilohm or greater) and limits the current flowing in the earth conductor of the coax to tens of amps rather than kiloamps. For low frequencies and DC, of course, the 'scope is still earthed. When this trick is used it is necessary to put 0.1 μF bypass condensers between both live and neutral to earth at both ends of the mains inductor. This is because equal pulse currents flow down the three cores of the mains lead.

For very high frequencies, the earth of the system is not definable anyway, and it radiates in a more or less balanced way before settling down with something tending to zero on the top plate.

The 'scope used in the experiments has a number of unusual characteristics and is a home-built system. The tube is an old Ferranti tube (made in limited numbers and no longer available) which has an accelerating potential of 20 kV, magnetic focusing and a twin line X deflection system consisting of two rods crossing the tube with the beam passing between them. The deflection sensitivity is very low (5 1/2 kV/cm on the film) which is a very useful characteristic when measuring high voltage pulses, because it makes the attenuators simple to build. This tube was the first one to write faster than the velocity of light about 15 years ago (1958). The active electronics in the 'scope is a single spark gap which provides the deflection sweep and brightening pulse. The sweep is approximately exponential, which we have found to be a useful feature in most of our work. The whole system is remarkably immune to pickup and, indeed, on one occasion we floated one of these 'scopes up to 1 million volts on a microsecond pulse and then recorded a high speed pulse, a situation that most 'scopes would not be happy in.

With the particular version of the 'scope we were using, the tube HT voltage was quite stable and hence the X deflection constant, but the sweep deflection system was somewhat time-dependent. However, whenever the time measurements were important, a calibration was done before and after the shots. The exponential nature of the sweep was also a minor inconvenience in some of this work, but, in general, the very fast response time of the 'scope (see the later section on output pulse rise time) and its immunity to pick up were very considerable advantages. The only reason that a brief description of the 'scope has been included is that it explains various unusual features of the waveforms obtained.

Any fast 'scope (better than or equal to 100 megacycles response) should be able to perform quite well, but more care will be needed with the attenuation, and with pick up suppression, than we had to use.

There was one additional feature added to the bank which has not been mentioned to date and that is the location of a 40 ohm resistor across the pulse output of the condensers. This consisted of some 40 large 1 watt resistors in a series parallel arrangement and had two functions. Firstly it provided a load to discharge the system, if the bank were fired and the start gap did not break, thus discharging the lines in a few microseconds and so avoiding a late time track around the edges of the lines. Secondly it added to the damping of the oscillations in the bank circuit in such circumstances to prevent it ringing on too long. This is not very important but as a general policy we do not like capacitors and gaps to ring through too many cycles. The resistive load across the capacitor could not be too low in value, as it would affect the ringing gain of the charging circuit, so that the value chosen was a compromise, on the high side.

HIGH VOLTAGE DESIGN CONSIDERATIONS

The last item to be mentioned in this section is the resistors used in the system. Except for the oil Blumlein output load/ current monitor, these were old-fashioned 1 watt ceramic cased composition carbon resistors (size 3 cm long by 1 cm in diameter). these resistors can be used as delivered up to 15 kV per resistance, for high speed pulses such as those encountered in these experiments. They will absorb up to 2 joules without any significant change in resistance, either pulse or DC, and can be used up to about 5 joules with some slight drift, which, however, can be cancelled out in a potential divider arrangement. Around 8 joules per pulse they blow up. Unfortunately, our Stores no longer stock these, but now use physically much smaller resistors and while we have some stocks of the old type, when we had to make up the 0.1 ohm resistor we had to use the more modern type. These we have not tested for high voltage pulse operation, but we have looked at a range of American resistors in the past, of very similar size. These all showed DC changes at significantly lower overall voltages and energy depositions (reasonable because of the small length and lower volume) and in general if a 10 per cent DC drop in resistance was found after a number of pulses, then the impedance was down 30 to 40 per cent during the high voltage pulse. This is indeed what happened to the output load and is discussed further towards the end of this note. Such resistors would still be useable in a voltage monitor providing the signal is taken off one of a chain via a high impedance feed, so the tap off resistance had essentially the same voltage pulse on it as would the others in the chain.

Figure 12d-12 shows how the voltage monitor was attached to the line. An essentially similar method of construction was used to make the attachment between the bank and the two lines. This attachment also formed the two inductors, but because it was required that these feeds should not have more than 40 nH or so,

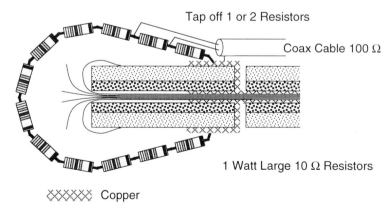

Fig. 12d-12 Monitor chain attachment.

the feeds were 8" wide and were attached to the lines through long slots in the plywood. The feeds were also only about 1 cm apart over most of their length and were wrapped in sheets of mylar and had curved up edges, to prevent flashover during the charging pulse.

60 CM LINE RESULTS

Ringing Gain

The ringing gain results will be given for the second version of the line, where the calculated capacity of the LHS was 30 nF and the observed 24 nF, while those of the RHS were 26 and 17 1/2 nF respectively. This version was the one with only one sheet of 1/32" polythene in the LHS line. Figure 12d-13 gives a tracing of the waveform for bank volts of 30 kV and gives a first peak gain of 1.13 and a half period of 150 ns. The reason tracings are given of the records is that it was intended to take a series of good records for this note at the end of the experiments, but this was not done as I had to go into hospital and was relatively immobile for a period of 5 weeks. During this time, various parts of the system had to be used for another investigation. This left only the records in the book and these were working records usually containing 4 or 5 traces on each print in order to save Polaroid film. As such, they are rather confusing and it was decided to trace representative records rather that have a lot of photographic reproductions done which would not have been of high quality and would also have been rather confusing.

The value of the total load capacity being rung from the bank was bounded by the values of 41 nF (no air breakdown) and 56 nF (total air breakdown). The voltage on the bank was not very high

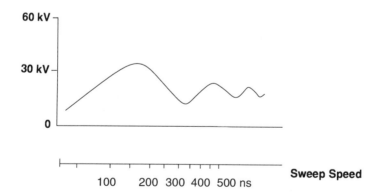

Fig. 12d-13 Ringing wavefront.

HIGH VOLTAGE DESIGN CONSIDERATIONS

(it could not be raised much higher because the start E/P would have broken later on in the wave form) and hence complete breakdown was unlikely and in addition the fact that some charge separation existed in the line after a series of firings, both suggest that the total air breakdown value would be too high. Thus a value about 50 nF for the load capacity is taken. This gives a value for the bank plus feed inductance of around 62 nH, based on a value of C_{eff} of 37 nF. This is reasonable, but perhaps a little low, since the bank inductance was 45 nH, giving an estimated total inductance of around 67 nH. However, neither of the above numbers is better than about 10 per cent, nor is there much point in obtaining greater accuracy, as was mentioned earlier.

To get the calculated first peak gain, using the approximate relation given in Figure 12d-9, a value of R_p is needed as well as the value of R_s, which is 0.2 Ω approx. There is besides the dt/dc term, a 100 Ω monitor and 40 Ω across the bank. The dt/dc is estimated to be a time of $\sqrt{LC_{eff}}$ equal to 100 ns and a capacity change of about 10 nF giving about 10 ohms. Thus $R_p \sim 7\ 1/2$ ohms, most of which comes from the dt/dc term, (which cannot strictly be treated as a plain resistance, anyway). The first peak gain becomes 0.74 (0.74 + 0.76) which comes out <u>very</u> fortuitously as 1.12. This degree of agreement is highly accidental and an agreement to 5 per cent would be much more usual, and worse would not worry me much. The main virtue of these calculations is to suggest how improvements can be made. For instance, if the damping had mainly been in the series resistor, efforts might have been made to decrease this. Alternatively it is seen that the 40 ohms across the bank output was not contributing much to the loss of ringing gain. Also, if more detailed analysis of the waveform is made (allowing for the RC drop), a damping factor of about 0.67 per half cycle is obtained, compared with the one suggested by the above calculation of 0.75. However, this too has its errors, both in measurement and theory, so the disagreement may not be real. The treatment also suggests that at much higher voltages the gain may drop to only a few per cent over one, as the air in the line breaks down more completely.

Out of Balance Signal

Figure 12d-14 shows a tracing of the out of balance signal across the sharpening gap, measured directly with the 'scope. The amplitude is about ± 1.6 kV for 30 kV on the bank, i.e. about ± 5 per cent. The record was the best obtained after a small amount of fiddling, this being done by changing the spacing of one of the feeds, i.e. altering the inductance. I do not know what prepulse voltage can be applied to the E/P gap without altering its performance, but I would not think that a 5 per cent prepulse would have much effect. However, this can easily be tested, although we did not have time to do this.

460   CHAPTER 12

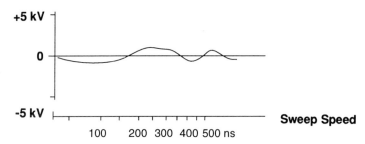

Fig. 12d-14 Out of balance signal.

Start E/P Gap

Figure 12d-15 is a tracing of a record of the operation of the start gap without the sharpening gap going. From this the breakdown field can be obtained for various gases and pressures. Most of the work so done was obtained with version A of the LHS and no RHS line, fairly early in the programme. This data, and also some derived from the sharpening gap, have been written up in "Results from Two Pressurised Edge Plane Gaps", J C Martin; I Grimson; SSWA/JCM/729/319. These results will not be covered again in this note. However, two points will be briefly covered. One is the apparent electrical length of the line and the degree of overswing shown in Figure 12d-15; the other is the number of channels.

The apparent electrical length of the line deduced from Figure 12d-15 is about 40 ns, considerably longer than the expected time of 25 ns. The explanation for this lies in the finite rise of the pulse from the start gap. The calculated rise time is dealt with more fully below, but can be taken as about 23 ns (maximum slope parameter) for the start gap working with about 10-15

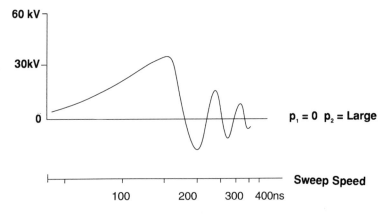

Fig. 12d-15 Start gap operating.

HIGH VOLTAGE DESIGN CONSIDERATIONS

channels in it. This pulse reaching the sharpening gap, which is open circuit, is reflected and returns to the start gap, at which it again reflects with a change of polarity, but in addition it is integrated by the inductance of the start gap, giving a pulse whose maximum slope is now about 40 ns, and this is delayed by about 10 ns more than the two-way transit time of 25 ns. The process repeats itself each time the maximum slope increases, the time at which this occurs being delayed as well. Figure 12d-16 shows three successive pulses and the resultant waveform is sketched. The effect in the case shown is to increase the apparent two-way transit to about 42 ns and also the first peak voltage, instead of doubling, rises only to 78 per cent or so of this.

To check the real electrical length of the line, a solid gap was substituted for the start gap and this gave a two way transit of 25 ns. The rise time of this pulse here was about 7 ns (maximum slope). Even in this case (which was done with version A of the LHS line) the overswing was about 90 per cent of voltage doubling. The two causes of this were the fact that the voltage at the gap was less than at the monitoring point, as mentioned early. This was the major effect, but there was still a residual effect of the finite rise of the pulse from the gap. In the record shown in Figure 12d-17, the lower impedance, version B, LHS lines were used and the rate of charge of the line was significantly less and about 5 per cent of the failure to double is due to this effect. Thus the first reflected pulse would be expected to be about 73 per cent

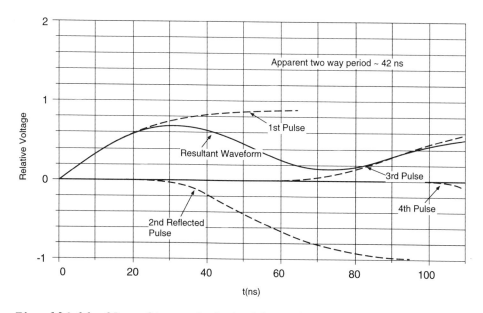

Fig. 12d-16 25 ns line switched with real life start gap.

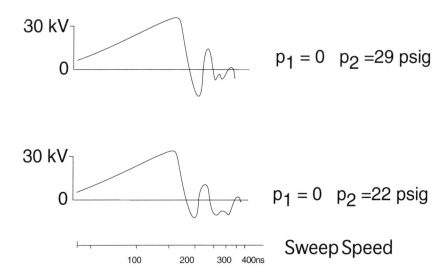

Figure 12d-17   Start and sharpening gaps operating.

of voltage doubling; alternatively the overswing is 50 per cent of the initial switching amplitude, as the record shows.

As the maximum current the sharpening gap can drive is directly related to peak voltage, the poor rise time of the switch has a significant effect on the short circuit current after sharpening. In addition, the rate of rise of voltage on the sharpening gap is directly related to that of the start gap output pulse and this is one of the factors which controls the number of channels in the sharpening gap, and hence the rise time of the sharpened pulse. The major factor in the rise time from the start gap is the "external" inductance of the spark gap body and if this had been made twice as long, the start gap rise time would have been significantly improved, leading to more overswing and more channels in the sharpening gap.

The start gap inductive and resistive phases are not the only things which can give poor pulse rise times, as some gases show fizzle at pressures not much over one atmosphere. This is covered in the edge plane note and, for instance, carbon dioxide is particularly bad in this respect. As a result of the data obtained in that note, it was decided to use nitrogen of the gases tested, although air was not much inferior. In addition, the best number of channels was obtained from both the edge plane gaps with the edges being driven negative. But, again, the results with the edges positive were scarcely inferior. However, the data to be covered from now on will be given for nitrogen and negative edge polarity.

HIGH VOLTAGE DESIGN CONSIDERATIONS

The number of channels in the start gap depends on the point on the waveform at which it is arranged to fire. In order to get as much volts on the line as possible, while still having a reasonable rate of change of voltage at the time of gap firing, it was arranged for the start gap to operate at about 90 per cent of the peak line volts. In fact, the number of channels was not a very rapidly varying function of this firing voltage, provided the gap closed before peak.

When the start gap closed at 90 per cent of peak volts, the number of channels was about 16, on average. An open photograph of two of the start gap firings is shown at the top of the page of photographs.* The camera is simply tilted between records and some additional, rather faint, channels can be seen in the original Polaroid prints, which have been lost in the reproduction.

An approximate calculation of the number of channels to be expected will now be given. Familiarity with the multichannel note will be assumed. The inductance of the various parts of the gap will now be considered, as was done for the bank gap.

Spark channel radius $\sim 3 \times 10^{-3}$ $L_{spark} \sim 2 \times 0.7 \times 5$

~ 7 nH/channel.

Inductance of feed to channel ($\omega = 15$ cm) at 6 nH/cm

$$L_{feed} \sim \frac{15 \times 6}{12} \sim 7\ 1/2 \text{ nH for 1 channel.}$$

Stub Inductance (4 stubs) ~ 1.5 nH
(each side)

Gap body Inductance $\sim 12.6 \times \frac{7}{20} \sim 4.5$ nH.

The first two terms depend on the number of channels and give

$$L_{in} \simeq \frac{6}{n} \text{ nH.}$$

Considering first the number of channels to be obtained,

thus $Z_{eff} \sim Z_{line} + \omega L_{out}$

$\sim 0.4 + 5 \times 10^7 \times 6 \times 10^{-9}$

~ 0.7 ohms.

---
*   Editor's Note: These photographs were not available.

Therefore $\tau_L \sim \dfrac{16}{0.7\,n} = \dfrac{23}{n}$ ns

where n is the number of channels.

Using representative breakdown fields, τ_r can be obtained and is

P psig	τ_r (ns)
0	$13.6/n^{1/3}$
14	$10.2/n^{1/3}$
22 1/2	$8.4/n^{1/3}$

Also $\tau_{transit} \sim 15/30n = 0.5/n$ ns.

This, depending as it does on 1/n, can be combined with 0.1 τ_L to give

$0.1\,\tau_L + 0.8\,\tau_{trans} = 2.7/n$

and Table 12d-II can now be obtained.

Table 12d-II
Multichannel Gap Analysis

n	$0.1\ \tau_L + 0.8\ \tau_{trans}$	$0.1\ \tau_r$			ΔT		
		p = 0	14 1/2	22	p = 0	14 1/2	22
1	2.7	1.4	1.0	0.8	4.1	3.7	3.5
2	1.3	1.1	0.8	0.6	2.4	2.1	1.9
4	0.65	0.9	0.6	0.5	1.55	1.25	1.15
8	0.3	0.7	0.5	0.4	1.0	0.8	0.7
16	0.15	0.55	0.4	0.3	0.7	0.55	0.45
32	0.07	0.45	0.3	0.25	0.52	0.37	0.32

Now T is required, which is related to the rate of rise of the pulse at firing time, and for breakdown at 90 per cent of peak volts T ~ 2.2 $\sqrt{LC}$ ~ 110 ns. Also σ is wanted. Figure 10c-3 of the Multichannel note gives it as 0.25 per cent for t_{eff} ~ $\sqrt{LC}$, which is for firing at 90 per cent of peak volts. To this must be added the jitter caused by the error in edge plane separation. In this gap the maximum change of gap was about 0.7 per cent, but this was a pretty uniform slope from one end to the other and the transit time helps to reduce the expected σ ~ 0.17 per cent (which would apply to a gap whose spacing varied randomly by up to 0.7 per cent) to more like 0.1 per cent in this case.

Thus $2T\sigma \approx$ 2 x 110 x (.0025 + .001)

$\approx$ 0.77 ns.

Using this value, the values in Table 12d-II suggest that the number of channels should be about 12 and only weakly dependent on pressure. This indeed was what was found after the gap had some conditioning shots.

Cranking this number back into the earlier expressions for the gap inductance, one gets a value for this, when feeding into the line, as being equal to

(16/12 + 6) nH
= 7.3 nH approximately

and not very dependent on the exact number of channels, most of the inductance coming from the external gap body term. Thus a two times longer gap would halve this term and also give more channels and thus essentially halve the above value. For the start gap as made

τ_L = 7.3/0.4 = 18 ns

and τ_r = 5 ns

giving an expected rise time of the order of 23 ns.

For a gap twice the width, the rise time (allowing for an increase in number of channels to about 20) would be about 14 ns (or a little more in real life, probably), a big improvement.

Once again the real value of the crude calculations is not to determine the exact number of channels to be expected so much as to locate the major terms in what controls this time, so that it can be predicted beforehand roughly, and also to suggest useful improvements in the construction of the system. In this case the external gap inductance is the controlling feature and this can be

HIGH VOLTAGE DESIGN CONSIDERATIONS

improved by using a smaller cylinder for the gap, or by increasing its length, the practically preferred route in a rebuild or new design being the latter.

The pressure in the start gap is denoted by p_1 and the approximate best value for this as a function of bank volts is given in Figure 12d-18, which also gives the optimum chosen value for p_2, the pressure in the sharpening gap. As such, Figure 12d-18 is a copy of the operating conditions graph for the 60 cm width line in its second version.

Sharpening Gap Performance

Once again the breakdown field data that were obtained for this gap are summarised in the edge plane note and will not be duplicated here.

Figure 12d-17 shows tracing of a couple of records for different values of p_2, the pressure in the sharpening gap. Again, these are for nitrogen with the edge negative. As the pressure in the gap is raised, the gap fires higher up the overswing, but at the time it fires the rate of change of volts is lower, hence a smaller number of channels results. Thus, while there are more volts to drive the load current with high values of p_2, the rise

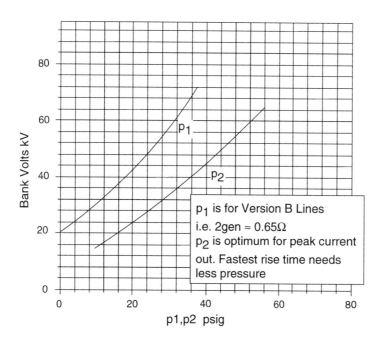

Fig. 12d-18 Optimum gap pressures.

time of the resultant pulse is poorer. As the RHS line is only some 10 ns long, the peak current reached in the 0.1 ohm load reaches a maximum value and then falls as the pressure is raised over the optimum. Thus the best condition obtained for these tests is where the sharpening gap is operating some half way down the overswing, but this is of course determined experimentally and will depend on the constancy of the edge plane separation, and also on the degree of overswing achieved, which depends in turn on the rise time of the start gap and the electrical length of the LHS line.

The second pair of photographs at the end of this note each shows two pairs of open shutter records of the start gap operation, the difference between the pairs in each photograph being the pressure in the sharpening gap p_2.* Here in particular there has been a noticeable loss of quality in the photographic reproduction (which was unavoidable) and many fainter channels have been lost and some of the brighter ones have coalesced. However, for the smaller value of p_2 in each case more channels can be seen. In addition, the bald spots where the edge plane gap was bigger, are clear, but with the greater rate of rise of the pulse at firing time these tend to fill in. The photographs were obtained with the aid of a mirror at 45° over the top of the sharpening gap and the camera was operated around f16, plus a neutral density of value 2.0 with Polaroid film speed 200. The start gap channels are significantly brighter, as they carry more current plus the ringing current of the bank. The aperture of the camera needs to be opened up as the number of channels increases, in order to keep the intensity of the brightest one roughly constant.

On the more routine of the shots, the number of channels was judged from the visual after-image and this was quite good for the start gap, but gave numbers like a half or less of the number recorded photographically in the case of the sharpening gap viewed at a distance of about 10 feet via the mirror. This is due to the eye lumping together, or not recording, those channels towards the edge of the field of view, I believe.

Table 12d-III lists the number of channels in the start gap and sharpening gap (recorded photographically) for a series of firings as functions of p_1 and p_2 for different bank volts, and also given is the peak deflection of the output current measurements. N_1 is the number of channels in the start gap and N_2 in the sharpening gap.

---

*   Editor's Note: These photos were not available.

HIGH VOLTAGE DESIGN CONSIDERATIONS

Table 12d-III
Number of Channels Versus Pressure

Bank Volts = 37 kV;			p_1 = 0 psig;	Average N_1 = 16
N_1	p_2 (psig)	N_2	N_2 average	Current Deflection (cm)
17	0	39, 32	35	.75
15	15	46, 48	47	.77
15	22	38, 40	39	.76
17	29	26, 21	23	.70
Bank Volts = 45 kV;			p_1 = 15 psig;	Average N_1 = 16
14	22	37, 30	33	.90
16	29	37, 37	37	.91
20	37	26, 25	25	.90
14	44	16, 13	14	.80
Bank Volts = 56 kV;			p_1 = 22 psig;	Average N_1 = 15
18	29	38, 35	36	1.10
18	37	35, 32	33	1.13
10	44	29, 27	28	1.14
16	52	18, 30	24	1.08

From this, an "optimum" operating set of conditions were obtained for this particular run, which are given in Table 12d-IV.

Table 12d-IV
"Optimum" Conditions

Bank Volts (kV)	P_1 (psig)	N_1	p_2 (psig)	N_2
37	0	~ 16	20	40
45	15	~ 16	30	37
56	22	~ 15	40	32

The crude multichannel calculations will now be given for these conditions.

Inductance of a Spark Channel

The spark channel radius is approximately 10^{-3} cm, giving

$$L_{feed} \sim \frac{55 \times 6}{12} \sim 27 \text{ nH for 1 channel}$$

$$L_{spark} \sim 2 \times 0.6 \times 6 \sim 7.2 \text{ nH/channel}$$

giving $\quad L_{in} \simeq \frac{34}{n}$ nH.

L_{stub} (10 stubs each side) ~ 0.35 nH.

Note the stubs were shorter and made a lower inductance connection to the square cross-section rod than in the start gap and bank gap cases.

$$L_{gap \ body} \sim \frac{12.6 \times 3}{55} = 0.69 \text{ nH},$$

giving $\quad L_{out} \simeq 1.0$ nH.

The impedance of the line as seen by the sharpening gap ~ 0.65 ohms plus 0.1 ohm load resistance ~ 0.75 ohms. For the multichannel calculations ωL_{out} has to be added, where $\omega \sim 3 \times 10^8$, i.e. $Z_{eff} \sim 1.05$ ohms.

Thus $\quad \tau_L = \frac{34}{1.05 \ n} = \frac{32}{n}$ ns.

The transit time is $\frac{55}{30}$ ns $= 1.8$ ns,

and the 0.1 τ_L and 0.8 τ_{trans} can be combined to give $4.7/n$ ns.

Using the breakdown field derived from the peak to peak voltage measured, the resistive phase can be obtained and is

p (psig)	τ_r (ns)
0	$12.3/n^{1/3}$
14 1/2	$9.3/n^{1/3}$
22	$8.3/n^{1/3}$

Using these values, Table 12d-V can be calculated.

HIGH VOLTAGE DESIGN CONSIDERATIONS

Table 12d-V
Multichannel Calculations for Optimum Conditions

(Times in ns)

n	$0.1\,\tau_L + 0.8\,\tau_{trans}$	0.1 τ_r			ΔT		
		p = 0	14 1/2	22	p = 0	14 1/2	22
10	.47	.57	.43	.36	1.04	.90	.83
20	.23	.46	.34	.31	.69	.57	.54
30	.16	.39	.30	.26	.55	.46	.42
40	.12	.36	.27	.24	.48	.39	.36
50	.10	.33	.25	.22	.43	.35	.32

Now the values of T and σ are required. The slope of the voltage waveform was obtained from the records and for the optimum operating conditions was about 40 ns for each pressure. For other values of p_2 this changes, of course, getting larger the greater the pressure is. The estimation of σ is unfortunately not very good because of the variation of gap length, which was up to 3 per cent in the sharpening gap. Judging from the distribution of the bald patches, the sparks only occurred where this was more like 2 per cent. With something like 40 channels this corresponds to a span of some 5 σ, i.e. $\sigma \sim 0.4$ per cent, due to gap variation. The intrinsic scatter from the MC note is about 0.15 per cent for a $t_{eff} \sim 15$ ns. Thus $\sigma \sim 0.55$ per cent, rather uncertainly. I realise that if the distributions are Gaussian, the straight addition of the two σ's is not valid, but if the distribution is otherwise (as it may well be), straight addition could apply. However, worries about this are overwhelmed by the basic assumptions as to the effective scatter of the gap width.

Using the above values

$$2\,T\,\sigma \sim 0.45 \text{ ns}$$

and from Table 12d-V the expected number of channels is in the range of 30, decreasing slightly with pressure. The numerical agreement here is, again, rather fortuitous and not to be taken very seriously: however, it is obviously in the right ball park. Moreover, it shows that if the gap had been better constructed (a maximum difference in spacing of some 1 per cent with a σ of the order of 0.2 per cent of the gap length), many more channels would have resulted. Again, if the start gap had been made twice as wide, T would have been considerably shorter and again more channels would have resulted. Indeed, so many that the impedance feeding each channel would have risen beyond the range of validity of the relations used in the MC note. However, it would be expected that at least 100 channels would have been obtained with the improvements mentioned above.

The calculated rise times of the pulse from the sharpening gap for the assumed optimum conditions are given in Table 12d-VI.

With the suggested improvements, the calculated rise time would be under 3.7 ns. This could be decreased further by reducing the edge plane separation and running at higher gas pressures; however, this could not be done very much.

Table 12d-VI shows that the rise time is improving as the pressure is increased and in the later series of shots this was done, as the system was operated at higher voltages.

A second point that the calculations show is that providing the sharpening gap has the same wave front impressed upon it along its length, the standard deviation of the output wave will be around ± 1/2 ns and with improvement this will probably halve. Indeed it may well do better than this, because with 20 channels in a length equal to the distance from start gap to load (10 cm), averaging will smooth the ripple down to something below 0.1 ns. However, there will be large-scale changes across the wave front on top of these small scale variations, caused by possible variations in amplitude across the width of the line in the wave hitting the sharpening gap. These are briefly mentioned again towards the end of this note.

Table 12d-VI
Calculated Risetimes

P_1 (psig)	P_2 (psig)	N_2	L_{gap} (nH)	T_L (ns)	T_r (ns)	T_{tot} (ns)
0	20	40	1.8	2.4	3.6	6.0
14 1/2	30	37	1.9	2.5	2.9	5.4
22	40	32	2.0	2.6	2.5	5.1

HIGH VOLTAGE DESIGN CONSIDERATIONS

Output Pulse

The output pulse was monitored by attaching the 'scope cable directly across the resistors that formed the nominal 0.1 ohm load. There is some small inductance in this load and this was cancelled out to the first order by arranging the length of the stripped inner of coax to have an inductance such that the integrating effect of this was approximately equal to the inductive spike time of the load resistor array.

Figure 12d-19 gives a tracing of a typical output pulse. The observed waveform from the 'scope has to be corrected for the 'scope response. This is given in Figure 12d-20. Most of the late rise time is in the delay cable, which allows the 'scope to trip before the signal is displayed and hence accounts for the shape of the 'scope response. Shown in Figure 12d-20 is the calculated response to a step waveform and also the observed one, which was measured with a strip line generator switched by a tin tack driven solid gap. The agreement is amply good enough for the present purposes.

Figure 12d-21 gives the pulse front corrected for the 'scope response. In addition to decreasing the rise time, the correction for the 'scope response also raises the peak voltage that an infinite bandwidth 'scope would have measured. Once again the measured two-way transit time is longer than the 12 ns that the RHS beyond the load represented. This is, as before, due to the inductance of the sharpening gap and also possibly to some integration of the pulse when it bounces off the open circuit end of the RHS line. At these voltages corona conditions must change over a centimetre or two at the end of the line and this represents a small integration effect.

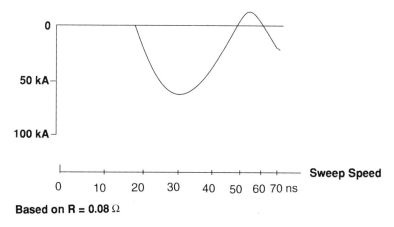

Fig. 12d-19 Load pulse.

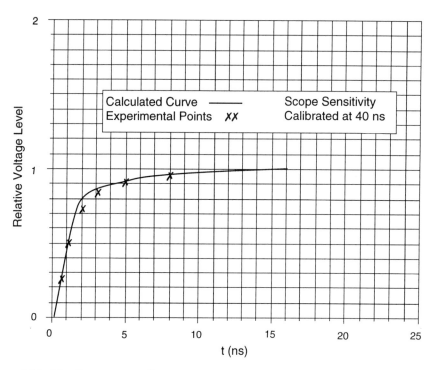

Fig. 12d-20 Response of scope to step input.

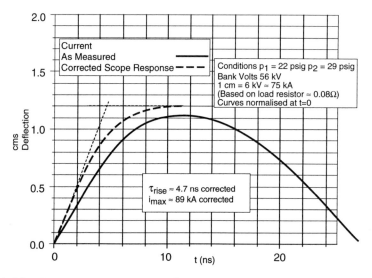

Fig. 12d-21 Output current waveform.

HIGH VOLTAGE DESIGN CONSIDERATIONS

The maximum rise time for the record shown in Figure 12d-21, after correction for the 'scope response, is 4.7 ns (maximum slope). The conditions for the record were $p_1 = 22$ and $p_2 = 29$, a bit below the optimum, but one which gives a faster rise than that calculated in Table 12d-VI, which gives the conditions for maximum peak current. Indeed, because of the faster rate of change of voltage at the sharpening gap breakdown point, the mean field was 10 per cent up on that used for the Table 12d-V and Table 12d-VI calculations. This reduces the resistive phase by 13 per cent, giving a calculated rise time of 4.8 ns. Again the agreement is too good, but I assure you, unfiddled.

Output Current

As was mentioned earlier, the smaller 10 ohm resistors were used to make the nominal 0.1 ohm. When originally assembled, the resistance was abut 10.1 Ω for each resistor. However, after the next series of high voltage shots to be described, a representative series of resistors was measured and found to have dropped their average DC value to 9.2 Ω. The energy dissipated in the 0.1 Ω resistor assembly during the high level shots can be roughly calculated. The peak current was 110 kA and the duration of the fast pulse was effectively about 20 ns. However it was followed by a chain of further pulses equivalent to a total of about 6 pulses (including the bank ringing current as well), giving an effective duration at full power of about 120 ns. Thus the energy dissipated for resistance would have been about 1.4 joules. The volume of carbon composite is a little smaller in the new resistors, so this might be equivalent to about 2 joules per resistor, which would not be expected to have been enough to produce a significant DC change of resistance in the old resistors. However, the old big resistors are mechanically enclosed in a ceramic body which supports them against any weak shocks caused by the heating. Other tests on USA resistors showed that they suffered DC resistance changes at levels of a joule a resistor even though the volumes of the resistor carbon composite material were similar. Also all the tests we have done show that when a DC change of about 10 per cent drop occurs in a few pulses, there is an additional 20 to 30 per cent drop in resistance during the pulse. This is caused by the fact that the current densities in the actual carbon filaments are very much greater than those obtained by taking the average cross-section of the composition rod, and heating and shock expansion cause a reduction of the resistance of these filaments. Anyway, as can be seen from Table 12d-VII, there is a progressive reduction of measured voltage across the carbon resistor loads compared with the voltage across the sharpening gap at closure time. The values agree well at low current levels but fall as the voltage and current are raised.

It is very difficult to see any reason why the system should not show a linear (or, rather, slightly super linear) relation between volts in the sharpening gap and the output current. The super linearity would be expected, because the rise time of the gap improves as the voltage is raised. Thus it is assumed that the load resistors are dropping their pulsed resistance, and the data obtained on other resistors is assumed to apply to this type. Thus in addition to the observed DC drop of the load resistor to 0.092 ohms, there would be a further pulse drop while the pulse volts were on, amounting to an extra 20 per cent or so at the top voltages recorded during the final current test shots. Obviously the best solution would have been to have repeated the measurements with a linear current monitor; however, as this was being planned the programme had to be stopped. In fact, in the absence of, say, 150 of the old 10 Ω 1 W resistors, it was not easy to make such a resistance with an adequately low inductance and an adequate heat-absorbing capacity. The one planned would have been a nichrome sheet one. Alternatively, and perhaps additionally, an integrated B dot coil would have been employed.

However, Table 12d-VII summarises the data obtained.

Table 12d-VII lists the operating conditions for the series of shots and then gives the voltage measured on the 'scope. This has to be corrected for the 'scope response (+ 8 per cent). The next column gives the estimated pulse resistance of the load based on a DC value of 0.092 ohms. From this a measured current is derived and then, using a total line impedance, the short circuit current is derived, taking out the load resistance. The next to last column lists the voltage swing (measured on the records) from the level at which the start gap had fired to when the sharpening gap

Table 12d-VII
Operating Data

Bank Volts (kV)	p_1 (psig)	p_2 (psig)	Load (kV)	Corrected 'scope Rise Time	Estimated Load R Ω	i (kA)	i S/C	V_{p-p} (kV)	i S/C Calc. (kA)
37	0	20	4.7	5.2	.087	60	68	44	68
45	15	30	5.5	6.0	.083	72	81	57	88
56	22	40	6.7	7.3	.078	93	104	70	107
64	32	54	7.3	7.9	.073	108	120	83	127

HIGH VOLTAGE DESIGN CONSIDERATIONS

operated. The final column gives an independently determined short circuit current based on this and the calculated likely line impedance of 0.65 ohms. As can be seen, the line impedance determined current is above that estimated from the load resistance measurements, but the agreement is quite good. Using the average values of the peak current, the output of the line at 64 kV on the bank would have been about 123 kA into a short circuit. For the bottom line of shots, the start gap was operating at 65 kV. A few shots were done at 70 kV operation of the start gap, with no signs of breakdown and, indeed, at no time did the line track; these shots would have corresponded to 135 kA short circuit current.

With a twice width start gap, the overswing after the operation of the start gap would have been 15 to 20 per cent higher, giving good sharpening gap operation at 90 kV, with 65 kV on the bank, which again would give short circuit currents of about 135 kA. Thus there is a very reasonable expectation that a start gap rebuilt line, or, at slightly higher voltage of operation, a line of 60 cm wide, would give 135 kA. Thus it is expected that the 140 cm wide line would be expected to run at 300 kA and possibly a bit higher.

The system was found to be easy to fire and reproducible in its characteristics, and could easily be fired once a minute, although the life tests which were planned unfortunately were not carried out. However, the start gap and LHS line had about 2000 firings at between 35 and 60 kV on them, and the full system some 1000 shots, mostly in the same voltage range.

This concludes the summary of some of the data obtained with the 60 cm wide experimental line. Before going on to the suggested 140 cm wide system, it may be worth briefly mentioning the effort and cost involved in the 60 cm experiments. The work was done by myself and a vacation student (Ian Grimson), and while it took about 7 calendar weeks, only some 6 weeks' time was actually available during period. The total construction time for the bank and the lines was about 3 weeks. During this time essentially everything was made, excluding the bank gap which happened to be around. However, the gas pressure and flow gear had to be assembled and if this were excluded, the time to make all of the bank plus lines, including all the gaps, would have been the 3 weeks. The power pack was to hand, as were the untabbed condensers. The remaining 3 weeks were roughly split between monitoring the performance of the system, including running it up to the maximum volts at which it was tested, and in making the measurements reported in the edge plane note. However, it must be stressed that we knew pretty well what we were doing and it would take someone inexperienced in the area considerably longer than this.

As regards costs, the local condensers (which were to hand and were, of course, reused afterwards) cost about £200 and the cost of the rest of the materials in the bank and line would add about another £200. If we had had to build a power pack (which we do with components ratted from a component store on site, or obtained secondhand), it would have cost us another £200 or so.

The only engineering effort used was in cutting the brass sheet accurately to the widths required for the edge plane gaps and in making the brass fittings for the pressure seals for the start gap. The sharpening gap seals were made by us.

To make the larger line discussed below would cost at most £300 in materials, I estimate, excluding the power pack and condensers, and would take us now, with what we know, about the same time as the 60 cm line did. A bit more time would be needed to do the monitoring, to ensure the long wave flatness of the wave front beyond the sharpening gap, but obviously we would not repeat the extensive series of measurements with different gases in the edge plane gaps. So, roughly, I would expect to spend about 10 man weeks to make and diagnose the operation of the full line. However, again it must be stressed that we would have all the materials to hand and would know pretty exactly what we were doing, so the time scale suggested applies only to us.

140 CM WIDE SYSTEM

Bank

There are two modifications I would suggest be made to the basic bank we slung together for the 60 cm wide line tests. These are to go plus and minus for the charging volts and also to split the bank either side of the uniform field DC switch. This latter change reduces the inductance of the switch and also reduces the performance required from the condensers.

With regard to the plus and minus charging, this eases any tracking problems across the gap (not that we had any trouble here) and also for the price of another set of rectifiers doubles the output of a power pack. In building our power packs, most of the cost is in variac, transformer, etc. and a ± 35 kV unit is significantly easier and cheaper to build than a 70 kV one. There is no need, of course, for the power pack to be stabilised, or anything fancy, and it only needs to supply a milliamp or two of current to charge adequately quickly, as far as I can see.

Figure 12d-22 shows a drawing of the gap and a sketch of the condensers and output lines. The suggested gap would still use a 2" OD perspex tube and have the brass rod and gap dimensions given

HIGH VOLTAGE DESIGN CONSIDERATIONS

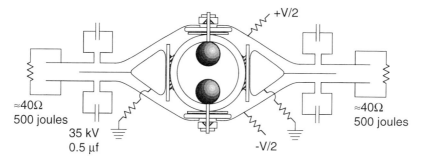

Fig. 12d-22 Sketch of gap and diagrammatic layout of capacitors. The Figure shows exaggerated insulation thickness.

for the 60 cm line bank gap. However, the rods would be made 10" long and be bowed up slightly over the centre 4" or so, so that the breakdowns occurred in the central region. I would suggest 6 connections to either rail, but you could get away with 5 if you wanted to. The inductance of such a rail gap, fed from both sides, would then be approximately

L_{spark} ~ 4 nH

L_{feed} (based on a 16 cm width) ≈ 5.3 nH

L_{stubs} ~ 1 nH

$L_{gap\ body}$ ~ $\frac{12.6 \times 14}{16} \cdot \frac{1}{2}$ ~ 5.5 nH.

The reason that 16 cm width is taken is that this makes the L_{feed} ~ $L_{gap\ body}$. However, it pays to make the rail and the external copper sheets wider than this, as some current flows out there and so reduces the inductance below that calculated in the above simplistic way. This gives a total gap inductance, fed from both sides, of about 16 nH, or maybe a bit less.

If the condensers have an inductance (suitably tagged, of course) of 10 nH each, the four capacitors add 10 nH, and if the two "inductor" feeds are arranged to have an inductance each of a 20 nH, this gives an all up inductance around 40 nH. If you can't get four 10 nH, 35 kV capacitors, then you surely can get eight 20 nH capacitors and put them in pairs. This may be necessary anyway, to get the required current rating (dealt with below).

The total line capacity would be about 120 nF and I would suggest a total bank capacity of 500 nF, to give you a bit more ringing gain than we had. Thus each capacitor (if you use 4) will need to be 0.5 microfarad and I would suggest you might get 40 kV condensers, to have a bit in hand. So a suggested requirement for the capacitors is four of 40 kV 0.5 microfarad, each of whose inductance is 10 nH or less and whose ohmic resistance should be less than 40 milliohms (not a very restrictive requirement). A further requirement for the capacitors is that they should be capable of giving 40 kiloamps peak current each (reached after the start gap closes) and they should have a good life with something like 60 per cent reversal. These requirements should be obtainable, but if they are difficult they can be halved (or doubled, depending on what parameter it is), by using eight capacitors.

In the charging circuit the lumped inductance will be 40 nH and the C_{eff} about 100 nF. This gives a $\sqrt{LC_{eff}}$ of about 63 ns, a little longer than that used in the above tests. However, I do not feel the increase will be important. However, the ringing gain should be significantly better than in our system and at 30 kV I reckon (if the wide lines are assembled no worse than ours) that you should get a peak gain of around 1.30, compared with 1.13. This should enable the system to work at rather less volts on the bank, or drive the line higher, or fire a bit earlier in the waveform, before peak and win back the difference between 63 and 50 ns in the charging time, if this proves necessary.

As is shown in Figure 12d-22, resistor chains are placed across the two outputs, in order to discharge the system reasonably quickly in the event of the start gap not going (very unlikely at high volts). These should have a value of about 40 ohms each and be made of chains of 8 resistors in series and absorb about 500 joules each, at worst. If you were using our old 1 watt resistors, you would need about 100 total on each side.

If the system is plus and minus charged, the flux excluder on either side of the gap can be earthed via a 10 kilohm sort of resistor chain and the earth plane can extend past the capacitors, as is shown in the sketch. This helps to isolate the HT tabs on the condensers from each other and makes the tracking problem a ± 35 kV not a 70 kV one.

The bank stores some 1.2 kilojoules at 70 kV and therefore is a dangerous one if contact is made with the DC connections (you can kill yourself with this energy). However, all the mylar wrapping tends to make it very difficult to get at the DC but an automatic dump system should be arranged to short out the bank and power pack when anyone approaches it. The energy that can be collected from the lines when these are operating is much less. The human body is about 200 ohms at this voltage and frequency and in normal operation at like 70 kV the volts are around for about 1 microsecond, so that it is possible to get about 30 joules from it. This will not kill anyone, but it would be a nasty shock which one would not wish to repeat.

HIGH VOLTAGE DESIGN CONSIDERATIONS

The feeds to the bank, which also act as the tuning inductors, need to be about 20 nH each and hence want to be of width about half their length, when they are separated by 1 cm. Figure 12d-23 shows a possible location of the bank and the feed points to the line. The output to the RHS line needs to be split in two, as shown, in order to feed the 10 cm of line between the output gap and the gas cell. In our experiments, the 0.1 ohm load resistor charged this capacity adequately quickly, but it will be necessary to balance this line too, to keep the prepulse on the gas cell down to whatever is necessary (presumably a few kV prepulse can be tolerated, especially as it is only on for about 100 ns). Figure 12d-24 is included to show the origional layout.

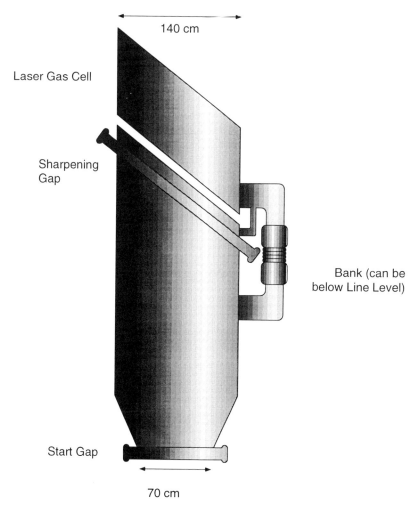

Fig. 12d-23 Top view of possible layout.

The Lines

The construction of these is as for the 60 cm line. The main differences are not that the sharpening gap is slanted and that the start gap is wider. Figure 12d-24 gives the original arrangement. Although I would expect the rise time of the output from the start gap to be better than in our tests, I would not rely upon this much, hence the lines are about 1 metre longer overall than the system we made. Anyone building this wide line might chance their arm and decrease this some, but if the pulse rise time from the start gap is no better than in our tests, they will then have to increase the thickness of the LHS line by using two sheets of polyethelene so as to raise the impedance at the switch and regain the rise time required.

Difficulty may be experienced in getting wide enough sheets of mylar; however, if the slimmer sheets can be joined by overlapping about 2" and twinsticking over, say, an inch, then they should be OK. Such joints should be staggered across the width. Hopefully it will be possible to get the 1/32" polythene in a wide enough roll, otherwise, rather less desirably, this can be overlapped. Clean the sheets with acetone or other solvent, otherwise the twinstick won't, and put a double layer of twinstick, i.e. one on each sheet.

Start Gap

I would make the rails of the start gap about 70 cm long, with the perspex body about 3 1/2 feet wide. The scaled up length of the rails of our gap would only be some 35 cm long over the uniform field part, so such a length should usefully reduce the gap's inductance. I would also decrease the gap from 0.72 cm to about 0.6 cm. This raises the pressure in it and avoids operation near atmospheric pressure, where the jitter is rather large (see edge plane note). It should perhaps be considered whether to make the

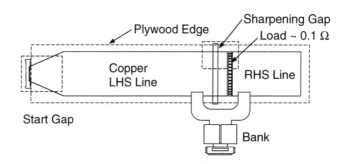

Fig. 12d-24 Top views of line showing original locations.

mylar over them, so that the edge can be removed for resharpening, should this become necessary after a lot of shots, but normally the mylar overwrap should cover the contact points. Because of the greater width of the start gap, the side angle of taper into the line is not very much bigger than in the 60 cm line, but again the line insulation should be reduced in thickness modestly, as you approach the gap, in order to keep an approximately constant line impedance.

Sharpening Gap

This will be about 2 metres long and inclined at roughly 45° to the axis of the line. I think the spacing should probably be kept at 0.6 cm or thereabouts. It would probably be simple to make the 1/4" OD rail in one piece, but the knife edge probably should be made in two bits with the ends rounded where they nearly join in the middle. The perspex tube we get has a very good constancy of diameter and is cast, which is probably better than extruded tube. However, if there is a slight taper in its diameter, packing behind the 1/4" rod can be used to maintain an adequately constant spacing. Because the start gap will have a better rise time, in all probability, and can certainly be better made, it is likely to have some 400 channels in it. It is, of course, to be slanted and hence a 55 cm length of it will be feeding not 0.65 ohms but more like an ohm. Using these numbers, it is calculated that for p_2 pressures greater than 40 psig, the rise time on the far side of the sharpening gap should be 3 ns, falling a little more with increasing pressure.

The small wavelength ripple should be less than 0.1 ns, but if there is a change of amplitude across the front of the pulse moving up the line, then there will be a time difference across the width, after the sharpening gap. Thus if there were a 5 per cent difference in amplitude of the pulse arriving at the sharpening gap, the time difference one side to the other would be of the order of 1 ns for the expected value of T at firing time of about 20 ns. Such a difference could come about as a second order effect of their being a voltage gradient down and across the line from the feed point. Alternatively the extra distance the wave has to travel to reach the late end of the sharpening gap might conceivably produce a few per cent difference. However, if this were to happen, it could be corrected by changing the angle that the sharpening gap makes to the axis of the line. In order to check whether this is necessary, the time of arrival of the pulse beyond the sharpening gap must be measured across its width. This measurement needs to be made to a fraction of a ns and is quite a challenging one, as it has probably to be made on a single pulse. It is possible that a signal taken from the start gap can be used to trip the 'scope with adequate constancy, so as to measure the pulse arrival time at various points across the line on successive shots, but I rather doubt it. If not, the polarity must be measured on a single shot and then mixing the signals from a fixed central point and a variable point across the width would be the best way, I think.

CONCLUDING COMMENTS

I am afraid that by this time most of the readers of this note will be rather tired of its detailed nature, purely apart from its other failings. I whole-heartedly agree with them. On the other hand, anyone who tries to build the 140 cm wide system without previous experience in the modern field of high speed pulse voltages, may wish it was twice as long (some people are masochistically inclined). To them I must apologise for any numerical or other inconsistencies and for all of the things that are either left out or opaquely explained.

The note gives a detailed account of the construction and operation of the test 60 cm wide line and suggests the outline design of a 140 cm line, which is pretty confidently expected to provide a short circuit current of 300 kA rising in about 3 ns. The very best of British luck to anyone who wishes to try.

ACKNOWLEDGMENTS

Ian Grimson very ably assisted me in the construction and testing of the 60 cm wide line and my warmest thanks are extended to him for this help.

I am, as usual, eternally indebted to Mrs. G.V. Horne (or at least until the next note) for her untiring labours in taking my terrible scrawl and turning it faultlessly and quickly into elegant typescript and English. My one real worry is that my unbelievably awful spelling will one day undermine hers, which is perfect, by some form of contagion.

HIGH VOLTAGE DESIGN CONSIDERATIONS

COVERING NOTE ON NITTY GRITTY DETAILS

Twinstik Double Sided Transfer Tape comes from Evode Ltd., Stafford, Staffordshire, and the reels are 60 feet long, 3" and 1" wide. The cost of the 3" used to be £3/4 reel. They also make Type 528 impact adhesive, but any impact adhesive will do.

Simplex (Rapid) clear is made by Dental Fillings Ltd., of London N.16. It comes in two components, a powder and a liquid. We get the powder in 5 lb cans and the liquid in 40 fluid oz cans. Do not get it in small quantities: it is much cheaper in bulk. The method of mixing I use is to add liquid to the powder until it becomes pretty fluid (a rapid transition), then stir for about 1 minute. At this time (dependent on temperature) the powder absorbs the liquid and it becomes much more viscous. A little more liquid is added until it is about motor car oil consistency. For pouring this can be carried out for the next 4 minutes or so (UK room temperature), after which the mixture gets more and more viscous. At 5 or so minutes it is good for daubing on surfaces, at around 10 minutes, puttying, and a couple of minutes before it goes off it can be worked like clay. It is exothermic when it sets and in bulk (> 100 cc) can foam as it goes off. Compact volumes greater than this can't really be cast. It is useful to make it up in a sturdy polythene beaker from which the set waste can be extracted. If the same beaker is to be used quickly again, wash it out with water and dry, otherwise the free radicals left accelerate the setting process.

In order to reduce the chances of crazing of the perspex during the setting (which looks unsightly but does not have much of a weakening effect), dampen the surfaces near the joint with a wet Kleenex; also in tubes blow air through the tube (gently at first) and put a damp Kleenex inside near the joint. To join a fence onto a gap body, wait 3 or 4 minutes after mixing and in a couple or three layers build up a ~ 1 mm, ~ 7 mm wide, layer on the perspex sheet, blowing between a little, to form a skin. Then press the tube onto the fence, gently, and smoothly. If you push hard down you will have to stand there holding it down for 15 minutes while it sets, otherwise bubbles will suck in along the joint. I usually file a slight flat along the cylinder where the fence is to join and roughen the joint area with emery paper. Incidentally, extruded perspex tube has a lot of strains in it and a vapour crazing may be very difficult or impossible to prevent. In this case the only way out is to anneal the tube, but this may well distort it. Heating in water at ~ 90°C for 1 hour may be the best way, if you have to do it, Simplex is a very useful tool for high voltage engineering in general.

Capacitors of the right sort of characteristics (but not necessarily completely filling the specifications) are made by the following firms, among others:

Maxwell Laboratories Ltd.,
P.O. Box 20508, 9244 Balboa Avenue, San Diego, Cal. 92123 USA

Also

Capacitor Specialists Inc.,
Escondido, Calif., USA

(UK Suppliers:
Hartley Measurements Ltd., Kent House, High St., Hartley Wintney, Hants, UK)

and

Hivotronics Ltd., Wella Rd., Basingstoke, Hants, UK

Capacitor Specialists quote £110 for a 0.6 μF 35 kV < 15 nH condenser. Unfortunately the cost is hardly dependent on capacity for such smallish condensers, so buying these and putting them in parallel is quite a bit more expensive. I would ask what is the charge for a capacitor with the required characteristics and what are the values for the same parameters of the nearest capacitor that they routinely make. They will all quote at least 12 weeks delivery, as none of they carry stocks, and may well take longer.

While I have specified a set of parameters, these can mostly be eased without too much trouble resulting. I would suggest the internal resistance can rise by a factor of 2 or maybe a bit more, but should easily be met. The capacity can be ± 20 to 30 per cent. The voltage preferably should be 35-40 kV. The maximum current rating should be close to or higher than that quoted, or trouble may well set in. The inductance is the major one; 15 nH is probably OK, but keep the tabbing good and broaden out the copper sheets as quickly as possible as they leave the condensers. Approximately 1 mm of mylar or polythene is all that is needed to hold off the volts, except for tracking troubles, but liberal use of Twinstik and folding of layers of mylar around the bent up copper edges and blotting paper should do OK. Test the set up (or a bit of it) without the capacitor (model the face of this) first and check that it doesn't track up to at least 50 kV and preferably 60 kV, for 35 kV use.

If trouble does set in, the bank can be put in a polythene bag and this filled with freon. This usually raised the tracking voltage by like 1 1/2 times, but it is difficult to make a gas-tight bag. Test freon level by either lowering a lighted match into it, or by floating a balloon on its interface. Left overnight, the balloon may sink to the bottom, even in the unlikely event that the freon is still there, as some sort of diffusion seems to occur.

Well, good luck if you go ahead. Please do not hesitate to write with questions and, if possible, a visit by someone for even a week here will save a lot of money and time.

J.C. Martin
19 February 1973

Section 12e

NOTES FOR REPORT ON THE GENERATOR 'TOM'*

J.C. Martin

INTRODUCTION

The system consists of a 12 stage triggered gap Marx generator which produces the main high voltage gain in the system. The inductance of this Marx is reasonably low, being 2.3 microhenries, but if it fed the 140 ohm load directly, the 10 to 90 rise time would be about 36 nanoseconds and far too slow. In order to speed up the rise time of the pulse, the normal peaking capacity circuit is used. This basically establishes the required current in the Marx inductance at the time the voltage on the condenser has reached approximately that of the erected Marx. An output gap then closes and the current diverts from the peaking capacitor to the load. In order to produce a fast rise time to the output pulse, the peaking capacitor and the output switch must have a low inductance and the feed from them to the load transmission line must match the impedance of this. The external conical transmission line is attached to the generator at the output face.

Figures 12e-1 and 12e-2 show photographs of the system. In the side view, the Marx generator is on the left, the peaking capacitor is in the middle and the output gap is in the top strip line and has the perspex fins attached to it. The output face is on the right of this view. Because of the fact that the generator was built and operated in an ordinary laboratory (floor area 23 feet x 14 feet), a wide angle lens had to be used in taking the pictures, which accounts for the distortion in them.

---

* SSWA/JCM/735/407 (Rev.)
   No Date

Fig. 12e-1 "Tom" generator.

Fig. 12e-2 "Tom" generator, side view.

HIGH VOLTAGE DESIGN CONSIDERATIONS

The requirement for the output pulse was a rise time of about 7 ns. The amplitude of the pulse was required to be about 650 kV maximum, but it was decided that, if possible, the system should be tested up to 1 MV. This was to ensure reasonable trouble-free use in operation. It was also considered desirable to make the generator capable of feeding lower impedance transmission lines than 140 ohms, since it was possible that these might be required at a later date. The result of this decision was that the generator feeding a 140 ohm load could produce pulses of full time width at half height of up to 600 ns and the system was designed to produce a range of pulse widths.

The design and performance of the individual main components will now be described and then a summary given of the system performance.

MARX GENERATOR (Circuit Considerations)

The Marx consists of 12 stages, each stage containing a 90 kV capacitor of 70 nanofarad capacity. These capacitors are plus and minus charged, so at each end of the Marx there is an untriggered half voltage gap. The other 11 gaps are all triggered and all the gaps are contained in a single pressurisable air-filled column. This, of course, enables the breakdown voltage of the gaps to be simply altered by changing the air pressure.

In order to have reliable operation, it is essential that the Marx operates over a wide range of voltages with a constant pressure of air in the gaps. This is achieved by coupling the trigger resistors across a number (m) of stages back down the Marx. See Figure 12e-3, where the schematic layout is shown for m = 3. When gaps n and n + 1 have fired, the electrodes of gap n + 2 have a large difference in potential from the trigger electrode of the gap. The trigger resistor discharges the stray capacity from the trigger electrode to the main gap electrodes as these erect past it, effectively producing a 2 1/2 V pulse on it, where V is the stage voltage. This causes vigorous triggering action of the bottom half of gap n + 2. In many applications this is sufficient to give an acceptable triggering range, as after the bottom half of the gap is fired, the full gap volts, V, appear across the top half of the gap from the sharp edge of the trigger electrode to the top electrode, causing this part of the gap to break rapidly. The time of breaking of this half and the bottom half of the gap can be calculated from the edge plane breakdown relations for air, which are

$$F_+ (dt)^{1/6} = 22 \, p^{0.4}$$

and

$$F_- (dt)^{1/6} = 22 \, p^{0.6}$$

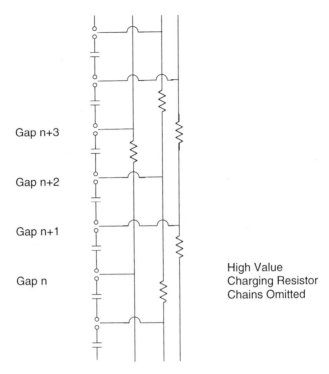

Fig. 12e-3 Schematic of triggered Marx coupling over three stages.

as given in "Nanosecond Pulse Techniques", J.C. Martin, SSWA/JCM/ 704/49, which is referred to in future as NPT. In the expressions, F is the voltage in kV between the point or edge and the plane divided by their spacing d in centimetres; t is the breakdown time in microseconds; and p is the pressure in atmospheres. As can be seen from these relations, as the air pressure is increased, the required breakdown mean field rises at a rate roughly equal to the square root of the pressure increase. The self break of the gap themselves, however, increases more like the pressure ratio (actually a little less than this ratio - see NPT). Thus as the gaps are pressurised, the time of triggering at constant fraction of breakdown volts decreases and also the triggering range increases somewhat. However, to obtain the fastest erection (and the largest triggering range - the two go hand in hand in a well designed system), it is desirable to ensure that more than V be applied to the top half of each gap after the bottom half has been fired with a 2 1/2 V trigger pulse. This can be achieved by the method schematically shown in Figure 12e-4 (again for a m = 3 layout). In this diagram the impedances shown between every third condenser can

HIGH VOLTAGE DESIGN CONSIDERATIONS

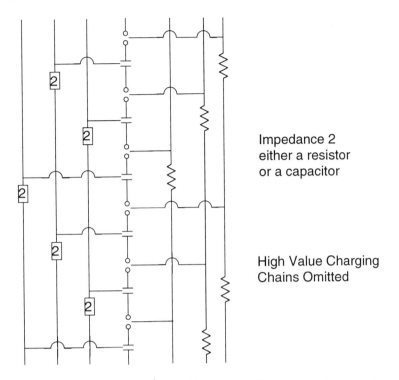

Fig. 12e-4 Wide range triggered Marx coupling over three stages.

be either stray capacities built into the structure or the charging resistors. When the capacities, condenser to fourth condenser up, are very much bigger than those across the gaps, the pulse voltage applied between the trigger and the top electrode has a value of 2 1/2 V; however, real life values mean the pulse is somewhat less than this, but it can be made comparable to the trigger pulse applied to the bottom half of the gap, and an optimised system with very good triggering range results.

The triggering method described above works regardless of strays to ground, which the original Marx depended on for its operation, and indeed works as a propagating system in an indefinitely long Marx. The method is obviously a general one and can be used with m having any value greater than 1: the higher the value of m, the greater the triggering range. However, the number of resistors and complexity of the system obviously increases, as well as the erected voltage along the trigger resistors and in general it is not necessary to use a value of m greater than 3.

In the system used in the TOM Marx, the layout favoured an m = 2 arrangement and it was also possible to build in good stray capacity coupling across two stages simply, so the specification of the Marx is m = 2 triggered, with capacity coupling. As will be described below, over most of the working range the Marx would trigger down to 35% of its self break and had erection times of less than 100 ns over eminently useable operating ranges.

The reason a large triggering range is very desireable in a reliable Marx is that with a large number of spark gaps in a big multistage generator, the statistics of gap breakdown mean that it is undesireable to operate at even 80% of self break of the collection of gaps. In addition, small fluctuations in voltage setting, pressure in the gaps, the presence of hot degraded gas in the column from previous firings, all combine with other factors to make it desirable to operate well below the nominal self-breakdown voltage of the set of gaps and to have a good range of gap gas pressure to operate the Marx in without significant alteration of its erection time. This the system described above, which was originated at AWRE, provides simply and economically.

The full circuit of the Marx is given in Figure 12e-5. The charging resistors which link across each gap tend to discharge the helpful strays and are made fairly large in value, while the trigger resistors have a more modest value to speed the triggering action. During a long decay time firing the trigger resistors take most of the energy (up to 3 kilojoules total at maximum test voltage) but in the case of a self break of one gap in the Marx which does not lead to full erection, the charging resistors absorb the energy. This deposits more energy into those resistors which are across the gap which self breaks and each of these can get about twice the energy in a single capacitor, i.e. about 1/2 kilojoule each. Normally the Marx fires and transfers all but a few percent of its energy into the output load, but during tests without the peaking capacitor it has to be designed to absorb this energy in the copper sulphate resistors. The discharge time of the Marx is then about 15 microseconds.

The use of plus and minus charging in the Marx has several very useful consequences. The first is that the power pack is smaller and cheaper. A second advantage is that the charging leads can be much smaller and the corona problems at plugs, etc. are much eased. A major advantage is that the midplane trigger electrodes are at earth and this avoids DC blocking trigger capacitors. There is only one disadvantage associated with this arrangement and this is the bottom half gap and its triggering. The trigger pulse was made large enough to trigger the bottom half of the first triggered gap up and to overvolt the bottom half gap adequately provided there is no attenuation in the circuit. In the system as built there is a large resistor in series with each output line of the power pack but there are fairly large capacities from the two HT cables to earth and these reduce the trigger pulse actually appearing across the bottom half gap. There is a simple solution to this problem and that is to add more resistance at the base of each charging column. However, in the particular system described here

HIGH VOLTAGE DESIGN CONSIDERATIONS

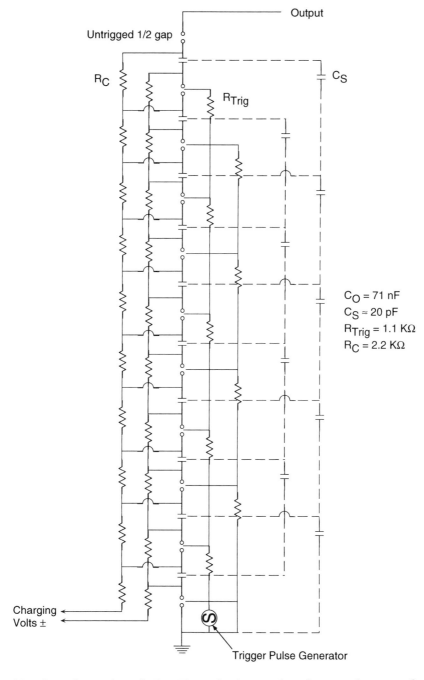

Fig. 12e-5 Schematic of Tom Marx (triggered m=2 capacity coupled).

it had been found necessary to locate the safety dump resistor back in the power pack. Because of this it was deemed unsafe to add carbon based high value resistors in the charging lines at these points. Consequently at high values of pressure in the spark gap column the trigger pulse plus the DC levels are not adequate to fire the bottom half gaps and hence the triggering range, after improving as the gap pressure is raised, starts to decrease again slightly at the highest pressure levels, due to too a small applied trigger pulse with the particular circuit evolved for this Marx.

Inductance of the Marx

The construction of the Marx was arranged to reduce the inductance as far as was practical without causing additional assembly problems. The capacitors (made at AWRE) have an inductance each of about 20 nanohenries at their output face, when properly tabbed. There is a 10 cm feed from the face of the capacitor to the spark gap column face, giving rise to a further 20 nH or so. The inductance of the spark gap column was approximately 30 nH for each gap. Thus the expected inductance of the 12 capacitors feeds and gaps was about 800 nH. To this has to be added the inductance of the central spark gap column viewed as a rod of about 5 cm radius to a plane up the output face of the Marx. This gives rise to an inductance of about 8 nH/cm, giving an additional 800 nH as the Marx is 100 cm tall. There is an additional 200 nH in strip connectors from the top and bottom of the Marx to the shielding bungs. Thus the inductance is roughly calculated to be 1.8 μH when referred to a wide plane down the output face of the Marx. In actual use the peaking capacitor is located some 20 cm from this face and is fed down converging 40 cm wide feeds. This gives rise to a further 400 nH or so. Measurements made with a shorting metal strap, copying the geometry of these feeds and the peaking capacitor, gave an inductance of the Marx of 2.3 micro-Henries, in better agreement than the rough calculations would be expected to give.

The reason for performing the crude estimations is that it shows that the capacitors account for only 10% of the total working inductance, and the spark gaps some 15%. Thus it is not worth improving these, since even if they could be reduced to nothing, the improvement would be small. However, the inductance of the central core is quite considerable and if this region could have been made wider or shorter, significant improvement would have resulted. However, the working gradient of the Marx under maximum voltage tests in air was already over 1 million volts per metre, so very considerable extra work would have been involved to, say, double this gradient. Equally, a completely new layout would have been necessary to increase the effective electrical width of the central region. Indeed, to get any real improvement, rail gaps nearly the full width of the Marx would have been necessary to produce a significant reduction by this approach.

HIGH VOLTAGE DESIGN CONSIDERATIONS

During the ringing tests which gave the Marx inductance, the internal resistance was measured from the damping of the waveform. The measured value was 6.3 ohms, which is only 20% greater than that to be expected from the 12 capacitors in series, which was 5.3 ohms. In view of the large number of plug and socket and other joints in series, this is a very fair value.

DC Tracking

The spark gap column and the feed from the face of the capacitor to it were mocked up and tested to 120 kV, where the maximum voltage to which the capacitors could be taken was 100 kV. The capacitors themselves could not be overtested significantly and after a series of experiments the DC tracking voltage was raised to over 100 kV. Each condenser, after the rather elaborate tabbing and insulation wrapping, was tested to 100 kV. Five of the 14 condensers tracked during these tests, all at a joint in a thin perspex sheet which had been damaged during an over-enthusiastic filing operation. These were easily repaired and retested, when they passed the high voltage tests without further trouble. The DC tracking across the capacitor output face probably represents the weakest link in the whole system, but apart from units suffering from mechanical damage, no unit tracked during the 100 kV test. However, under conditions of high humidity, trouble might be experienced with corona currents at the test voltage, but at normal maximum operating voltages no trouble should be experienced.

Spark Gap, DC Breakdown

The spark gap electrodes were 1" diameter spheres set in a 7.5 cm OD perspex tube of wall thickness 0.5 cm. The physical gap between the electrodes was 1.22 cm for the triggered gaps and the half voltage untriggered gaps had a spacing of 0.53 cm. This latter spacing was arranged to give a breakdown voltage for a half gap of 0.6 of that of the triggered gaps.

For the triggered gaps the field enhancement factor for 1" OD gaps at 1.22 cm spacing is 1.35, giving an effective uniform field gap separation of 0.90 cm. Using the treatment outlined in the NPT note, the breakdown field and hence voltage of a single gap was calculated as a function of pressure and the values are given in Table 12e-I. The breakdown voltage of a set of 11 gaps is of course less than that of a single gap and taking a slightly optimistic value for the standard deviation of a single gap of ± 3%, the breakdown voltage of the set comes out at about 92% of that of a single gap. Also given in Table 12e-I is the observed values of the self break of the Marx at 1, 2 and 3 atmospheres pressure absolute. The self breakdown voltage was not determined at higher

Tabel 12e-I
Breakdown Voltage

p (atmospheres)	F (kV/cm)	$V_{single\ gap}$ (kV)	$V_{self\ break}$ 11 gaps (kV)	$V_{observed}$ (kV)
1	40	36	33	33
2	71	64	59	58
3	101	91	84	81
4	129	116	107	

pressures as it was felt undesirable to do this intentionally at over the working voltage.

The spark gap column and connectors were pressure tested to 88 psig, i.e. 6 atmospheres, and could certainly be used to higher pressures if necessary. In other Marx generators at pressures over 60 or so psig the standard deviation of breakdown of a single gap begins to increase significantly, so the breakdown of a set would not necessarily be expected to increase completely in line with the calculations. In addition, the breakdown values are those obtained after a few conditioning self breaks at 1 atmosphere pressure when the system has been left unused for a time: this is to clean off the whiskers which grow on the ball bearing electrodes.

Pulse Breakdown

The detailed breakdown results will be given later in this report after the section dealing with the mechanical construction of the Marx. However, a few words now may be of use in explaining some of the physical features built into the Marx.

Pulse breakdown and tracking within the Marx is a major preoccupation of a designer building a high gradient system. In particular, tracking down the copper sulphate resistors can cause trouble. These can be made longer by spiralling them around in space but this means effectively that they have to be made of flexible tubing and the water diffuses through the relatively thin walls, even when the many joints necessary are made so that they do not leak. Solid carbon composite or metallic film resistances can obviously be used but these have to be specially developed to take the energy which may be dumped in them, not to mention the high voltage pulses they have to survive. They also in general have to be potted to resist the pulse voltages which during the erection phase can reach significantly more than the stage voltage. As

HIGH VOLTAGE DESIGN CONSIDERATIONS

such, it was decided that it was much cheaper and quicker to stay with copper sulphate liquid resistors in perspex tubes, with which a lot of experience had been obtained in the past.

In order to minimise the pulse tracking of these components, it was decided to pass these through sheets of perspex which were located between the capacitor stages and to seal the penetrations with a cold setting acrylic cement. Indeed, in the case of the trigger resistors, these were formed on the sides of the spark gap column to reduce the number of connections. The obvious disadvantage of this process was that it would make maintenance of the spark gaps and resistors very difficult. However, it was felt that the necessary gas and liquid seals could be made very long-lasting. In addition, a six stage version of the Marx had been crudely and quickly made, nearly a year before, as a test bed and this had been left lying around unattended and unloved. On examination at the end of the period, the liquid resistors had essentially remained as they were originally and the spark gap performance was unaffected. Thus the decision was taken to make the Marx main frame largely unmaintainable. The capacitors could be simply unplugged and changed and if the Marx were to become damaged, it could possibly be repaired but at worst could be rebuilt in about two weeks.

The subsequent breakdown overvoltage tests showed that the solution adopted had indeed suppressed internal tracking completely and hence the chance of damage to the Marx from this cause was essentially zero at normal operating voltages. Mechanically the resulting structure was also very robust, while still being easily manhandable.

MARX GENERATOR (Mechanical Construction)

Figures 12e-6 and 12e-7 show the side and top views of the Marx. The most important part of the structure is the central core, comprising the spark gap columns with their attached trigger resistor columns, and the two high impedance charging resistors. Figure 12e-8 shows a cross-section view of this region.

The pressure vessel for the spark gap column is made of 7.5 cm OD perspex tube with 0.5 cm wall thickness. The spherical electrodes are made from 1" OD phosphor bronze ball bearings and are spaced 3" apart up the column. These are brazed onto countersunk 2BA bolts. These penetrate through the pressure column walls and are tightened down so that the sloping faces of the countersunk heads seal against the edges of the 2BA clear holes drilled in the wall. The sharp edge of the drilled hole is filed off on the inside of the perspex tube, to avoid too high a stress cracking the perspex locally. A second pressure seal is obtained by 'simplexing' (Simplex is the two-component cold setting acrylic cement

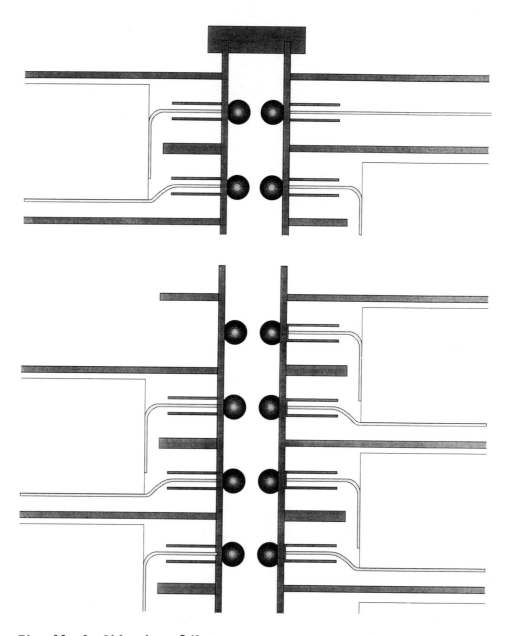

Fig. 12e-6 Side view of Marx.

HIGH VOLTAGE DESIGN CONSIDERATIONS 501

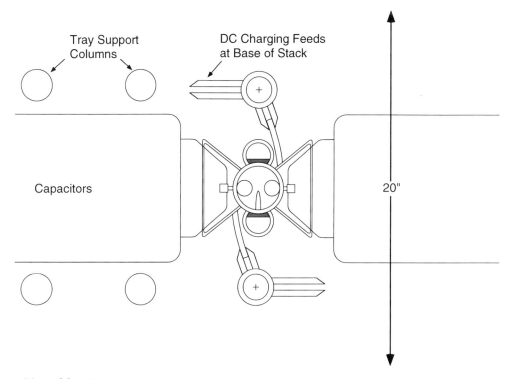

Fig. 12e-7 Top view of Marx.

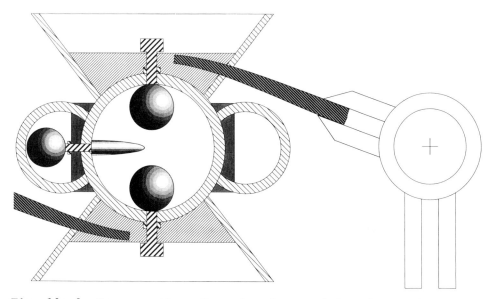

Fig. 12e-8 Cross section of spark column and charging resistors.

mentioned above, and has become a verb locally because of its very extensive use in all the AWRE high voltage generators) a seal on the outside of the tubes. The same simplex joint holds on the wedge shaped boxes into which the capacitors plug. The electrical contact between the capacitor feeds and the spark gaps is made by wander plugs and sockets, the plugs being 2BA tapped and screwed onto the ends of the bolts holding the ball bearing spark gap electrodes. The charging leads are stripped 50 ohm cable inner sheathed in PVC tubes and are also simplexed at both ends. The trigger electrode is made from 3/8" rod which at the triggering end is cut to a wedge shape and then brazed onto a countersunk 2BA bolt and sealed as the gap electrodes are. All the copper sulphate liquid resistor electrodes are made of copper sheet which is roughly cut into flattened circular discs and brazed to 2BA nuts, in the case of the trigger resistors, and 2BA countersunk bolts in the case of the charging columns. The trigger resistor columns are cut from 4.0 cm OD perspex tube and simplexed all along the length of the spark gap column, as is indicated in Figure 12e-8. Where the copper electrodes are brazed to the brass nuts or bolts, the alloy is covered with cold setting Araldite to protect against weakening at the points by local electrolytic action.

The electrodes are introduced from both ends of the charging columns, which are made out of 2" OD perspex tube, by a simple rod tool and of course have to be made so that this can be done. There is a charging column electrode of each polarity at each spark gap position, but the trigger electrodes are introduced alternately from each side, each trigger resistor linking every other gap. The spark gap column and the two charging resistors penetrate all the 13 perspex sheets forming the trays on which the capacitors are placed. At each penetration a simplex seal is made by casting the cement around the columns. The trays are made out of 3/8" perspex and are approximately 20" wide by 22" long, as is shown in Figure 12e-7.

In addition to being held at the spark gap column and charging resistors, these trays are supported by 4 columns of 4 cm OD perspex tube, simplexed in place in pairs on either side of the capacitors.

The capacitors consist of 2 parallel sets of condenser windings, each set comprising 9 pads in series, each pad being about 0.3 μF capacity and rated at 10 kV. This assembly of windings is cast in an epoxy block of dimensions 10" long, 8" wide, and about 4 1/2" tall. The tabs of the pads are joined near the bottom surface of the capacitor and the strip connector is returned close to this, to keep the inductance low. The output feeds of the capacitors are wrapped in several layers of Mylar and the output connection nearest the spark gap column is cast in perspex, to increase its tracking voltage. A detailed description of the insulation and

HIGH VOLTAGE DESIGN CONSIDERATIONS

connections of these capacitors would take several pages and probably be incomprehensible without having a unit in front of the reader. Also these capacitors are made at AWRE only, so it is unlikely that anyone else would use them.

Figure 12e-9 gives a sketch of the end facing the spark gap column and shows the area of greatest difficulty, where the strip line connectors separate and where 90 kV has to be supported across a small distance without tracking. Two sheets of Mylar are folded and break this region into a number of separate spaces. These Mylar sheets are stuck to themselves and the perspex-covered face of the capacitor with Twinstik, a double sided sticky tape made by Evode Products. This, too is very extensively used at AWRE and has given rise to the locally used verb "twin stuck".

The trigger pulse transformer is located under the second capacitor in the Marx, the output of the trigger pulse generator feeding the bottom of one of the trigger resistor chains. The bottom of the other trigger resistor is joined directly to the earthy output of the Marx.

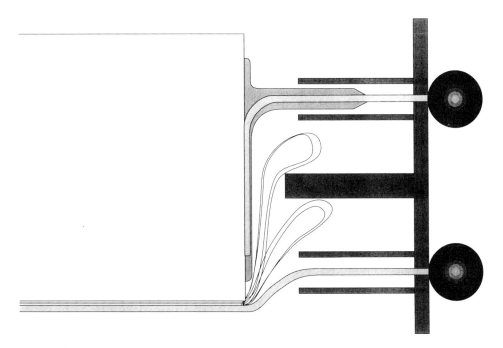

Fig. 12e-9 End of capacitor showing some of the insulation.

There are voltage grading rings at either end of the Marx column, the one at the top being much fatter than the bottom one, as the lower one is largely shielded by the laboratory floor. The form of these high voltage shielding structures, or bungs, is racetrack in plan and toroidal in cross-section. The actual form of the cross sections is fairly complex but is based on the curves given in "Electrostatic Grading Structures", J C Martin, SSWA/JCM/706/66. The complex shapes can be quickly and cheaply made from polyurethane foam of about 3 lb per cubic foot density, which is then covered with 2 thou. aluminium foil which is twin stuck onto the polyurethane foam. The note referred to above outlines the mode of manufacture and, as an example, the two complex bungs for the Marx were made in less than a day by one person.

The top bung is some 9" thick and has overall dimensions of 34" by 50" and weighs about 20 lb. A hemispherical grading structure made traditionally would have cost a great deal and would have reached the ceiling of the laboratory. The top electrode was designed to reach breakdown at about 900 kV for a long pulse, but the presence of the laboratory complicates the design slightly. The connection from the output half gap was taken to the top grading electrode and then via an aluminium foil covered foam strip connector to the peaking capacitor.

MARX GENERATOR (Test Results)

Erection Time

A series of tests was performed with the first two capacitors in place. These were designed to measure the time from the firing of the triggered pulse gap, which switches a small capacitor into the primary of the auto-transformer (see section on trigger pulse generator), to the firing of the first part of the first triggered gap. The trigger pulse has a polarity the reverse of that of the erected Marx and initially fires the bottom half of the first triggered gap. It was intended that the pulse should then fire the bottom half gap, but, as was explained earlier, at high pressures the pulse amplitude was too low to do this at the bottom of the triggering range. The time interval between trigger generator gap to first half of a gap going was roughly constant at 60 ns and this time has been taken off the Marx erection time measurements, which, again, were referenced back to the trigger pulse gap time. This is because the trigger pulse rise time is a function of the design of the trigger pulse generator and could be decreased at will. The erection times given below are therefore the time from first part of a gap firing in the Marx to closure of the last gap.

HIGH VOLTAGE DESIGN CONSIDERATIONS

Figures 12e-10 and 12e-11 give this erection time as a function of gap volts over self break voltage of the set of gaps for various air pressures in the column. Figure 12e-10 shows the delays measured when the Marx is erecting for positive output (i.e. negative trigger) and Figure 12e-11 shows the performance for the opposite polarities. The hiccup on the curves for negative Marx output voltages has been observed in other generators and is believed to be connected with the firing of the first couple of gaps rather than with the later stage of the erection. As the curves show, there is a large triggering range and at 65% of self break, the erection times lie in the range 100 to 200 ns, which is a healthy state of affairs. In general it has been observed that jitter in the erection time is less than ± 5% of this time and hence there are perfectly usable operation regions where the jitter in erection time of the Marx is likely to be ± 3 ns, or less. Later overall system delay and jitter times were measured and will be mentioned towards the end of this report, but these times also include the jitter of the trigger circuit and hence are not as low as those estimated above. The trigger pulse used in the erection time tests was about 60 kV out of the auto transformer, for 11 kV on the capacitor which feeds it.

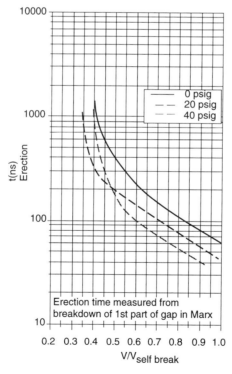

Fig. 12e-10 Positive Marx output.

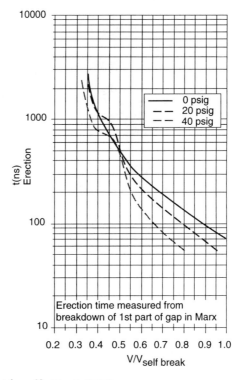

Fig. 12e-11 Negative Marx output.

Long Pulse Brealdown Tests

The long pulse breakdown of the Marx was investigated by firing the Marx into its own internal resistor chains, which gave an exponential decay to the pulse with a time constant of about 15 microseconds. These tests were more severe than the operation in its final form when the decay time would be more like 300 ns.

At about 820 kV the Marx tracked down, after a microsecond or so, and in addition, open shutter photography showed the track ran down the trays at the curved ends of the Marx and disappeared into the Marx, tracking a stage or two on the opposite side of the Marx before returning again to go to ground. While the level was close to that at which it had been designed, the mode of flash-over was considered unsatisfactory. It was considered possible that corona from the edges of the capacitors farthest from the spark gap column was charging up the perspex tray immediately under this point. This would relieve the field on the outer edge of the capacitor during the DC charging, but if it happened during the erection would enhance the pulse fields at this point, leading to local breakdowns which would then link up, propagate, and lead to final complete tracking. In order to test this hypothesis, 1/32" polythene sheet corona guards were wrapped around the ends of the capacitors and taken back along the top and bottom of the blocks. This indeed had a decidedly beneficial effect, not so much in raising the long pulse breakdown voltage as in changing the track location. With these polythene corona guards in place the Marx broke down at about 830 kV after some 4 microseconds, but now the discharge path was between the grading structures, completely missing the Marx between them. The actual location of the breakdown was where it had been expected - at the centre of the curved ends of the bungs. Although it could not be definitely established from the photographs, it is very likely that the streamer was leaving the top shielding electrode and travelling towards ground.

The curved length of the arc was 110 cm approximately, giving a mean field of 8 kV/cm. Hence the value of $F(dt)^{1/6}$ was almost exactly 22 kV/cm, which is what would have resulted from a simplistic application of the edge plane breakdown formula given above. It is in fact not unreasonable that this relation should apply, because the incipient streamers rapidly leave the shielding effect of the top electrode and then travel as if they had originated from an edge. If the edge plane relationship is assumed to apply, then the breakdown voltage for the much shorter proper pulse would be some $(50)^{1/6}$ higher - i.e. about 1.5 MV. Indeed, the Marx was subsequently tested up to 1.05 MV with the shorter proper decay time and no signs of breakdown were observed.

The case for the applicability of the streamer transit relationship is strengthened when the flash-over data from the tests on the peaking capacitor are included. As is mentioned below, this component tracked at an effective pulse voltage of a little over 600 kV with the long pulse tests, but went to 1.05 MV with the proper pulse with no trouble.

Thus there is some ground for believing that the pulse breakdown voltage of the Marx structure may be as high as 1.5 MV, but it is certainly very adequately safe for operation at 650 kV. The Marx was tracked a few times at about 830 kV and showed no damage; indeed, by visual inspection it was not possible to see where the arc had joined the two bungs and the Marx itself was unharmed by testing in this fault mode.

PEAKING CAPACITOR

Circuit Considerations

Figure 12e-12 shows the simplified circuit which will initially be used to explain the working of the peaking capacitor circuit. In the circuit it is assumed that $C_p \ll C_o$ and hence the voltage on it will ring up to approximately 2 V, where V is the voltage on C_o. Figure 12e-12 shows this and also the current in the circuit which reaches a peak value of $V/(L/C_p)^{1/2}$. If C_p is now chosen so that $(L/C_p)^{1/2} = Z$ (the output load), then at the time the voltage on the peaking capacitor has reached V, the current flowing in the circuit equals that necessary to provide V across the resistive load. So if the output gap is arranged to close at this voltage, a fast rising output pulse will result.

Another and physically more enlightening way of looking at the circuit is that current flows into C_p through the inductance and it is so arranged by varying C_p that when the voltage on it has reached V, the current in the inductance is that which is required to provide V across the load resistor Z. At this point the output switch is closed and the current ceases flowing into the peaking capacitor and diverts into the load. This picture enables a number of extensions to be made to the simplified peaking capacitor circuit.

Firstly, if the output switch is closed too soon, the current in the inductance will be nearly equal to required value, but the voltage will be too low. Thus the voltage across Z will rise rapidly to the switch closure voltage V_1 and then ring up with a period $2\pi \sqrt{LC_{eff}}$. This ringing will continue damped by the effect of Z across C_p. This condition is shown in Figure 12e-13. Also shown is the case where the output switch closes at a voltage greater than V.

A second extension is to the case where C_p is not very much less than C_o. It is clear from the above discussion that a

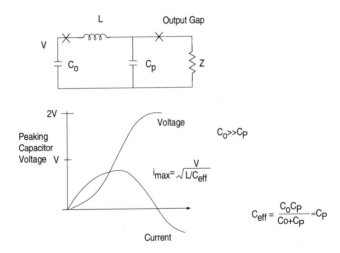

Fig. 12e-12 Simple Peaking capacitor circuit.

HIGH VOLTAGE DESIGN CONSIDERATIONS

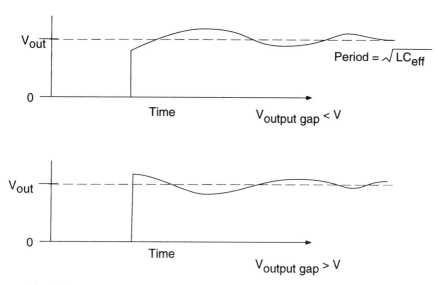

Fig. 12e-13 Load voltage waveform for output gap operation at other than V.

condition can still be produced where the output switch can be closed at voltage V_2 (less than V) such that the required current is flowing in the inductance. It is true that there is a L di/dt voltage across the inductance now, but the term is small and leads to only very slight oscillations on the tail of the waveform. Thus the peaking capacitor circuit is not restricted to negligible values of C_p compared with C_o: indeed quite satisfactory operation has been reported by Physics International, at values of $C_p \sim 1/3\ C_o$.

A further extension is to the case where there is a finite resistance R in series with the inductance (a situation which always applies in practice). In this case $i_{max} \sim V/((L/C)^{1/2} + R)$ and if the output switch is closed at a voltage of $V(Z/(R+Z))$ proper operation is obtained at a reduced output pulse level.

The peaking capacitor circuit can be extended to further cases, but in general can be made to work quite well despite finite load capacitors, extra series, parallel resistances, etc. in the Marx circuit, providing these do not take too large a fraction of the current or voltage in the circuit.

It is now necessary to consider what determines the rise time of the output pulse when the output gap is closed. This depends on the inductances of the peaking capacitor and the output switch and these must be made much less than the original L if a substantial decrease in pulse rise time is to be obtained. Basically, if L_1 is the sum of the switch and peaking capacitor inductances, the resulting output rise time is L_1/Z and providing $(L_1/C_p)^{1/2} < Z$, ringing will not occur significantly. for TOM, the 'output' inductance was like 60 nH and the peaking capacitor typically of the order of 300 pF. Thus $(L_1/C_p)^{1/2} \sim 14\ \Omega$, much lower than Z. The rise time caused by the lumped constant peaking capacitor circuit would then be fractions of a nanosecond. However, there is the resistive phase of the output switch to be included and, more importantly, the finite velocity of light and its transit time effects in the real generator. These effects are considered in more detail in a later section.

It is probably worth mentioning that while the treatment of the peaking circuit given above provides a very good description of its operation, in a real life system other effects enter. One of these was found to be the capacity of the top terminal of the Marx to ground and the circuit had to be modified to minimise the effect of this. In addition, to obtain the shortest decay times called for, the output load was not a simple resistance but a combination of inductances and resistance. This in its turn had effects on the waveform, which had to be worked out on the generator.

A further point is that in real life there is a range of values for the closing voltage of the output switch which produce acceptable waveforms. Referring to Figure 12e-13, it can be seen that if the output switch closes at a higher voltage than the matching current condition dictates, a bigger output pulse is obtained at the expense of some minor oscillations on the tail of the waveform. Consequently by adjusting the voltage at which the output gap closes, a faster 10 to 90% rise time can be obtained for a very small oscillation on the waveform and so compensated for the loss of gain introduced because of the series resistance in the Marx. Thus the final waveform is obtained by tuning up the system by practical alterations, adjustments, and experimentation dictated by the theoretical analysis outlined above, combined with monitoring of the waveform.

The final point in this section is the inductance and resistance of the condenser. These parameters were not measured but can be adequately calculated. The inductance of the capacitor depends either on the location of the external feeds or, as in this case, on its mode of use. If 20" wide straps were run down one face and a pulse injected into the condenser, it would exhibit an internal inductance of about 6 nH, calculated from the total Mylar thickness, width, and length. The same figure would apply if a very sharp rising pulse were sent down a parallel or conical strip transmission line whose axes were orthogonal to the output face of the capacitor. This is the figure that applies to its use in the present system. The internal resistance in the same method of use is about 0.01 ohm.

HIGH VOLTAGE DESIGN CONSIDERATIONS

Peaking Capacitor Constructional Details

The production of a 1 MV pulse charged adjustable capacitor of several hundred nanofarads, whose inductance must be only ten nanohenries or so, would seem to those not versed in modern high voltage techniques to be a major undertaking. It is pleasant to record that it took 3 people only 1 day to make it.

Normal high voltage capacitors are made of a large number of sub-capacitors or pads in series, each pad being rated for 10 kV or less. Previous work at AWRE had shown the possibility of making capacitors with 100 kV per pad, when working in air, and significantly higher that this working under freon. The basic dielectric used is Mylar in a number of thin sheets. There are two flash-over problems to be overcome. The first one is the overall voltage across the stacked capacitor and this is dealt with by using the cheap field grading techniques described in the section on the Marx generator. The second area of tracking problems is where the Mylar sheets, which have fields of up to 2 MV/cm in them, leave the edges of the metal covered blocks which form the spacers between the pads. This problem is overcome by splitting the layers of Mylar apart so that the 100 kV pulse volts appear across 6 isolated air-filled voids. Tracking occurs in each of these voids but because tracking is a powerful function of the voltage applied, the distance the streamers go is only a centimetre or two, as the voltage driving each streamer is only about 15 kV.

Figures 12e-14 and 12e-15 give side and plan view of the generator and show the location and general form of the peaking capacitor. The spacing metal covered blocks are 20" by 10" and are 1 1/2" in thickness. There are 11 full spacers and 2 half thickness, attached to the field shaping plastic foam top and bottom electrodes. As the total thickness of Mylar dielectric is less than 1 cm, the overall height of the stack of pads is 18", while the field shapers add another 12" or so in total.

In order for the capacitor to have a value close to that of the stacked Mylar sheets, the air films between the many layers of Mylar (~ 60 layers) must be very thin. This is most easily achieved by making the metal spacers pliant and not rigid as might be expected. The form of the spacers is 1" of plywood faced on either side by 1/4" sponge rubber. This in turn is covered by 2 thou aluminium foil which is wrapped right around the spacer block and twin stuck on. The Mylar sheets (5 to a layer, in general) and the blocks are then stacked up on the bottom field shaping electrode and the top electrode put in place and 4 lead bricks added to compress the stack. The plywood acts as rigid members to distribute the load, while the compressible layers of sponge rubber help to take up any non-uniformity in the gaps between the blocks. An additional force helping to assemble the stack of Mylar sheets

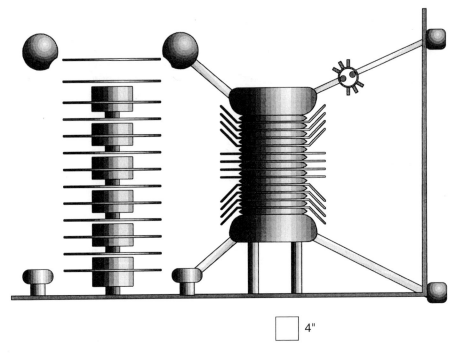

Fig 12e-14 Side view of generator.

occurs after the condenser has been pulse charged a few times. The thin films of air break down after a fashion and charge separation occurs, leaving patches of charge of opposite signs on the sheets of Mylar, and the attraction of these significantly helps to assemble the condenser. This effect has a mildly adverse aspect to it, in that when the capacity of the condenser is to be changed (something that takes about 10 minutes), the stack is disassembled prior to reassembly with a set of mylar sheets with different thickness. The separation of the localised patches of charge now acts like a static induction high voltage generator and as the sheets are peeled apart, a series of small but sharp discharges occur, usually to the fingers of the devoted experimentalists. Because of the very high surface resistivity of Mylar, it is not much help to leave the peaking capacitor for a day or two before disassembling. However, if the polarity of the Marx is reversed and a couple of low voltage shots done, the effect is much reduced and life made decidely more comfortable for the research worker.

As an example of the degree of assembly that can be achieved, one arrangement with 12 stages, each of 4 sheets of 5 thou Mylar, would have had a capacity of 500 pF when completely assembled. After weighting down and firing a few times the capacity produced by the Mylar was 420 pF after taking off the capacity of the bungs.

HIGH VOLTAGE DESIGN CONSIDERATIONS 513

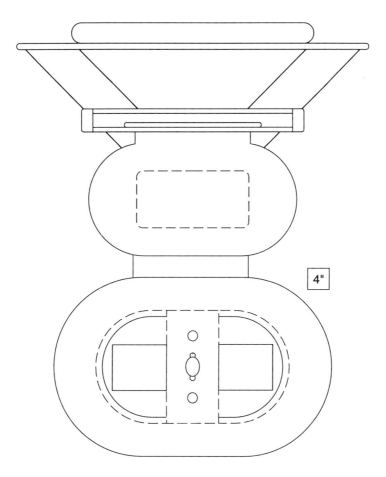

Fig. 12e-15 Plan of generator.

This capacity was measured with a standard low voltage bridge; when a high voltage pulse was applied, the capacity would have been higher, because of air breakdown. These values show that the total air layers had a capacity of some 2500 pF and indicates that the total average thickness of the air films was about 0.05 cm. This means that each air film had an average thickness of 0.3 of a thou, something that would have been extremely difficult or impossible to achieve with rigid metal spacers.

Various thicknesses of Mylar were used to adjust the capacity, but the final version had 4 sheets of 3 thou Mylar and 1 sheet of 5 thou Mylar in each stage. The top 2 sheets were 3 thou and were bent upwards, while the bottom 2 sheets were again 3 thou, bent downwards. Figure 12e-16 shows a cross-section of the edge of a couple of stages of the peaking capacitor.

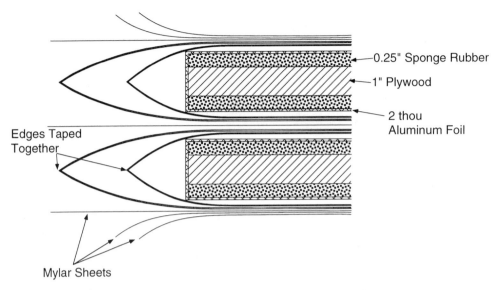

Fig. 12e-16 Edge grading for two stages of peaking capacitor.

Peaking Capacitor Test Results

Pulse Voltage Flash Over

   Once again, long pulse flash over tests were performed on the peaking capacitor. There were 5 layers of 5 thou Mylar in each layer and the middle layer stuck out very little more than the others. This combination flashed over at about 520 kV. The track occurred away from the rounded ends, where it had been expected, and went down the side nearest the Marx. Various aspects of the photographs of the flash over suggested that it was originating in the middle of the capacity and then going both ways. In order to counteract this, the central 5 thou Mylar sheet was extended a couple of inches all round and the top 4 central sheets were dished by folding at the edges to produce a tray with edges that sloped at about 45°. The 4 bottom central sheets were likewise treated downwards and Figure 12e-17 shows a cross-section of the final stack. In this configuration the flash over voltage was 570 kV with the long tail pulse. However, on the wave form there was also the peaking capacity ringing wave form which went up to 780 kV for a very short while. Using streamer transit theory, it was estimated that the complex combined wave form was equivalent to a smooth long tail pulse of about 620 kV, which was only 10% less than the grading electrodes had been designed for. The track was now at the curved ends of the bungs, where it was expected, and ran out over the bent up sheets of Mylar and did not get anywhere near the capacitor pads. The condenser was tracked several times, the discharge

HIGH VOLTAGE DESIGN CONSIDERATIONS

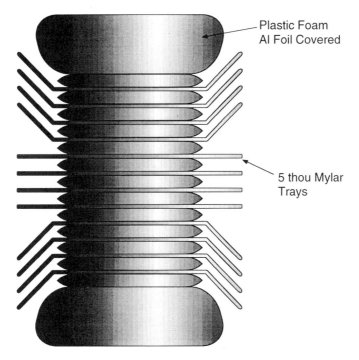

Fig. 12e-17 Cross section of peaking capacitor.

going in a different place each time and not damaging the Mylar trays at all. Indeed, the same Mylar trays were used in all the subsequent experiments.

Using the streamer relation, it was estimated that with the proper wave form the flash over voltage would be 1.2 MV, and indeed the capacitor survived an overtest shot at 1.05 MV with no signs of tracking at all. It is considered that with a re-contouring of the bungs and making them rather wider and fatter, the high speed pulse breakdown could have been made equal to that of the Marx. As it was, the capacitor was obviously entirely safe from flash over at the nominal maximum working voltage of the system, so this was not done.

Ringing Gain and Waveform

When the voltage across the peaking capacitor was first monitored, the wave form was just about what was expected - that is, a damped oscillation of the correct period. The peak of the first voltage peak was about 1.4 times the erected Marx voltage, which is significantly less than the expected value of 1.65. The latter figure allows for the finite capacity ratio between the Marx and the peaking capacity and also for the measured late time Marx resistance of a little over 6 ohm. However, this discrepancy was not unexpected, as the internal resistance of the Marx was measured from the damping of the ringing waveform over several cycles. During the first ring up cycle there is some excess resistance, due to the resistive phase of the spark channels in the gaps. In addition, air films in the peaking capacitor break down and corona occurs around the edges of the stacked capacitor stages, leading to a dC/dt term in the circuit. This acts like a resistance in parallel with the peaking capacitor and adds to the damping of the first peak. These two effects, it was estimated, could easily account for the extra damping which caused the first peak to be lower than the simple calculation would give. This meant that the peak current in the peak capacitor circuit would be too low to provide the output volts into the load resistance. However, as was explained earlier, this could be compensated for by increasing the peaking capacitor over the simple case theoretical value, which would then increase the current at the time of output switch closure.

As was stated above, the first peaking capacitor set up gave essentially the expected waveform; however, when it was rebuilt with a different capacity it became apparent that this was a coincidence, because with the different value for the peaking capacity, the ringing waveform had two periods, one of the periods being the expected one, the other being close to the original test set-up period. This second period at first sight did not appear very important, except from the point of understanding what was going on. However, it was expected that it might reappear on the output waveform, so two days were devoted to trying to locate its source. Some twenty significantly different alterations were made to the system and the monitoring in attempts to show conclusively what was causing it. These tests eliminated a number of possibilities but, unfortunately, it was not possible to show conclusively that the explanation developed was the correct one. The only suggestion left viable after these tests was the one that the capacity of the top of the Marx to the rest of the laboratory was ringing. This capacity was measured to be about 80 pF and an inductance of about 5 μH would be necessary in the ground return to the Marx: this appeared to be about what was involved. Attempts were made to move sheets of metal close to the Marx and to change the ground inductance, but the changes in these quantities that could be achieved were calculated to be too small to be seen and indeed so it proved experimentally. The approximate lumped constant equivalent circuit is shown in Figure 12e-18 if this explanation is indeed the correct one.

After two days' hard work, the problem had to be left, probably solved but not certainly proved, in order to move on to the output switch and the full system tests.

HIGH VOLTAGE DESIGN CONSIDERATIONS

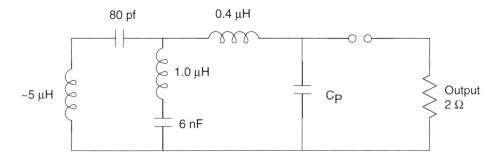

Fig. 12e-18 Approximate circuit including Marx bung strays.

OUTPUT GAP

Circuit Considerations

Two aspects of the gap performance will be briefly considered in this section. Firstly the pulse breakdown of the gap and its tracking will be considered. Secondly the component of the rise time due to the output gap will be approximately treated.

Pulse Breakdown

The pulse breakdown of pressurised air and SF_6 have been covered by J.C. Martin in "High Speed Breakdown of Pressurised Sulphur Hexfluoride and Air in Nearly Uniform Gaps", SSWA/JCM/732/380. This shows that there is a weak time dependency of the breakdown field (a 1/6th power) and, for air, a weak gap length dependency of the breakdown field. For SF_6 there is some evidence of a large area low jitter conditioning process. For more details the reader is referred to the above note. However, to summarise, for times appropriate to TOM ($\sqrt{LC} \sim 40$ ns) the uniform breakdown fields of about 270 kV/cm for air and 560 kV/cm for SF_6 can be achieved for pressures of 80 psig. The actual gap spacing for the TOM full scale gap is 3.05 cm, with a field enhancement factor (FEF) of about 1.24 giving a uniform field equivalent spacing of 2.42 cm. Thus at 80 psig the pulse charged breakdown voltages are 660 kV for air and would be 1.35 MV for SF_6. The gap can be pressurised to 100 psig, so that air can be used up to about 700 kV, at which pulse level SF_6 can be substituted to go higher than the generator has been tested.

With regard to the external flash over, because the gap is pulse charged very rapidly, it was expected that simple short anti-tracking barriers around the gap would prevent the streamers in the air completing. This was indeed the case and judging from the results obtained with a half scale gap, the flash over voltage of the full scale gap would be around 1.6 MV in air.

Rise Time Caused by Gap

Rigorously the rise time component due to the gap cannot be separated from the effect of the feed electrodes which join the peaking capacitor to the output face and, indeed, the treatment should proceed some way down the conical transmission line. A proper treatment would be a full 3D electromagnetic treatment over this region, which would also have to include the physics of the gas plasma channel. However, to an adequate accuracy, it is considered that simplified treatments of the individual stages of the problem are applicable and useful and at least a help in assessing the relative effect of the various components contributing to the pulse rise time. It turns out that at high pressures, the spark gap does not contribute much to the rise time, most of this coming from the pulse rattling around in the output feed line.

The times discussed below will be e-folding or maximum slope times, not 10% to 90% rise times, and reference to the NPT note should be made for a review of the two parameters and their relation. The note should also be consulted for the relations covering the resistive and inductive rise times.

At the nominal maximum output the mean field in the gap is about 220 kV at 80 psig and this leads to a resistive phase rise time of about 0.7 ns. The more complicated question relates to the inductive rise time. The gap consists of two long cylinders separated by about their diameter. The spark channel is arranged to close in the middle region of the gap and the two roughly equal lengths of parallel cylinders act as pressurised gas insulated low inductance lines to feed current into the spark channel. The inductance per cm of such an arrangement is about 5 nH/cm. The length of the cylindrical electrodes is about 50 cm, as is the width of the lines which feed them at a number of places along the electrodes. The inductance of the long cylinders with current flowing uniformly into them along their length is then $50 \times 5 \times 1/12 \approx 21$ nH. The inductance of the spark channel at the time of interest is about $2 \times 3 \times 5 \approx 30$ nH. Allowing for the inductance of the multiple feeds through the wall of the perspex pressure vessel, the gap inductance is about 55 nH.

Providing the impedance R and spacing of the line in which the gap is placed are small enough, the inductive rise time is given by L/R and this would suggest an inductive rise of about 0.4 ns. However, R = 140 ohm is too high for this simple treatment to be applicable, because the pulse front from the spark channel could only travel 12 cm outwards in this time. Thus the limit of applicability for this treatment is $R \leq 30$ ohm or thereabouts.

However, another way of dealing with the issue is to consider the problem of a 25 cm long transmission line of impedance about

160 ohm, fed along its length by resistors which total 140 x 2 ohm. This produces a series of output pulses along the direction of the feed line. Averaging across the width of this line, an estimate of the e-fold rise time can be obtained for this case and this gives an effective inductive/transit rise time of about 0.8 ns. Calculations with lower impedance feed lines show that this treatment converges to the answer given above for low enough impedance lines. If the impedance of the feed line rises much above 140 ohm, the problem has to be treated as that of a radiating antenna; however, fortunately this is not necessary in the present case.

Thus the expected rise time of the gap in a 140 ohm line of small spacing would be expected to be about 1.5 ns maximum slope parameter, or about 2.2 ns, 10 to 90%, with 80 psig air in the gap. As the gap pressure is lowered, the rise time increases slowly because the resistive phase increases, but the effect is not large and calculations similar to the above can be used to obtain the answer.

Incidentally, it should be mentioned that a strip line of 140 ohm cannot have both a small separation and a width of about 50 cm in real life, because it requires a dielectric constant considerably less than 1. However, what is meant by the above statement is that velocity of light transit effects across the separation of the lines are ignored: the effect of these will be covered very approximately in the section on the output feeds, where the final rise time of the pulse is obtained.

Output Gap Mechanical Construction and Tracking Results

It was found to be impossible to track the full scale TOM gap, so details of a "half scale" gap, made and initially tested by Messrs George Herbert, Mike Hutchinson and Denis Akers, will be included along with tracking results obtained with this gap.

The half scale gap consisted of 5/8" cylindrical electrodes mounted inside a 2" OD perspex body. These electrodes had had approximately 1 mm removed from their inward-facing surfaces and then the cylinders contoured to give an effective radius of curvature of the order of 1.2 cm. The actual electrode separation was 1.25 cm and it is estimated the FEF was of the order of 1.20, giving a uniform field equivalent spacing of about 1.04 cm.

The 1 foot wide scaled transmission line feeds which were attached to the gap were made from 3/4" wood contoured into "uniform field" profiles at the edges and wrapped in aluminium baking foil. Approximately 3 mm perspex sheet was wrapped around the lines and simplexed on to the 2" OD body of the gap. In addition, 3 fins of 3 mm perspex were bent and simplexed on to the body of the gap on each side of the gap. these fins stuck out 4" from the axis of the gap and were curved in the planes containing the gap axis. Figure 12e-19 shows a sketch of the gap. In the half-scale tests, the gap was taken up to about 450 kV in air, with the bottom plane of the transmission line 18" away, with no breakdowns across the gap or between the lines observed.

Subsequently the same gap was installed in TOM and the half-scale gap went up to a little under 800 kV in the tests described here. At this level a track occurred which flashed over the perspex wrapped feeds and punched through the base of the 3 fins where these were simplexed to the body. When the gap was examined, it was observed that there were some small bubbles in one of the joints and also that it was upside down. The small holes made by the track were drilled out, patches simplexed on, and the gap reinstalled the right way up. Subsequently the same gap went to 800 kV without further trouble. Such a voltage on an air-insulated gap is progressive and was partly achieved because of the rapid charging in the peaking capacity circuit ($\sqrt{LC} \sim 28$ ns). The inductance of the half scale gap, in a one foot line, is about 22 nH and in a 140 ohm transmission line would add 0.6 ns to the rise time (it was working at 100 psig SF_6). Of course the rise time of the output

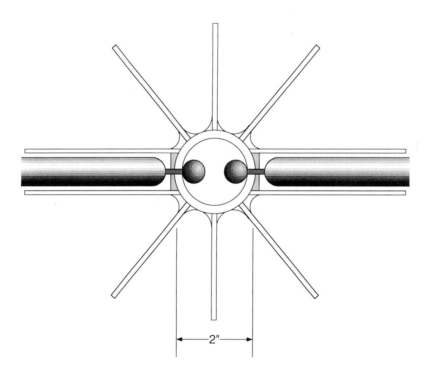

Fig. 12e-19 1/2 scale Tom gap, 800 kV in air.

HIGH VOLTAGE DESIGN CONSIDERATIONS

pulse is mainly controlled by the wave fronts rattling around across the depth and width of the line, but an 800 kV air-insulated gap with these characteristics, which is simple to make, is quite useful.

The full-scale TOM gap was similar in design but had a 4" OD perspex pressure vessel. The electrodes were about 50 cm over the uniform field part and were made out of 1 1/4" tube with solid end caps added at the ends where the contouring was done. The physical gap was 3.05 cm but, just to make the problem of the FEF factor even more complex, brass was removed from both outwards and inwards facing surfaces of the rods. However the field enhancement factor was estimated to be about 1.24, giving a uniform field equivalent spacing of about 2.42 cm.

The full-scale TOM gap never showed any signs of tracking and performed well, with no signs of wear at all. Initially the spark channels were located at the joints between the solid end caps and the hollow tube. However, as a few sharp points at these joints were removed, the discharges moved towards the centre of the electrodes, where these had intentionally been very slightly bowed together.

The relationship between output gap breakdown voltage and pressure is not a unique curve, for two reasons. Firstly the breakdown field depends weakly on the peaking capacity period, going as $(LC_{eff})^{1/12}$. Secondly there is a range of output gap breakdown voltages at which an acceptable waveform is obtained. However, Figure 12e-20 gives a representative curve relating the output gap pressure to the system output voltage, where the gap is pressurised with air. The curve is for negative output, as there is a small polarity dependency which decreases as the pressure is raised.

OUTPUT FEED LINES AND FACE

Circuit Considerations

The two output conical transmission lines between the peaking capacitor and the output face are not symmetric because the conical transmission line to which the generator is attached is not a symmetrical one on account of the influence of the ground plane.

At the peaking capacitor, the two lines are of equal width (20") but the earth plane broadens towards 5 feet width at the output face. The top line (which contains the output gap located close to the peaking capacity) diverges to a width of 3 1/2 feet at the output face, where it is separated from the earth plane by 3 1/2 feet. The length of the feed between the peaking capacitor and

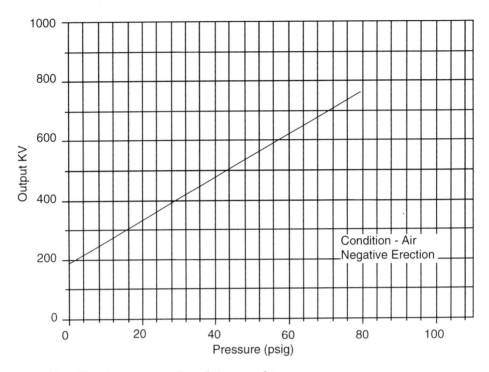

Fig. 12e-20 Output gap breakdown voltage.

the output face is 1 1/2 feet. The impedance at the output face is 140 ohm, but at the peaking capacitor the impedance is a little over 200 ohm. Thus this is a mismatch in impedance in this region. This has a significant effect on the rise time, even with an output impedance of 140, and when output lines or resistive terminations of 100 ohm or lower are used, the effect on the rise becomes important. While there is plenty of capacity in the Marx to drive a line of 50 ohms or so, in order to preserve the rise time the feed lines would have to be widened and also shortened a little. With these changes, lines of impedance down to 50 ohms could be fed with little adverse effect on the rise time. With the 140 ohm output line, the extra rise time due to this effect is about 1.5 ns (maximum slope parameter).

The final effect to be considered which increases the rise time of the output pulse is the location of the output gap in the feed line, in particular its position with regard to the peaking capacitor face and the separation of the output feeds. An approximate treatment of the effect of the wave fronts rattling around in the output diverging feed line was obtained by considering a 2D case where an infinitely fast rise time generator is assumed to be located at the position of the spark gap. The images of this source were located by simple geometrical considerations and some

HIGH VOLTAGE DESIGN CONSIDERATIONS

allowance was made for the finite width of the feed lines and 3D nature of the wave fronts by diminishing the strength of the image sources by a constant factor f for each reflection the wave front had undergone. The resulting wave vectors were calculated at three heights across the output face and the average of these at late times normalised to the known uniform field, thus giving the factor f. In order to obtain the effect on the rise time, both the transit time at the velocity of light and the image source effects were combined, and the resulting terms combined across the whole height of the face to give an effective rise time due to the geometry of the output feeds and the location of the spark gap. As is fairly obvious, it is helpful to locate the gap as close as possible to the output face of the peaking capacity (which represents essentially a dead short from the wave front point of view). The resulting crudely calculated rise time effect of the output feed can be roughly represented by a linear rise from 0% to 100% of about 4 1/2 ns.

The combined effect of the spark gap rise time (maximum slope rise time 1.5 ns), the feed impedance mismatch (maximum slope rise time 1.5 ns), and the output feed geometry (a linear ramp of 4 1/2 ns), and the output feed geometry (a linear ramp of 4 1/2 ns duration) together gave a maximum slope rise time of 6 ns and a 10% - 90% rise time only a little more - about 6 1/2 ns.

It is obvious from the above calculations that no very great accuracy can be claimed for the rise times obtained. However it is considered that they are in the right ballpark and in particular they show the relative importance of the various factors. The actual design of the output section was based on these calculations but naturally had to take into account the limitations imposed by high voltage flashover considerations.

The output face of the generator gave no trouble and never showed any signs of tracking, despite on occasions having some horrible things done to it by way of having incorrect length copper sulphate terminating resistors draped across it. Some work with a scaled model interface suggested that the tracking voltage of the interface would be about 1.4 MV in air, with the continuing line in place. Based on previous tracking studies at AWRE, it is considered that with flash guards simplexed on to the face, this breakdown voltage could be raised significantly, probably to 2 MV or thereabouts.

FULL SYSTEM PERFORMANCE

Monitoring

Two different types of oscilloscopes were used in the measures, a Mark 2C made at AWRE which had a rise time of about 5 ns (maximum slope parameter) and one developed in SSWA by Mr T. Storr which had a rise time of 1.8 ns. The Mark 2C was used in a doubly screened box and was mainly employed to obtain the peak amplitude and decay of the pulse. The T 'scope has a number of unusual characteristics. Firstly it is very immune to pick up and can be operated without any external screening. Its deflection sensitivity is 5 1/2 kV/cm on the film and therefore needs less attenuation of the pulse, which helps to prevent distortion in this stage. It has a near-exponential sweep, which is of use in many applications where the first pulse of a high speed pulse is of major importance and later pulses of less interest. In this application the exponential sweep was a mild disadvantage and the T 'scope was used to determine the rise time and peak amplitude of the pulse. The constancy of the sweep speed is not very high for this 'scope, there being some drift in it, but time calibrations were interleaved with measurements whenever rise time measurements were being made.

The monitor chains were made out of at least 100 1 W resistors and disposed in spaces as is described in the NPT note, so as to cancel out stray capacity effects. Thus when monitoring at the output face, where the field is non-uniform because the top plate is narrower than the ground plate, the resistors were more closely spaced as they neared the top plate terminal.

The impedance of the voltage divider for most of the rise time measurements was 3500 ohm and it was estimated that the residual uncorrected stray capacity had less than 1 ns effect on the rise time.

All the monitoring, charging, trigger leads, etc. were wired and, as is mentioned in the section on the trigger pulse generator, resistive breaks were introduced into all the feeds to the generator and inductors into the mains cables of the 'scopes. Thus the system was also single point earthed. These techniques were sufficient to give a very satisfactory immunity to pick up with both 'scopes.

As an example of the degree of pick up exclusion obtained, the Mark 2C records (the Mark 2C has a film deflection sensitivity of 70 volts/cm) were quite clean, even though the 'scope screen box was only made of gauze. At one time the 'scope box was within 2 feet of the hot plate of a transmission line operating at 800 kV and was in a free field gradient of about 300 kV/metre, but still gave pick-up free records, providing both doors of the 'scope were

HIGH VOLTAGE DESIGN CONSIDERATIONS

shut. With one door open, there was a few percent high frequency ripple visible.

The terminating resistors used were copper sulphate solution ones and were distributed across the end of the extension line, or across the output face, to match the fringing field to a first approximation. As the resistors were flexible, they could be curved along field lines and this was done in the case of those at the edges of the line, or on the output face. In the case of measurements at the generator face without a continuing line attached, the capacity of the solution resistors was such that $(L/C)^{1/2}$ was about equal to the impedance of the resistor.

The high voltage of the power pack was calibrated against a standard 0 to 80 kV voltmeter which had been originally calibrated to better than 1/2%. Time calibrations of the 'scope were performed with oscillators and cable pulse generators and were good to a couple of percent or better. The stability of the 'scope's deflection sensitivity was good and checked three times during the month or so that the measurements took, with agreement again to a couple of percent between calibrations.

Full System Performance Pulse Amplitude Measurements

As mentioned earlier, the amplitude of the output pulse can be controlled over small limits by altering the voltage at which the output gap fires with a constant DC voltage on the Marx. In the initial experiments with the 600 ns tail and 140 ohm termination on the output face, the output switch could be fired at a slightly higher voltage than the exact peaking condition without introducing significant oscillations on the tail of the waveform. This more than compensated for the internal impedance of the Marx and an output pulse closely equal, or slightly greater than, 12 times the capacitor voltage was obtained. Table 12e-II lists the results for a series of firings under these conditions.

This table also shows good agreement between the two 'scopes, which record at very different voltage levels, and, in this case, with different resistive attenuators as well.

In later measurements, particularly those with the line across the laboratory, the impedance was down to a little under 100 ohm. In addition an inductance bypassed by a resistance has been added, to remove a small oscillation. As well as this, more resistance and an inductive resistance had been added across the Marx to reduce the waveform tail to 200 ns full time width at half height. All these changes increased the effect of the internal resistance of the Marx and the output pulse eventually was nearer 10 times the capacitor voltage, most of the remaining 17% of the stacked capacitor voltage being across the internal Marx impedance.

Table 12e-II
Pulse Amplitude Measurements

Capacitor Volts (kV)	Output Mk 2C (kV)	Output T 'scope (kV)	Mean Observed Output (kV)	12X Capacitor Volts (kV)
24	282	294	288	288
36	420	448	434	432
42	520	520	520	504
54	660	640	650	648
66	800	810	805	792

The variation in amplitude was checked in a couple of series of runs and in both cases the standard deviation of the output pulse voltage was less than ± 1%. In one series, the system was fired 8 times in 2 minutes, which is as fast as the power pack would charge the system, but the deviation was still less than ± 1%.

There is a small prepulse inherent in the peaking capacitor approach and this was about 8% of the main pulse amplitude for a line impedance of 140 ohm. For 100 ohm termination the amplitude is more like 6%. The prepulse was present for about 70 ns before the main pulse, for most of the generator arrangements used, but this depends on the value of the peaking capacitor. The prepulse level is a function of the rate of charging of the peaking capacity, the output switch capacity and the load impedance, and agreed with the calculated values.

Full System Performance Pulse Rise Time

Two main series of rise time measurements were performed, although a number of other determinations were made, which agreed with the results of the main investigations. The first series (6/12/72) was made at the output face of the Marx. This was terminated with 4 solution resistors of total impedance 120 ohm, slightly lower than the expected impedance of the line to be attached. It was appreciated that such measurements might give slightly too long a rise time because the fringing field beyond the output has to be established. A second effect which might influence the observation is that the field lines at the output face, without the line connected, are more complex in shape than would be the case when the generator is in use. In an attempt to check this, a parallel plate nominal 140 ohm line was built across the laboratory for 13 feet, after the generator was repositioned in a corner. The monitor was then placed 2 feet down the line from the output face, on the centre line of the electrodes. The nominal free time before any reflection can get back from the end of the line is 22 ns. However, life is not as simple as theory and, of course, the line was very far from being like its idealised theoretical equivalent. Because of the size of the laboratory, the end of the line was within 2 feet of the Mk 2C 'scope in its box and not much more than 3 feet from the operators. The Marx was also within about 2 1/2 feet from the walls, on which were mounted a ring earthing strip and 2 large iron clad switch outlets. Thus it was not too surprising that when the matching load to the line was determined experimentally it was found to be a little under 90 ohm. This means that something like an extra 250 ohm was in parallel with the normal fringing fields because of the presence of the room and its contents. The pulse would not immediately see this extra load but it would appear over a time of the order of 10 ns. Thus it was not possible to define the impedance into which the pulse was being fed, but it is very likely that this was lower than the loading in the previous case.

One incidental result from this line was that it confirmed the streamer breakdown transit relation, in that the workers survived unharmed. Indeed, despite the small clearances, at no time was breakdown recorded, even though the line was taken up to 840 kV.

Thus both sets of rise time measurements can be criticised, but it is believed not in any very substantial way. Certainly it is true that the 2 sets of measurements were obtained under very different conditions. Table 12e-III lists the 'scope response corrected rise times (both parameters) obtained, and also the calculated values for the 140 ohm load referred to earlier. In the table the observed values are only good to ± 1 ns absolutely and both cases are for load impedances significantly lower than 140 ohm. With regard to the maximum slope times for the case with line attached (9/1/73) results, there were signs of a small amplitude osciallation on the records which may have increased the maximum rate of rise of the pulse and so affected maximum slope parameter rise time, making the observed value too low by a nanosecond or so. This oscillation could have been generated at the discontinuity between the conical feed lines inside the generator and the parallel line which was attached to the output face. In its proper use, a conical transmission line is attached to the output face of the generator, matching smoothly to the one inside it.

Table 12e-III
Corrected Rise Times

Set Up	Output Volts (kV)	τ_{max} Slope (ns)	τ_{10-90} (ns)
6/12/72 Output Face Measurements Load - 120 Ω	360 500 660	8.1 7.1 5.4	7.8 7.3 6.8
9/1/73 Line Measurements Load ~ 100 Ω	210 300 450 600	7.1 7.1 5.9 4.4	11.0 8.7 8.0 7.2
Calculated Values Load = 140 Ω	350 700	7 6	7 1/2 6 1/2

The results are graphed in Figures 12e-21 and 12e-22. As these show the 10 to 90 rise time, results are in good agreement and almost too good agreement with the approximately calculated values. The agreement would be even better if a lower load impedance had been used in the calculations, but, as was explained earlier, these calculations are approximations to a very complex

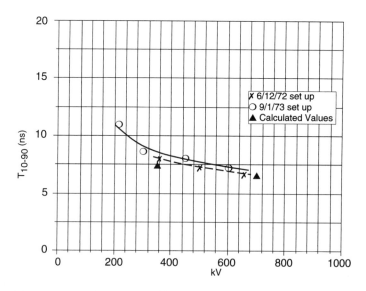

Fig. 12e-21 Pulse voltage 10-90% Risetime.

HIGH VOLTAGE DESIGN CONSIDERATIONS

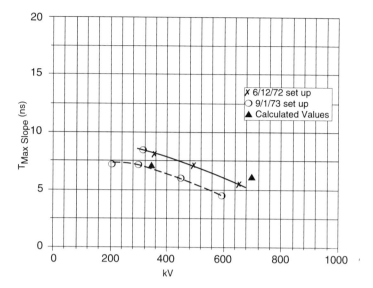

Fig. 12e-22   Pulse voltage max slope risetime.

situation and are surely not good enough to justify such refinement. However, the agreement is very satisfactory and suggests that for 140 ohm line, the rise time will be in the range 7 ns to 6 ns for output voltages between 300 and 700 kV.

The maximum slope parameter agreement is not so good, but still well within what would be expected, especially if the 9/1/73 results are optimistic by 1 ns, as was suggested above. The results shown in the graph suggest that for 140 ohm line, the maximum slope rise time will be in the range 7 to 5 ns for output voltage between 300 and 700 kV.

Full System Performance General Remarks

The system is very easy to operate and tolerant of intentional and accidental fault mode operation.

The Marx was fired some 1400 times and, apart from the high overvoltage tests (where the air flow through the system was very low and operating pressures had not been properly established), the Marx may have prefired twice, but even in these cases the trigger pulse generator gap may have operated unintentionally.

The complete generator was fired some 1200 times and was operated in a number of fault conditions without damage of any kind.

The only component which was damaged was a capacitor in the Marx, during the long tail flash over tests before the polythene anti-corona guards were installed. During the tests, as was mentioned, the breakdown track dived across inside the Marx twice, on occasions, and after one such track a condenser started to flash over near its output feeds, on recharging. The breakdown streamer in its path across the trays had punctured the mylar insulation, leading to subsequent DC tracking. The condenser was removed and repaired without trouble. Apart from this understandable failure, no component was damaged at all.

The system was intentionally self broken at high level: it was also fired with the output switch heavily over-pressurised. In one unintentional fault mode test it was fired with the added inductance between the Marx and the peaking capacitor missing. It operated quite correctly, apart from producing a loud bang.

Of the 1200 complete system firings, some 100 shots were over the nominal output of the system. However, as the life of the system is a very powerful function of the voltage (something like an inverse eight power), these shots represent a very much larger number of firing under normal operating conditions, besides establishing that the system was very safe from the point of breakdown and tracking.

The generator weighs about 300 lb and is mechanically reasonably robust. Apart from the Marx frame, it is completely disassembleable and, in particular, the peaking capacitor is very simple to change. Thus with the ease of altering both the capacity and the inductance of the system, a wide range of parameters can be covered, a fact that was of considerable help in performing the experiments on the high speed breakdown of pressurised air and SF_6 mentioned earlier.

FUTURE CAPABILITIES OF THE DESIGN

The Marx over its capacitors was 1 metre tall, almost exactly, and was thus operated at a gradient of 1 MV/m in air. To make a 1.5 MV user system (which should be tested for a few shots at 2 MV) operating in air, would require a height of some 2 metres, excluding the field shaping bungs. Including these, the Marx overall would be around 7 1/2 feet tall, which would mean it would be uncomfortably close to the ceiling of a standard laboratory. It would, of course, also need some 20 to 24 condensers in it rather than the 12 in the present Marx. However, by placing the whole system in a freon-tight perspex box it is considered that the height of the Marx (the tallest component) need be no more than 5 feet overall. Some tests performed with another Marx showed that putting the Marx under freon raises the overall voltage gradient that can be supported by a factor of at least 1.6. Freon also enables the peaking capacitor to be operated at twice the gradient. The present 4" OD output gap will not track up to 2 MV easily under freon and when pressurised with SF_6 will operate up to this voltage. With the addition of flash guards, the air side of the output face will go to 1.5 MV, according to small case tests on a model, so an increase in the separation of the plates of 1 foot should enable the output face to be tested to 2 MV. Thus, putting the system under freon, increasing the number of condensers in the Marx, and possibly enlarging the peaking capacitor bungs, should give a very good chance of operating a new system at 2 MV for test purposes and enable regular use at 1.5 MV. The size of the pulser will be about a foot extra in all dimensions.

CONCLUSIONS

The mechanical construction of TOM is described and its performance is compared with the theoretical basis of its design. In general, quite reasonable agreement was observed.

The system is simple to operate and has a very safe margin against breakdown and tracking.

It has a wide operating range and has easily adjusted main parameters and has been operated a considerable number of times under test very satisfactorily. It is tolerant of fault mode operation and was never damaged, even under a high degree of overtest.

The design is considered to be capable of extension to operation at 1.5 MV, with a 2 MV overtest level, while still being capable of being operated in a normal size laboratory.

APPENDIX A*

TRIGGER SYSTEM

This consisted of a power pack unit delivering a DC voltage of 11 kV from a voltage doubling circuit and a small mechanically operated spark gap which shorted out the HT at the input to the cable leading to the unit at the base of the Marx. The cable from this power pack to the pulse generator thus fulfilled two functions: that of DC charging and also in providing the trigger pulse that fired the rail gap in the trigger pulse generator. This cable was made 100 metres long so that the power pack and firing button could be positioned remotely from the Marx. The mechanically operated ball bearing closure gap could easily be replaced by a triggered gap of the corona type, a thyratron, or solid state device, if command triggering were required. The polarity of the trigger power pack could be easily reversed by changing around a small panel on which the rectifiers were mounted.

The unit at the Marx consisted of a small 75 nF low inductance capacitor with a rail gap mounted on it. When this gap was fired the capacitor was placed across the primary turns of an air cored, oil immersed, auto-transformer. The capacitor and spark gap were contained in a small perspex box located near the base of the Marx, which can be seen to the right in the side view photograph of the generator. A low inductance strip line fed the output of the condenser to the auto-transformer, which was located in the space under the second condenser in the Marx.

The rail gap on top of the small capacitor had two 3/8" OD brass main electrodes with a sharp edged trigger electrode mounted off centre between them. The spacing of the edge of the trigger electrode from the main rail electrodes was 2 mm and 4 mm. The trigger pulse from the power pack doubled on reflection at the end of the cable and was arranged to break the 4 mm gap first. The gap volts were then applied to the 2 mm gap, causing it to break. Figure 12e-23 shows a schematic of the circuit. It was observed that the 2 mm portion of the gap frequently broke in multichannels, which is interesting, because this is a very low voltage gap and low voltage gaps are harder to break quickly than gaps with a decent voltage on them.

The DC breakdown of the gap was 15 kV and it would trigger down to 6 kV on the gap. This, too, is somewhat surprising because, of course, the trigger pulse from the power pack also decreased as the voltage was reduced. As is shown in the schematic, resistors are included in the trigger pulse connection and also in

---

* SSWA/JCM/735/407

HIGH VOLTAGE DESIGN CONSIDERATIONS

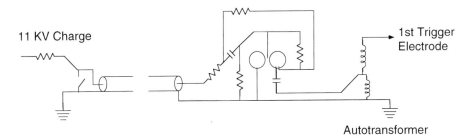

Fig. 12e-23 Schematic of pulse generator.

the earth connection. These are included to break the earth loop and so reduce the circulating earth currents which might otherwise give rise to pick up problems. This process is carried out in the main Marx which is earthed only at the bottom of its output face to the bottom of the transmission line. Resistors in the main Marx charging pack isolate these leads, as do resistances in the base of the charging column, so that there are low impedance earth loops in the system.

The pulse transformer is similar to those described in NPT notes, only here as the output voltage is only some 60 kV, the transformer is oil impregnated. The auto-transformer is wound on a 4" OD cylinder of perspex. The primary winding consists of 3 turns, while the secondary has 15 turns. The copper of the winding is 3 thou thick and tapers from 6 1/2 cm down to 1 1/2 cm at the output end. Figure 12e-24 gives the equivalent primary circuit parameters when the transformer has fired the first trigger gap, at which time there is a load of about 450 ohm across the transformer. From this circuit the gain can be calculated as about 5.2, the rise of the pulse 40 ns, and the fall time about 60 ns, to the first zero crossing.

Under test with a 450 ohm load on the auto-transformer, a gain of 5 was measured, with a rise time of 45 ns and a time to first zero volts crossing of 400 ns, in good agreement with the calculated values. Thus in normal operation the trigger unit gives out a 60 kV pulse.

The triggering system gave no trouble at all and required no adjustment. During the 1400 or so firings of the system, it

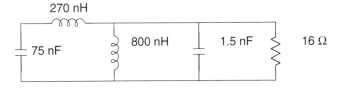

Fig. 12e-24 Primary circuit for autotransformer.

definitely prefired three times, just after some work had been done on the perspex box and dust had got into the gap. It may have prefired on two other occasions, or the Marx may have self-erected: either is a possibility.

The overall time from arrival of the trigger pulse from the mechanical switch to generator output time was measured under typical conditions as about 290 ns, with a jitter of ± 17 ns. Most of this jitter came in the trigger system, it is believed, as well as about 150 ns of the delay mentioned above. By redesign of the rail gap and the auto-transformer, the total delay could be reduced to 200 ns and the jitter reduced easily to ± 5 ns, and maybe better.

APPENDIX B*

POWER PACK AND DUMP UNIT

The load which when added to either side of the power pack reduces the output voltage to one half, is 4.5 Megs. Thus the total internal impedance of the power pack is of the order of 9 Megs and when the charging current is 2 mA, the total output is about 18 kV less than the open circuit output. A constant charging current of 2 mA charges the Marx condensers for a nominal maximum output pulse in 25 seconds, but the charging can be quicker than this, if necessary.

The power pack was tested to ± 60 kV without any corona at all. The dump was tested to over 2 kJ absorbed successfully.

The open circuit output of the Marx was linear with Variac input, 215 volts from the Variac being necessary to get a total output voltage of 120 kV. The dump could also be operated and not fire the Marx when this was at 90% or less of its self-break voltage.

I suggest a circuit diagram should be included, as well as possibly a photograph of the power pack. Also a description of the dump system. (Good luck!).*

---

* SSWA/JCM/735/407
** Editor's Note: This was not included in the note.

INDEX

"TOM", 490, 105,116, 133, 335, 337
1 MV EMP generator, 337
140 cm wide system, 478, 484
60 cm wide system
60 cm wide line, 443,478, 484
AERE, 22

Air
  cored transformer(s), 4, 28, 29, 37, 115
  gap(s), 139, 169, 172, 173, 264, 296, 326, 345, 347, 349, 356, 358, 366
AK impedance, 369
Alcohol(s), 50
Annealing, 53, 79, 80, 235, 237
Anode, 3, 7, 8, 10, 11, 24, 30, 119, 120, 212, 368, 369, 370, 375, 376, 377, 380, 381, 394, 395, 398, 402-405, 407-411, 432, 436, 437
Anode cathode voltage, 380
Aurora, 225, 259, 373, 375
Autobiographical details, 17
Avalanche, 48, 83, 84, 89, 90
AWE, 8, 9, 16, 31
AWRE, 1, 15, 16, 21, 24-26, 30, 32, 35, 40, 62, 71-73, 75, 82, 102, 104, 105, 110, 115, 119, 196, 228, 230, 258, 261-263, 282, 287, 300,

AWRE (cont'd)
  303, 311, 331, 335, 367, 373, 376, 421, 428, 429, 443, 494, 496, 502, 503, 511, 523, 524

Backfiring, 198, 305, 309, 310, 313
Beam
  energy, 1, 396, 401, 408, 409
  transport, 2, 376
Biography, 15
Blumlein, 6, 9, 10-12, 21, 31, 44, 45, 104, 117, 118, 185, 261, 262, 331, 335, 440, 441-443, 448, 457
Brass, 59, 66, 102, 103, 125, 212, 264, 265, 331, 338, 348, 350, 359, 443, 451-453, 478, 502, 521, 532
Breakdown
  field(s), 45-50, 52, 53, 79, 80, 81, 83, 84, 96, 127, 135, 140, 142, 145, 147, 148, 150-153, 170. 171, 173, 174, 198, 199, 201, 204, 211, 212, 214, 228, 230, 232, 238, 240, 243, 245, 248, 252, 253, 256, 266, 275, 312, 340, 342-349, 357, 358, 428, 438, 464, 467, 470, 517

537

Breakdown (cont'd)
  strength, 4, 45, 52, 77-79,
    232, 233, 236, 239, 245,
    256
  voltage(s) 46-48, 60, 62, 78,
    79, 96, 97, 99, 103, 127,
    135-137, 139, 142, 144,
    145, 171-173, 177, 206,
    227-229, 231, 232, 235,
    236, 239, 241-243, 245,
    249, 250, 263, 264, 275,
    281, 283, 283, 307, 310,
    311, 327, 329, 335, 340-
    343, 348, 354, 356, 358,
    365, 428, 445, 451, 491,
    494, 497, 498, 507, 517,
    522, 524
Bridge, 17, 60, 174, 273, 413,
  414, 513
Bubble growth phase, 205, 210

Capacitor store, 2, 3
Carbon tetrachloride, 185,
  186, 311, 321-323,
Cascade gap(s) 58, 276
Castor oil, 50, 179
Cathode(s), 2, 3, 7, 8, 10,
  12, 23, 30, 91, 114, 117-
  122, 146, 155, 194, 218,
  262, 360, 367-370, 373, 375,
  380, 394, 395, 402, 405,
  407-411, 425, 429, 431-434,
  436-438
Charging resistors, 4, 129,
  132, 493, 494, 499, 501, 502
Child-Langmuir, 120, 373
Conditioning, 47, 96, 153,
  227, 257, 258, 341, 348,
  466, 498, 517
Conducting channel, 54, 267,
  300
Construction(al) techniques,
  18, 102, 130
Copper sulphate, 4, 5, 54, 89,
  114, 115, 129, 228, 246,
  248, 413, 414, 494, 498,
  499, 502, 523, 525
Core members, 33
Cornell, 13, 32, 73, 116, 117,
  259, 375

Corona gap, 103, 283, 352, 364
Cost, 2, 4, 32, 39, 58, 60,
  61, 71, 103, 132, 276, 376,
  433, 435, 477, 478, 485,
  486, 504
Crazing, 102, 485
Current monitoring, 125
$CuSO_4$, 38, 42, 65, 66, 67-70

DC breakdown, 47, 49, 50, 82,
  96, 140, 152, 171, 172, 235,
  238, 274, 352, 354-356, 358,
  359, 351, 362, 445, 497, 532
  dc charged gaps, 264, 347
  dc tracking, 497
Deionized water, 185, 414
Diaphragm tube(s), 119, 256,
  257, 259, 262
Dielectric breakdown, 15, 71,
  75, 82, 95, 236
Diode(s), 30, 117-120, 122,
  262, 368, 375, 377-381, 398,
  400, 401, 405-412, 425-421,
  433, 436-438
  insulator sections, 430
  pinch, 120, 377, 398, 429
Divergent field(s), 49
Dose on axis
Drop outs, 78, 265
Dump resistor(s), 5, 36, 69,
  496

E-beams, 117
E-folding time, 55, 56, 267,
  268, 298
Earthing, 67, 126, 527
Edge
  control, 449, -451
  grading, 53, 449, 514
  plane, 63, 82, 84, 94, 97-
  99, 101, 112, 169, 263, 266,
  271, 274, 279, 282, 299,
  306, 307, 310-314, 329, 342,
  343, 441, 442, 454, 460,
  462, 466-468, 472, 477, 479,
  482, 491, 507
  plane gap(s), 63, 64, 266,
  271, 299, 311, 313, 441,
  460, 462, 478

INDEX

Effective time(s), 63, 170, 177, 185, 257, 322, 343, 454
Electrode(s), 11, 41, 42, 46-50, 51, 54, 56, 58-60,
  erosion, 296, 364
  field grading, 418 2, 63, 69, 74, 77, 78, 81, 82, 83, 85, 86, 88, 89-93, 96-100, 102, 103, 109, 110, 129, 136, 140, 142, 146, 152, 156-160, 166, 168, 169, 173, 182, 185, 196, 213, 228, 231, 237-239, 245, 246, 249-251, 257, 259, 264-266, 270-281, 283, 291, 293, 296-298, 304, 307, 313, 322, 329-333, 336-338, 340-342, 348, 350-352, 354, 355, 357, 359, 361, 364, 365, 413, 414, 417-419, 421, 422, 430, 444, 452-454, 491, 493, 494, 497-499, 502, 504, 507, 511, 514, 518, 519, 521, 532
Electron
  beam diagnostics, 375
  flux, 375, 377, 378, 380
Electrostatic grading structure(s) 133, 417, 504
Erection time, 40, 111, 112, 494, 504, 505
Ethylene glycol, 179
Evostick, 130, 239, 240

Faraday cup, 380, 381, 401, 408
Fast charging, 2, 215, 313, 327, 329, 425
Fast circuits, 117
Fault modes, 132
Fef, 47, 48, 173, 182, 194, 336-338, 345, 350, 418, 419, 421, 519, 521
Field
  distortion, 58-60, 97-101, 276-279, 351
  emission, 30, 71, 81, 82, 274, 277, 351

Field (cont'd)
  enhancement factor, 141, 173, 201, 336, 338, 418, 420, 421, 497, 521
Field Emission Corporation, 30, 71
Fizzle, 169, 342, 354, 364, 441, 462
Flash over field, 429
Flux excluders, 428, 433, 434, 444
Four element gap(s), 283, 352, 353
Freon, 7, 21, 48, 49, 61, 88, 89, 114, 136-138, 487, 511, 530, 531

Gap
  breakdown voltage, 263, 522
  pressure(s), 111, 447, 467, 496, 519
Gas
  channel(s), 287, 289
  discharge laser, 274
  gap(s), 58, 59, 263, 265, 266, 269, 272, 276-278, 292, 297
  lasers, 32, 155
Gas(eous) dielectric breakdown, 82
Gaseous switches, 261
Glass, 7-9, 24, 26, 30, 104, 256
Glycerine, 50, 81, 178, 179, 185, 186
Grading
  ring(s), 256, 504,
  structure(s), 129, 133, 504, 507

H36, 21, 367
Helium, 158, 159, 163-167, 273, 288
Hermes 2, 370, 372
Hermes II, 368
High current, 25, 62, 63, 74, 95, 102, 118, 119, 121, 126, 264, 271, 285, 296, 331, 352
High speed pulse breakdown, 94, 145, 515

Hull, 75-77, 89, 95, 107, 117, 125, 133
Hydrogen, 98, 273, 279, 284, 288, 302
IB gap, 153, 312
ICF, 23, 31
ICSE, 9, 10, 235
Inductance, 3, 4, 6, 28, 36-43, 45, 54-56, 58, 60-63, 65-67, 104, 70, 74, 85, 92, 97, 100-102, 107, 109-115, 118, 122, 125, 126, 128, 159, 169, 236, 237, 262, 267, 270, 272, 276-278, 283-285, 287, 289-292, 296-298, 300, 301, 306, 307, 310, 313, 314, 321, 325, 327, 332, 337, 341, 354, 359, 364, 365, 410, 425-431, 433-437, 441, 443, 444, 446-448, 455, 559, 461-463, 466, 470, 473, 478-480, 482, 486, 489, 496, 497, 508-511, 516, 518, 520, 525, 530, 532
  of a spark channel, 470
  of the Marx, 496
Inductive fall time, 300
Interfaces, 38, 53, 54, 69, 85, 89, 859, 449
Intrinsic breakdown, 47, 62, 77, 78, 96, 204, 210, 212, 245, 273, 282
Ion Physics Corporation, 3, 30, 35, 71, 73, 278
Irradiated, 83, 103, 169, 170-172, 274, 285, 351, 352, 355, 356, 358, 365, 366
IT, 431

Jerry, 29, 337
Jitter, 58, 60-64, 84, 96, 98, 99, 103, 111, 171, 271, 275, 276, 279, 282, 284, 295, 296, 297, 305-313, 318, 319, 322, 323, 326, 327, 329, 340, 342, 347, 348, 355, 356, 363-365, 482, 505, 517, 534

Klystron, 23

L.C. Marx, 41
L/R Marx output circuit, 113
Large area water breakdown, 191, 198
Large area(s), 198, 211, 215, 245, 438
Lark, 29
Laser(s), 32, 33, 58, 84, 121, 155, 169, 274, 276, 278, 304, 355, 440
  pumping, 169
  trigger(ed), 58, 84, 274, 276, 278, 304
Lexan, 256, 258
Life measurement(s), 239
Light source, 91, 92, 104, 105
Linear accelerator, 1, 22-24
Lines, 6, 9, 10, 12, 21, 28, 43-45, 48, 50, 53, 54, 58, 62, 65, 69, 88, 117-119, 122, 128, 149, 151, 198, 240, 262, 264-266, 283, 288, 289, 293, 313, 314, 319, 327, 337, 341, 354, 385, 400, 428, 439, 440, 448-450, 454, 456, 457, 458, 461, 477, 478, 480, 482, 491, 496, 504, 518, 519-523, 525, 527
Liquid(s) 2, 15, 36, 46, 47, 50, 51, 55, 59, 61, 63, 69, 72, 77, 82, 85, 89, 93, 94, 96, 98, 101, 129, 177, 178, 181, 185, 191, 196, 228, 231, 240, 242, 252, 261, 263, 266, 267, 270, 271, 273, 275, 277, 282, 287, 291, 292, 295, 299, 300, 305, 306, 310, 311, 315, 321, 322-326, 425, 427, 428, 431-433, 485, 499, 539
  gap(s), 61, 96, 98, 101, 271, 282, 292, 299, 310, 311, 323, 324, 539
  pulse breakdown, 80, 539
Long pulse, 255-258, 309, 337, 370, 372, 373, 504, 506, 507, 514
Low voltage irradiated spark gap, 285, 351

INDEX

Management, 17, 18, 27, 262
Marx
  generator(s), 3, 10, 15, 26,
    28, 39, 40-42, 71, 88, 107,
    109-112, 114, 263, 264,
    306, 313, 489, 491, 498,
    599, 504, 511
  award, 15
  generator (mechanical construction), 494
  inductance, 112, 497
Maxwell Laboratories, 42, 118,
    302, 428, 486
Mean angle, 121, 369, 375-380,
    392, 395, 396, 403, 404,
    405, 408
Mean breakdown field, 40, 47,
    201, 228, 238, 348
Mean electron angle, 378, 379,
    403, 404, 408, 412
MINI 'A', 10, 12
MINI 'B', 116, 122
Mogul, 11, 12, 26, 72, 367,
    370, 372, 408
Mogul C, 370
Mogul D, 367, 370
Monitor(ed)(ing), 2, 29, 30,
    36, 68-68, 70, 112, 125,
    127, 128, 169, 186, 259,
    262, 265, 270, 325, 326,
    340, 341, 356, 362, 367,
    380, 405, 407, 408, 410,
    412, 454, 455, 457, 459,
    461, 473, 476-478, 510, 515,
    516, 524, 527
Multichannel, 57, 61-63, 93,
    95, 96, 98, 100, 101, 110,
    118, 169, 171, 232, 262,
    263, 266, 269, 270-272, 278,
    282-285, 295-300, 304, 306,
    310-313, 320, 321, 323, 326,
    363, 425, 439, 441, 446,
    451, 453, 463, 466, 469, 471
  gap(s), 63, 101, 271, 295,
    296, 298, 313, 326, 331,
    465
Multistage, 8, 108, 262, 426,
    429, 494
Mylar, 6, 21, 28, 31, 52, 53,
    61, 62, 66, 70, 77, 80, 85-

Mylar, (cont'd)
    88, 89, 92, 93, 103,
    118, 127, 130, 133, 156-159,
    161-164, 167, 173, 174, 231,
    235-240, 242, 245-253, 283,
    307, 330, 331, 352, 359,
    363, 365, 408, 411, 427,
    444, 448-450, 480, 482, 483,
    486, 502, 503, 410-415

N.R.L., 245, 253
Naval Research Laboratory, 65,
    145-148, 158, 159, 163, 165,
    167, 169, 343, 462, 467
Nitrogen, 83, 91, 145
Non-uniform gaps, 139, 140,
    144, 173, 445
NPT, 77, 82, 84, 95, 97, 104,
    112, 115, 117, 125, 492,
    497, 518, 524, 533
NRL, 13, 30, 73, 116, 119,
    121, 255, 439

Oil, 2-6, 10, 11, 24, 26, 28,
    29, 31, 38, 42, 50, 51, 53,
    54, 61, 62, 78, 79, 81, 82,
    93, 98, 102, 114, 115, 128,
    136, 179, 181, 185-187, 225,
    226, 266, 282, 283, 289,
    303, 312, 322, 335, 336,
    427, 432, 457, 485, 532,
    533,
Open circuit, 8, 11, 44, 45,
    112, 113, 115, 127, 434,
    435, 473, 535
Oscilloscope(s), 64, 67, 68,
    70, 305, 401, 524
Output gap, 152, 348, 481,
    489, 509, 510, 517, 519,
    521, 522, 531

P.I., 225, 259
Paper, 23, 54, 76, 151, 159,
    235, 237, 239, 249, 305,
    309, 333, 438, 449, 450,
    485, 486
Parallel plates, 177, 187
Parapotential theory, 120
Peak to peak voltage, 235,
    236, 470

Peaking capacitor(s), 88, 112-114, 152, 337, 338, 341 489,
Peaking capacitor(s)(cont'd) 496,504, 507, 508-516, 521, 522, 526, 530, 531
Perspex, 3, 4, 6, 8, 21, 29, 38, 40, 52, 53, 60, 69, 70, 79, 82, 87, 102, 118, 129, 130, 136, 245, 246, 255, 256, 336-338, 359, 405, 429, 430, 443, 445, 446, 451, 453, 454, 478, 482, 483, 485, 489, 497, 499, 502, 503, 519-521, 530, 532-534
Photo diode, 379,401, 406, 407. 409-411
Physics International, 11, 30, 45, 61, 80, 82, 96, 118-120, 257, 258, 282, 303, 509
Pick up, 30, 66, 67, 70, 398, 455, 456, 524, 533
Plasma, 3, 8, 25, 28, 31, 32, 54-56, 58, 62, 83, 84, 88, 90-93, 99, 119-122, 158, 171, 173, 174, 259, 262, 267-269, 273, 274, 276, 279, 287-289, 292, 279, 301, 302, 304, 310, 314, 315, 326, 327, 354, 355, 369, 370, 373, 405, 408, 411, 433, 441, 446
 blob(s), 8, 119, 120, 411, 433
 channel(s)54, 56, 62, 83, 84, 90, 91, 158, 171, 173, 267-269, 273, 289, 297, 301, 304, 310, 315, 326, 327, 354, 355, 441, 446, 518
Plastic(s), 7, 8, 30, 52, 53, 70, 79, 80, 89, 93, 119, 128, 130, 158, 227-229, 231, 232, 236, 238, 241, 243, 245, 246, 249, 252, 257, 278, 283, 307, 352, 377, 379, 401, 455, 511
PLATO, 8, 10, 72, 335, 336, 340, 343, 344, 345, 348, 350
 gap, 336, 340

Plus and minus charging, 9, 38, 66, 327, 444, 478, 480, 491
Point plane breakdown, 211, 225, 309, 311
Point to plane experiment, 136, 182, 195, 195, 207 266, 274, 309, 310
Point-plane gap, 201-204, 205
Polar diagram, 375, 377-386, 388, 391, 393-397, 399, 400, 402-406, 408, 410
Polished stainless steel, 428
Polythene, 6, 8, 9, 11, 28, 29, 52, 53, 55, 62, 78, 79, 89, 97, 161, 187, 228, 230-233, 235-237, 243, 247, 247, 252, 253, 262, 269, 278, 282, 289, 303, 313, 400, 448, 451, 458, 482, 485, 507, 530
Positive bush(es), 79, 309
Poynting vector, 427
Prefire, 265, 529, 534
Prepulse, 54, 120, 122, 225, 338, 341, 342, 367, 369, 370, 404, 405, 443, 455, 459, 481, 526
Pressurised uniform gap(s), 94, 145
Protonic breakdown, 210, 213
Pulse
 breakdown, 13, 50, 72, 73, 78, 80, 82, 88, 93, 94, 135, 145, 199, 227, 233, 235, 239, 241, 242, 245, 285, 307, 335, 498, 506, 507, 515, 517
 charged gap(s), 97, 100, 103, 266, 355
 life, 80, 235, 239
 rise time6, 38, 44, 96, 262, 266, 270, 299, 300, 306, 363, 456, 504, 510, 526
 transformer, 5, 10, 13, 37, 39, 73, 89, 104, 115, 136, 185, 262, 284, 503, 533
Pulsed forming section, 2, 6
PVC, 69, 129, 130, 502

INDEX

Queen Elizabeth, 16

Radiogra(phic)(phs), 1, 10-12, 17, 18, 22, 24, 30, 31, 32, 93, 377, 398, 400, 409
film, 377
Radiography, 17, 18, 24, 30-32
Rail gap(s), 61, 98, 101, 102, 118, 171, 331, 335, 336, 348, 441, 443-445, 479, 496, 532, 534
Rayleigh-Taylor instability, 209
Razor blade(s), 8, 114, 120, 121, 194, 399, 400, 402, 405, 407. 411
Resistive
coupling, 108, 110, 111
phase, 6, 54-56, 62, 83, 91, 95, 104, 166, 236, 267, 268, 272-274, 283, 285, 287, 289, 290, 291, 293, 296-299, 302-304, 314, 317, 323, 354, 462, 470, 516, 518, 519
Ringing gain, 448, 450, 451, 458, 459, 479, 480, 515
Rise time (s), 6, 8, 21, 38, 44, 45, 54-57, 61, 70, 95-97, 107, 114, 115, 128, 151, 203, 236, 262, 264, 266, 268, 270, 272, 290, 292, 295, 296, 299, 300, 304, 306, 337, 341, 342, 361, 363, 379, 425, 427, 436, 437, 440, 455, 456, 460, 461, 462, 466, 468, 472, 473, 475, 476, 482, 483, 489, 491, 504, 509, 510, 517-529, 533
Rogowski, 54, 66, 125, 126
Rubber, 8, 237, 313, 359, 448-450, 511

S.M.O.G. (SMOG), 2, 3, 6, 8, 9, 10, 233
Safety, 5, 23, 30, 46, 67, 68, 126, 131, 455, 496
Sandia, 13, 30, 71, 73, 83, 98, 118, 119, 122, 154, 182,

Sandia (cont'd)
196, 266, 278, 343, 370, 372, 373, 375, 389, 421, 435
Scaled model(s), 127, 128
Scope, 37, 54, 56, 64, 67, 68, 70, 104, 128, 162, 169-171, 269, 305, 327, 341, 379, 401, 455, 456, 459, 473-476, 483, 524-527
Self absorption correction, 393
Self breakdown voltage, 78, 99
Self replicating curve, 230
SF_6, 4, 7, 10, 29, 40, 48, 49, 59, 60, 61, 82, 84, 88, 96-99, 102, 103, 118, 137, 138, 153, 265, 266, 274, 275, 278, 280-282, 327, 335-337, 340-349, 517, 520, 530, 531
Sharpening gap, 97, 103, 439, 440-443, 448, 451, 453-456, 459-462, 467, 468, 470-478, 482
Shellac, 202, 210-212, 218
Short pulse, 10, 17, 21, 24, 48, 49, 66, 182, 198, 199, 256-259, 263, 266, 270, 278, 306, 309, 375, 425, 427, 431, 433, 438
Silver paint, 159
Simplex (ed) 8, 60, 87, 102, 130, 337, 444, 454, 485, 499, 502, 519
Single channel gap(s), 61, 63, 271, 272, 296, 326
Snail, 29, 262
Solid
dielectric switch, 21, 29, 45, 62, 88, 96, 232, 269, 270, 283, 290, 295, 331, 439
dielectric(s), 56, 57, 77, 78, 81, 122, 269, 293, 296, 331, 433
gap(s), 55, 59, 61-63, 97, 267, 271, 282, 299, 300, 331, 461
Spark
channel(s), 54, 56, 58, 97, 157, 166, 232, 236, 266-

Spark
  channel(s) (cont'd)
    268, 273, 274, 277, 278,
    287-290, 292, 293, 295,
    296, 298, 301-304, 309, 315,
    317, 325, 470, 516, 518,
    521
  gap DC breakdown, 47, 49,
    50, 82, 96, 140, 152, 171,
    235, 274, 352, 354-356,
    358, 359, 361, 362, 445,
    532
  gaps, 4, 39, 46, 59, 60, 70,
    77, 78, 84, 88, 97, 100,
    101, 113, 132, 237, 239,
    241, 263, 265, 270, 274,
    274, 276, 283, 304, 320,
    325, 354, 361, 443, 444,
    447, 456, 462, 496-499,
    502, 503, 506, 518, 523,
    532
Standard deviation, 46, 52,
  62, 63, 77, 80, 142, 160,
  161, 196, 198228, 230, 238,
  241, 242, 243, 246, 247,
  251, 264, 266, 271, 283,
  284, 299, 319, 364, 370,
  432, 454, 497
Start gap, 440-442, 448, 451,
  452, 454, 455, 460-463, 466-
  468, 470, 472, 476-478, 480,
  482, 483
Stickout(s), 153, 183, 195,
  211, 212
Streamer(s), 31, 36, 48, 49-
  52, 71, 78, 79, 81, 83-87,
  89, 91, 94, 112, 140, 145,
  146, 151-154, 157, 171, 181-
  187, 191-199, 201, 202, 204,
  206, 208, 210, 211, 213-215,
  218, 219, 221, 226, 232,
  240, 252, 270, 273, 274,
  290, 304, 307-311, 318, 321-
  323, 325, 328-330, 342, 349,
  419, 507, 511, 514, 515,
  517, 527, 530
Strip transmission line(s),
  43, 53, 87, 446
Surface
  charge, 86, 156, 157

Surface (cont'd)
  guided sparks, 92
  sparks, 91, 155, 156
  tension, 418
  track(ing), 28, 127, 131,
    155-158, 160, 161, 166, 167,
    169, 173, 366, 444
Surpressed point plane breakdown, 211
Swarf, 367, 373
Switching, 15, 29, 38, 45, 54,
  57, 61, 70, 95-98, 104, 117,
  169, 170, 181, 261, 263,
  269, 271, 272, 276, 282,
  286, 287, 290, 295, 298,
  305, 326, 331, 335, 351,
  366, 425, 440, 462
Systems, 1-3, 8, 10, 22, 28,
  35-42, 45-47, 53, 58, 60,
  62, 63, 69, 71, 72, 78, 86,
  87, 94, 97, 104, 111, 113,
  115, 117-120, 122, 126-133,
  153, 157, 186, 232, 237,
  239, 241, 266, 271, 272,
  283, 292, 296, 305, 326,
  343, 375, 399, 421, 438,
  441, 451,

Terawatt diodes, 425
Thin sheet(s), 52, 77, 80, 93,
  130, 245, 252, 253, 511
Three element gap(s), 97, 275,
  278
Three metre edge gap, 326
Threshold voltage, 204-208,
  214, 272
TLD(s), 377, 379, 392, 393,
  399-402, 406, 408
Tom gap(s), 337, 338, 341,
  345, 347, 519-521
Tom Martin Cathode, 367, 368,
  373
Tom Marx, 493, 495
Townsend
  avalanche, 83, 90, 91, 140,
    145, 149, 150, 152, 153,
    252, 273, 348
  coefficient, 147
Tracking, 21, 28, 38, 39, 53,
  59, 85-90, 92, 97, 102, 115,

INDEX

Tracking, (cont'd)
127, 128, 130, 131, 155-162,
164-166, 169, 173, 174, 237,
239, 242, 243, 275, 278,
290, 307, 366, 441, 442,
444, 449, 450, 454, 478,
480, 486, 497-499, 502, 503,
506, 507, 511, 515, 517,
519, 521, 523, 530, 531
Transformer, 4-6, 10, 13, 28,
29, 37-39, 50, 51, 53, 54,
60, 62, 67, 68, 73, 78, 81,
82, 89, 93, 98, 103, 104,
115, 116, 128, 135, 136,
179, 185, 263, 283, 284,
303, 322, 427, 478, 503-505,
532-534
Transit time isolation, 63,
271, 296, 298, 299
Transmission line, 6, 36, 43-
45, 53, 69, 85-88, 118, 156,
240, 288, 292, 299, 327,
331, 336, 337, 341, 446,
489, 491, 518, 521, 524, 527
Trigatron(s), 58, 98, 153,
276, 278
Trigger
  pulse, 58, 60, 61, 63, 97,
    99, 100, 103, 153, 271,
    279, 280, 282, 284, 296,
    297, 299, 331, 332, 351,
    352, 354, 356, 359, 361-
    365, 493, 496, 503-505,
    524, 529, 532, 534,
  spacing, 99, 280-282
  system, 532, 534
Triggered
  Marx, 109, 110, 493
  uniform field rail gap, 331
Triggering 3, 11, 39, 41, 60,
84, 98-101, 107-110, 162,
274, 276, 278-280, 282, 304,
354, 362, 491-494, 496, 502,
504, 505, 512, 533
  range, 11, 39, 41, 60, 99,
    108, 109, 279, 491-494, 496
Triple points(s), 256, 257
Triton, 29, 262
Twinstik, 485, 486, 503
Two element gaps, 96, 277

U.V., 83, 84, 99, 155, 158,
210, 352, 355, 356, 359,
363-365, 434, 435
Uniform
  field, 46-48, 50, 53, 65,
    82, 88, 91, 140, 146-148,
    152, 169, 196, 212, 263,
    272, 274
  field breakdown, 196
  field gap, 47, 140, 146-148,
    274, 278, 338, 419, 441,
    497
Untriggered Marx, 42, 107
UV irradiated, 103, 274

Vacuum
  feed, 426, 428, 434, 436
  flashover, 122, 255, 258,
    429
  interface, 118, 119, 408,
    426
Virtual cathode, 120
Voltage
  divider, 65, 69, 125, 142,
    262, 524
  gain, 3, 4, 6, 44, 263, 444,
    489
  monitoring, 125, 262
Volume effect, 77-80, 127,
227, 230, 233, 243, 245,
248, 307

Water, 2, 6, 9, 10, 12, 28,
29, 31, 36, 38, 50, 51, 53,
58, 61, 69, 79, 81, 82, 93,
97, 98, 118, 128, 129, 177,
179, 181-183, 185, 186, 191,
196-199, 201-203, 205-210,
212, 213, 215, 225, 231,
232, 246, 248-251, 262, 266,
270, 282, 285, 289, 293,
303, 310-312, 314, 422, 427-
429, 433-435, 437, 438, 485,
498
  breakdown, 182, 191, 196-
    197, 199, 201, 210, 232,
    428, 437
  resistivity, 206, 210, 422
Web gap, 336, 341, 348
WEWOBL, 8, 10

Whisker(s), 78, 81, 119, 153, 196, 211-215, 265, 273, 342, 348, 351, 445, 498

X-ray, 1-3, 6-11, 22, 24, 25, 30, 32, 44, 54, 71, 118, 119, 121, 122, 300, 367, 369, 375-382, 386, 389-391, 393, 395, 396, 398-400, 402, 408, 410
  spatial distribution, 375
  tube, 2, 3, 6-8, 11, 30, 44, 54

Z pinch, 31